THE MIND OF THE HORSE

THE MIND OF THE HORSE

An Introduction to Equine Cognition

Michel-Antoine Leblanc

Translated by Giselle Weiss

Harvard University Press

Cambridge, Massachusetts
London, England
2013

Copyright © 2013 by the President and Fellows of Harvard College
All rights reserved
Printed in the United States of America

First published as *L'esprit du cheval: Introduction à l'éthologie cognitive du cheval,*
© 2010 by Éditions Belin

Library of Congress Cataloging-in-Publication Data

Leblanc, Michel-Antoine, 1941-
[Esprit du cheval. English]
The mind of the horse : an introduction to equine cognition / Michel-Antoine Leblanc ;
translated by Giselle Weiss.
pages cm
Includes bibliographical references and index.
ISBN 978-0-674-72496-9 (alk. paper)
1. Horses--Behavior. 2. Horses--Psychology. 3. Animal intelligence. I. Title.
SF281.L4313 2013
636.1--dc23
2013007726

For Catherine

CONTENTS

Color plates follow page 218

Michel-Antoine Leblanc has gotten us used to books that are both comprehensive and precise, relying closely on the scientific literature, which is otherwise largely inaccessible to the public. This book is no exception. It constitutes an impressive and singular review of the major research carried out in the area of equine cognition. Like any review, it reflects a state of affairs at a given moment, which the author himself points out in highlighting persisting gaps and the need for continuing research.

This perspective is important, especially at a time when (too) much information, at times contradictory, is being generated on horses' capacities for understanding, their modes of communication, their intelligence, or quite simply how they perceive the world. The author shares with us both his deep scientific and general knowledge, helping us to place the most recent work in a historical context and showing us how science advances little by little, through periods of regression, contradiction, and sometimes a step in the right direction. Consequently, nothing is ever definitively settled. Rather, an effort is made to develop questions precise enough that experiments can answer them. He conceals none of the challenges intrinsic to the species: The horse is a large, cumbersome domestic animal that is not easy to test. His observation is apt: We still know very little about the perceptual and cognitive capacities of the horse. His conclusion is simple: Let us remain humble and accept that we cannot know everything about this creature that is so close to us and yet so distant.

Humility notwithstanding, this book makes an essential point. Even if we do not possess all the anatomical, physiological, and behavioral information required for fully comprehending the perceptual world of the horse, we do know one thing: Its world is very different from ours! Horses do not see, hear, or smell the way we do. They differ in all these modalities. Although some scientists are tempted by images purporting to show what a horse sees, based on anatomical and physiological data, these are only images filtered by our brain. Today, we know that horses see some colors, see better than we do in the dark, hear ultrasounds to which we are deaf, and have a better sense of smell than we do. The present volume dutifully takes into account this body of information. It also cautions us—we riders and owners of horses, so ready to project our own vision of the world or to shape that of others to our own notions: The horse is a fundamentally different creature, and anything we can "say" on its behalf necessarily passes through the screen of our own cognitive apparatus! Only the horse, by its behavior, can give us some idea of what it perceives, what it does or does not like, and finally what can enhance or alter its well-being. Accepting and taking into account a difference that we can perceive only indirectly is not always very easy for the human brain!

Which brings me to another important point raised by this book: Perception is not a copy of reality, because the brain transforms the information it receives. Indeed, the brain—this formidable apparatus for perceiving, learning, and producing—is a product of evolution that differs between species and also as a result of experience. Social interactions, stress, and the physical environment, among others, are all factors that, even before birth, influence the specialization of different areas of the brain and their organization. All horses do not perceive the world in exactly the same way: Their genetic makeup and experiences contribute to the emergence of individuality. Although these processes are not well known in the horse, examples found in other species do show that it is less the size of the brain—an old idea that refuses to die—than its organization that determines its abilities. The horse has a fairly large ratio of brain to body size: Does that make it more intelligent?

Leblanc's restraint in this matter is worthy of praise: What do we mean by intelligence? Adaptability to novel situations? Adaptability to the needs of the species? Each case calls into play different processes. One horse may learn very quickly to open a box to find food but may never figure out how to take a detour; for another horse, the reverse may be true. Is one more intelligent than the other?

The author's proposal on this point is worth following: Let us ask what horses can or cannot do—their so-called cognitive capacities—as individuals as well as a species. For example, they obtain social information simply by listening to whinnies, but they can also use crossed information (here, images and sounds must be congruent). Like all species, they rely on multimodality (different senses) to construct their representations. In this domain, there is still much to be done, starting with how horses learn to represent the different humans around them. One thing is certain: Horses do discriminate humans and remember them, and in a more or less positive and generalizable way. Indeed, recent findings confirm the quality of equine memory, which enables the animals to reexecute experimental tasks learned a decade previously.

So, yes, this book is a substantial contribution both for those who share Leblanc's desire to understand as well as those who simply wish to know horses better to enhance their coexistence. This creature is different, and this difference must be accepted and taken into account, not demanding more than the animal is capable of within the "limits of its species" (see Breland, cited by the author). At the same time, it is capable of learning our words and our gestures, for example, and building representations of the world around it, all accompanied by a formidable memory. It is up to us to see that the positive aspects of its nature outweigh the negative—but that is another debate.

Martine Hausberger
Director of Research, CNRS
Director, Laboratory of Animal and
Human Ethology, University of Rennes 1

FOREWORD TO THE ENGLISH EDITION

The Mind of the Horse: An Introduction to Equine Cognition is a translated and updated adaptation of Michel-Antoine Leblanc's successful book, *L'esprit du cheval*. This English version, similar to the original French book, is an in-depth review of the scientific literature available to date on specific aspects of equine cognition and perception. While both scientific and popular books on the behavior and/or training of horses abound, accurate portrayal of the perceptual capacities of horses—and how those abilities relate to cognition and behavior—in a compiled book format has been a relatively rare occurrence. With *The Mind of the Horse*, we now have such a resource.

This well-organized book will hold appeal for both academics and the general horse community. Leblanc tackles complex subjects, such as the anatomy of the equine brain and the intricacies of equine perception, in a clear and approachable manner. Topics that could easily be overwhelming are written in a style that will hold the interest of most readers. The book is divided in a logical manner, starting with an overview of the nature of the horse and what is meant by intelligence and continuing with a discussion of the organization of the equine brain, with highly comprehensive chapters on equine perception. Whereas comparable books on equine behavior discuss the scientific literature to some degree, Leblanc systematically reviews individual studies with enough detail that the reader gains an excellent understanding of how experiments were conducted and how the findings of those studies have furthered scientific knowledge.

Horses and humans have interacted for millennia. The years have seen remarkable advances in horse management, handling, and training, as well as dismal failures. Successes—from both the human and the horse's point of view—most often resulted when humans attempted to understand the nature of the horse and, consequently, worked with the horse to achieve certain goals. Failures, on the other hand, frequently reflected a lack of knowledge coupled with an anthropocentric approach to dealing with these animals. In the distant past, of course, scientific data on the cognitive and perceptual world of horses were unavailable; nevertheless, astute trainers and handlers made good use of their observations of how horses naturally behaved. Yet, more often than not, humans mishandled horses. This is still an unfortunate truth today, in large part because of misconceptions and misunderstandings of what makes a horse a horse.

Readers of this book will come to understand that, while certain cognitive or perceptual abilities are indeed similar between horses and some other mammalian species, other abilities are strikingly different. Take, for example, equine vision. Impressive advances have been made with regard to research into the visual capabilities of horses, and much more is now known about equine color vision, depth perception, visual acuity, scotopic vision, lateral vision, interocular transfer, and so forth. Some of these studies confirmed common beliefs; others yielded unexpected results. For instance, horses are not color-blind but most likely dichromats with red-green color deficiencies. Visual acuity, depth perception, and night vision capabilities are substantially better than was thought previously. Although horses have an extensive field of vision and can detect stimuli appearing within a nearly fully encompassing circle, they can only identify and discriminate stimuli within a more limited lateral field of view. Finally, contrary to a popular myth, horses can certainly transfer visual information from one eye to the other.

Leblanc also discusses research findings on other sensory abilities, for example, hearing, smell, taste, and touch. These chapters are not as extensive as those on vision, mainly due to the fact that these topics have only been investigated to a limited degree compared with vision. Nonetheless, taken as a whole, the body of research described in this book has provided scientific answers to questions that have been asked throughout the ages; questions that, until relatively recently, were only addressed anecdotally or through folklore and often answered incorrectly. With this book, readers will now be better able to appreciate that horses do not perceive the world in the same manner as humans and may, therefore, react to their surroundings in unexpected ways. Horse behaviors previously viewed as negative by owners and trainers suddenly make more sense once physiological, cognitive, perceptual, and evolutionary aspects are taken into account. Likewise, greater knowledge about the basic and advanced learning skills of the horse and its impressive memory will enable the horse industry to tailor training and management more effectively and, thus, enrich the human-horse relationship as well as improve equine care and well-being.

Evelyn B. Hanggi, M.S., Ph.D.
Director of Research,
Equine Research Foundation,
Aptos, California

PREFACE

It is about forty years ago now since scientific studies of the behavior of horses started. Numerous studies have been devoted to the subject, and it is also true that book-length compilations are quite numerous. However, research studies exploring the horse's mental capacities are still few and far between.

It is mainly in the last ten to fifteen years that cognitive ethology has addressed the horse. This is a relatively new field with a double perspective: that of comparative psychology and that of ethology. It takes into account the study of the horse's cognition together with all the specific mental capacities of this animal, as built up by evolution. It is therefore not surprising that many aspects in this new subject are still unclear.

Nonetheless, the reader of this book will find that major breakthroughs have been made. In the past—and still today—a good deal of literature on the so-called intelligence of the horse was based solely on more or less speculative interpretations of reports that remained largely anecdotal, or to say the least, fragmentary. These have been gradually replaced by a whole corpus of varied knowledge, which though incomplete is predominantly based on rigorous experimental investigation, conducted in controlled situations that not only allow the advancement of knowledge but also the identification of the remaining gaps. This explains why, when considering the current state of our knowledge of the horse's mental capacities, we are confronted with a contrasting picture, some aspects of which are now quite clear whereas some others still remain in the dark.

This is where this book fits in. In the large field of the cognitive ethology of the horse, where vast areas of our knowledge remain to be explored but where research is on the move, it seemed the right time to me to systematically review the existing scientific literature on the subject and to take stock of all we do and do not yet know. I have approached this task according to the following plan:

The first chapter is devoted to a panoramic overview not only of all the knowledge gained during the last decades but also of all the uncertainties that remain regarding everything we know about the horse's nature and behavior: the way it adapts to its environment, what determines its diverse behaviors, and, more particularly, what its mental capacities are.

The second chapter is a brief retrospective of the way equine intelligence has commonly been assessed. The different opinions—including those of renowned riding masters—will be reviewed, as well as the controversies that stirred the equestrian and the scientific worlds on the theme of the alleged extraordinary powers of the horse's mind.

The third chapter is an introduction to the theoretical and conceptual framework within which cognitive ethology has its place, that is, on the themes of intelligence, cognition, and animal representation—all in the context of the development of comparative psychology, ethology, and cognitive psychology.

The fourth chapter is devoted to the main anatomical and neurophysiological characteristics of the horse's brain. This chapter also discusses perceptual and motor laterality in conjunction with the concomitant specializations of the brain hemispheres, which are more fully elaborated with respect to the horse in subsequent chapters, and the relation between the horse's brain and its mind, from an evolutionary perspective.

Before tackling the subject of the horse's perception of its own social and physical world through different perceptual modalities, I deal, in the fifth chapter, with the very nature of perception and a number of questions posed by its study in animals, including the horse.

Chapters 6 through 11 comprise detailed accounts of past research on equine perception. For each and every perceptual modality under study, I have compiled as thoroughly as possible all the existing experiments that testify to the horse's actual perceptual experience. All these results are situated within the framework of the present knowledge of the anatomical structures and underlying physiological processes.

Finally, I offer a few comments on the advances and gaps in the field in the form of a conclusion.

As the reader will note, the developments on the various perceptual modalities are fairly detailed. This is particularly the case with visual perception, which alone is the subject of three relatively extensive chapters, whereas, in contrast, only one rather short chapter is devoted to tactile perception. This should not be considered an implicit indication of any ranking of the importance of the perceptual modalities, especially given that the horse's perceptual experience is fundamentally multimodal. Rather, this presentation reflects the current state of our knowledge, the complexity of the subject, as well as the degrees of attention applied by researchers to the different aspects of the subject.

Let us hope that, beyond whatever interest readers will take in the information the book contains (and despite all of its weaknesses), it will be a valuable contribution and stimulate research to bring new knowledge to the field of horse cognition.

THE MIND OF THE HORSE

WHAT WE KNOW ABOUT THE NATURE OF THE HORSE

Millennia have passed since people looked on horses solely as a source of meat. Indeed, they were probably domesticated around 6000 years ago (Anthony 1996; Olsen 1996; Jansen et al. 2002; Levine 2005; Outram et al. 2009) in the western part of the grassy Eurasian steppe (Warmuth 2012) that extended 7,000 kilometers from the Carpathians to Mongolia. It was probably the only region in the world where the horse survived in appreciable numbers at the end of the Ice Age.

Then and Now

The earliest writings and depictions attest to the use of the horse going back around 4,000 years and refer initially to pulling chariots then military wagons. These practices appear to have been quite widespread, particularly in Anatolia (now Turkey). Moreover, representations in Iran and in Afghanistan show that equitation was practiced 3,500 to 4,500 years ago. At Ur, in Mesopotamia (in today's Iraq)—a site near Uruk where, coincidentally, writing was invented 5,300 years ago—a 4,150-year-old seal was found depicting a rider. Although it does seem that equitation as a military function, or a leisure activity for the elite, came along a bit later, by 1000 BC it was well known. That makes twenty-five centuries, beginning with Simon of Athens, then Xenophon, that people have been writing about the art of riding.

Does that mean that we now know everything there is to know about the nature of the horse? Hardly.

In *Questions équestres* (1997[1]), French general Alexis L'Hotte wrote: "The vast equestrian truths have come to the fore in all ages and belong to all schools . . . All horsemen, endowed with the spirit of observation and with considerable practical experience, have been able to make certain statements which have not been made by their predecessors or which escaped them; for the knowledge and use of the horse presents an endless field of research and observation" (200–201).

Moreover, the knowledge accumulated through the centuries has left incomplete traces, the fruit of an essentially oral tradition, with the exception of those relating to riding.

The breadth and importance of scientific work on animal and human learning conducted throughout the twentieth century classified, formulated, and validated the empirical observations accumulated over time by trainers. For example: "Sobriety of words must guide the *écuyer* . . . He must point out goals with clarity and never pursue two goals simultaneously" (L'Hotte 1997, 153). Or: "Regardless of the method employed, progress will advance all the more rapidly if errors are promptly eradicated" (199). Or: "Progression in the schooling of the horse must, quite definitely, be graduated, because gradual progress is the main road to success" (200).

Such guidelines, and many others, clearly showed early on that "the art of co-ordinating the muscular forces of the horse" (157)—L'Hotte's definition of equitation—rests on principles that have nothing to do with the locomotor apparatus. Moreover, in his well-known aphorism, "Calm, forward, straight" (153), which is

---

1. Date of the English translation. The original French work, written in 1895, was first published in 1906, after the general's death. The English translation kept the French title.

basic to proper equitation, the first two terms refer to aspects related to what could be called the psychology of the horse: that is, ensuring the animal's emotional stability and mobilizing attentional and motivational processes.

It seems to me altogether curious that generally, until very recently, in contrast to training for other animal species, instruction in equitation made very little explicit reference to the general principles of animal learning and their application. This may be due in part to the fact that in equitation, the very term *dressage* comprises two dimensions that are complementary but fundamentally distinct. The first of these we will call behavioral. It corresponds to what is commonly known in French as dressage for other animal species, defined by the *Petit Robert* (a standard French dictionary) as "the action of training an animal with the goal of habituating it to respond to human expectations." Such training fulfills one of the tasks of the general principles of learning as set down by the field of psychology over the last century. The second dimension is biodynamic and effectively corresponds to the definition given by L'Hotte. Its implementation consists in tasks that can loosely be called equine gymnastics and the "school of aids." This dimension so completely dominated the writings of the grand *écuyers* that it has more or less eclipsed the behavioral aspect.

Moreover, in the twentieth century, the development of ethology, that is, the biological study of behavior—and specifically, since the 1970s, equine ethology—has clearly shown how much remains to be learned about the true nature of the horse.

In fact, up to that point, anyone wishing to examine the life of free-ranging horses had to rely mostly on anecdotal accounts, which were frequently anthropomorphic and inclined to baseless clichés such as "the proud stallion defending his territory and reigning over a vast herd." In his seminal work published at the end of the 1960s, Irenäus Eibl-Eibesfeldt (1970), a student and later collaborator of Konrad Lorenz, compiled an extensive bibliography of 1,351 references. Yet, only 3 had to do with horses, and none among them with the conditions of life in the wild. The paucity of source material no doubt explains why the only reference to feral horses in this major reference work is, unfortunately, the following: "Among horses, there are large herds where a single stallion acts as leader throughout the year" (376)!

Another example is James Feist (1971), a young researcher studying the behavior of feral horses running free in the Pryor Mountain reserve on the border of Wyoming and Montana. One of the very first to devote himself to systematic observation in the field, Feist began his doctoral thesis by stating: "There are no published studies concerned with behavior of the feral or 'wild' horse in North America" (2).

Although the literature on the behavior of horses in the wild grew markedly in the 1970s, for a long time the contributions benefited only a small group of scientists. When I published my first book on equine behavior (Leblanc 1984), I could still write without exaggeration: "The scarcity of previous years has now given way to relative abundance. Nonetheless, the data accumulated have, for the most part, circulated mainly within the community of researchers" (13), and more specifically: "In the last seven or eight years, of all the articles published by *La Recherche* and *Pour la Science,* only one has dealt with this subject. Moreover, it was somewhat peripheral, as it was an article by Hans Klingel (1978) on the social life of zebras and antelopes!" (14).

This last point illustrates a rather strange situation, for far more was already known about the life and habits of zebras than about horses!

The profusion of magazine articles and popular books on the ethology of the horse over the last several years is enough to show that the situation is radically reversed. But now we face another paradox. These days, everyone (or nearly everyone) who is at all interested in our companion knows that horses live in harems and that stallions are not territorial. But to invoke the ethology of the horse[2] appears to have become sufficiently fashionable to condone

---

2. In the same way that people once carried on about horse "psychology."

and to label as "science" publications, practices, and training that, whatever their ultimate quality and intrinsic interest, generally have nothing ethological about them in terms of dressage or equitation aside from the name!

In any event, the ethology of the horse has now acquired legitimacy, which from a certain vantage is a source of satisfaction. "Ethological equitation" may be a misnomer and may regrettably be confused with the genuine ethology of the horse, but at least it has the virtue of leading more and more riders to think of their horse as an individual member of a species with its own characteristics and neither a machine nor a human clone! In other words, whether or not it is science, at least the horse benefits.

Discovering the Real Life of Free-Ranging Horses

A wealth of knowledge now exists on the behavior of the horse, both specifically with regard to individual and social behavior in the wild and what determines this behavior and to the construction of the individual during development and its relations with the environment.

Examples include recent detailed reviews aimed at specialized audiences of, for instance, veterinarians and ethologists, even though, unfortunately, at present these publications are available only in English (Waring 2003; McGreevy 2004; Mills and McDonnell 2005).

A further, still relevant illustration of the vitality of research on the subject is the work of Françoise Clément et al. (2002), which contains a list drawn up by Frédéric Lévy and Léa Lansade of twenty-eight teams of scientists throughout the world working on equine behavior.

It would be difficult to provide an overview of the knowledge thus far accumulated in just a few words. There is too much material. For a fuller account, refer to my recently published volume with Marie-France Bouissou and Frédéric Chéhu (Leblanc et al. 2004), in which I attempt to present a summary of French achievements in the field for a broad readership. The bibliography, which itself testifies to the substantial body of research, includes 446 references devoted to horses and related animals, though it still represents only a selection of "classical" work and up-to-date reports on the themes treated.

Here, I will simply mention:

- first, some aspects of life in the wild for the companion that we have taken to be a familiar animal for thousands of years, and some practical lessons that we can draw on its fundamental needs for use better adapted to its potential; and
- second, several observations on more experimental studies from a body of research in the field, in particular certain gaps in knowledge that persist, which will naturally bring us to the point of this book.

A preliminary aspect emphasized by field studies is the especially adaptable character of the horse, for feral horses are found living in very diverse ecologies. Some of these regions are temperate, of course, but they present varied features, such as the vast prairies of the Argentinean pampas (Scorolli et al. 2006), the swamp estuaries of the Camargue (Duncan 1992), or Cape Toï in Japan (Kaseda and Khalil 1996), as well as areas of brushwood and the New Forest in southern England (Tyler 1972) and the Kaimanawa range in New Zealand (Linklater et al. 2000). Other regions are arid, semidesert, or quasi-desert: Feral horses are found in much less mild climates such as certain areas of the western United States, for example, Wyoming (Feist and McCullough 1976), Nevada, and Arizona (Berger 1986), or islands such as Sable Island, off the coast of Nova Scotia (Welsh 1973), and even totally arid regions such as the Namibian desert (Stoffel-Willame and Stoffel-Willame 1999).

To thrive in such heterogeneous environments, the horse must possess a fairly impressive behavioral plasticity (Boyd and Keiper 2005). For example, it adapts the time it spends daily in each of its activities, grazing, resting, moving, socializing, and so on, known as its time budget, to the quantity and quality of the resources at its disposal. In particular, the time it spends each day eating may vary from 13 to 17 hours in a 24-hour period. Take the fairly extreme case of Namibia (Stoffel-Willame and Stoffel-Willame 1999), where dryness is particularly palpable and where a horse may have to travel up to 25 kilometers to drink. In summer, when the temperature may rise to 40 degrees centigrade in the shade during the day and tumble to 0 at night, the average time interval between visits to a waterhole reaches 30 hours. In winter, when temperatures still hover around 25 degrees centigrade during the day, the average interval increases to 72 hours.

All of these studies, carried out in very diverse environments, show that the ecological conditions were such that they also affected the social organization of the horses by perceptibly influencing the distribution and interactions of the groups, at the same time as the groups themselves were liable to have a nonnegligible impact on their milieu (Fleurance, Leblanc, and Duncan 2004).

The basic structure of equine social organization is now well known. It is observed not only in so-called domesticated horses (*Equus ferus caballus*) but also in Przewalski horses (*Equus ferus przewalskii*) and is shared with plains zebras (*Equus burchellii*) and mountain zebras (*Equus zebra*).

Horse society is structured as a harem, in which a male (though sometimes several) takes up with several females, who are accompanied by their offspring (until they are around three years old). The structure is stable, thanks as much to the relationship between the mares of the harem as to the presence of the stallion, because if he disappears—either replaced by another stallion or due to accidental death—the mares do not necessarily separate. Thus, for example, Ronald Keiper (1985), who studied ponies on Assateague Island, situated off the coast of Maryland, reported that two harems whose stallion had died guarded their integrity for several months, until another stallion replaced him. Similarly, in various cases of switches among stallions observed by Hans Klingel (1967) in plains zebras, the group formed by the mares and their offspring also remained stable.

Unlike most species of social mammals, where only males migrate, in horses, the young of both sexes leave the natal group to be part of a new family (Rutberg and Keiper 1993). Young females leave the natal group to rejoin harems, either already existing or being formed, whose males are unfamiliar. In this way, as shown by Anne-Marie Monard and Patrick Duncan (1996), they help to avoid inbreeding. The young males go through an intermediary phase lasting several years during which they live within small bands of bachelors (McDonnell and Haviland 1995).

Species that share this harem structure are not territorial: Bands that form in this manner, whether a harem or a bachelor band, live within more or less extensive home ranges of from one to a few dozen square kilometers that may overlap depending on the ecology of the habitat (Waring 2003). Typically, these bands interact very little.

In some semidesert regions, where horse harems are naturally scattered due to sparse vegetation, the bands actually avoid approaching each other, even when they meet on the way to a waterhole. This active avoidance of intergroup relations is not found in other habitats where food is more abundant and where the horse population is denser. What seems to be at play is rather a kind of indifference between the bands that sometimes come together in large gatherings. This is the case in The Netherlands, where hundreds of Konik Polski ponies on the Oostvaardersplassen reserve assemble in large herds (Vulink 2001).

This twofold observation—a basic social structure founded on harems and bands of young bachelor males that interact only rarely and the existence of vast herds—necessarily raises questions about the gregarious character of the horse. The answer appears to depend on the interplay between two social tendencies with different functions that sometimes converge and sometimes diverge:

- The first tendency is to band together, which is expressed quite naturally when food resources are adequate. It may have originated as an adaptive defense in response to predators (Treisman 1975). The gathering of groups into

herds may also, rather obviously, be related to behavior that is observed in ungulate species living in the open, in habitats where food resources are abundant and well distributed, for example, plains zebras or gnus in the Serengeti, in East Africa (Klingel 1969). Indeed, these proclivities correspond to defensive adaptive behavior on encountering predators (lions, say, or feral dogs), whose hunts end more successfully when they chase after isolated animals. Consequently, isolated individuals targeted by predators show a natural tendency to rejoin the herd to make themselves less conspicuous. Of course, these days, horses are only rarely exposed to such predators. Nonetheless, they may still manifest a phylogenetic inertia evolved and developed over hundreds of thousands of years of exposure to predators, which disappeared from their environment only recently. The walls of the caves at Lascaux, for example, which are only seventeen thousand years old, depict around a dozen felines. Moreover, today horses are still the target of predators such as biting flies, less "evident" pests, perhaps, but nonetheless real with respect to the large amount of blood that they suck (Keiper and Berger 1982). In the Camargue, formation of herds coincides with a reduction in the number of biting flies per horse, thus constituting an effective defense strategy against aggressors (Duncan and Vigne 1979).
  • The second tendency is for groups to disperse when predatory pressure eases and food resources are more meager and unevenly distributed. The demands of survival that these hardships imply take priority, and the "least risk" forces groups to disperse (Feist and McCullough 1976; Miller and Denniston 1979).

Grouping and group dispersal are thus adaptive phenomena whose balance varies as a function of the environment.

In the case of seasonal variation in food resources, groups may also (geographic situation permitting) temporarily migrate as a whole, without segregation by sex, toward more promising sites. This behavior is notably the case with Serengeti plains zebras. According to Olsen (1996), it was also typical fifteen thousand years ago of horses who left the green pastures of the Massif Central at the end of the hot season to spend winter in the alluvial plain of the Saône, using the route passing at the base of the Rock of Solutré,[3] where they were massacred. To some extent, the same strategy is still followed, for example, by the horses of the Great Desert Basin (Berger 1986). Depending on the season, the animals alternately occupy home ranges at high and low altitudes.

Still, however great the adaptability of the horse to the ecology, one must not confuse adaptability of the species with that of the individual. Taking the example of the survival of feral horse populations living in particularly harsh and geographically isolated environments, natural selection will exact a heavy toll on survival through merciless elimination of the weakest. The horses of Sable Island provide a further example: Counts done over a period of a dozen years (Welsh 1973) showed that the population there varied from more than 300 individuals to fewer than 150, which indicates the high mortality of some years! The average length of life on the island is barely six years for males and around four and a half years for females. Individuals living beyond eighteen years are rare: For every hundred births, only 7 or 8 individuals will reach eighteen.

Another very essential point is that, whatever the circumstances, the horse is a supremely social animal: In the wild, it never lives alone (McCort 1984; Feh 2005). It is either part of a harem, whose core persists even when it is temporarily deprived of its stallion, as indi-

---

3. The Rock of Solutré is located in eastern France, near Mâcon. At its base, which forms a cul-de-sac, is a major hoard of fossilized horse bones, spear tips, and flint tools greater than a hectare in size and 9 meters deep. Nearly a hundred thousand horses were killed there, between 32,000 and 12,000 years ago. The horses passing through were driven there to be massacred, by using branches to deviate their route and by, for example, setting fires to prevent them from turning back.

cated above (Keiper 1985), or it belongs for a time to a band of bachelors. Even in domestic groups, whose composition is determined by humans, the fundamentally social nature of the horse is just as evident (Hartmann, Søndergaard, and Keeling 2012; VanDierendonck and Spruijt 2012).

During the 1970s, horses of the Camargue, which had been domesticated for many generations, were released into nearly complete liberty over a space of 335 hectares at the Tour du Valat research center (Duncan 1992) for a ten-year study. The horses required relatively little time to reestablish a social structure similar to that of feral horses. The result shows that what determines many individual and social behaviors can be profoundly anchored in the genome of members of a species. It also testifies to the still somewhat arbitrary character of the constraints imposed on domesticated horses (such as confinement, social isolation, and relative inactivity) that can ultimately overwhelm their natural capacity to adapt.

Given these aspects, and especially taking into account the time budget of feral horses, the prevalence of stereotypies or repetitive movements (between 10 and 15 percent, on average) in stables is hardly surprising. In France, a survey carried out on a sample of nearly three thousand horses (Burgaud 2001) revealed a rate of 10.1 percent. In certain cases, increasing the possibilities of social interaction among the horses showing such symptoms of impaired well-being was shown to ameliorate their state, including artificial means such as installing a mirror (McAfee, Mills, and Cooper 2002) or even the simple image of another horse in their stall (Mills and Riezebos 2005). Nonetheless, experts agree that preventive action consisting of better taking into account the natural needs of the horse would be a more effective approach to the problem.

Equine Ethology Studies to Pursue

Apart from the basic characteristics of horse behavior—adaptability and its social component—the work of ethologists touches on many other, more specific aspects. By way of example, I will cite just three, which themselves represent fairly broad categories:

- behavioral ontogenesis, that is, development of the individual, beginning with birth;
- sociosexual development, in other words, how young males and females are differentiated in terms of their diverse individual and social behaviors—play, for instance—and what types of approaches they borrow, beginning with their natal group, to form a new family; and
- everything that has to do with reproduction in the wild (reproductive behavior of the stallion and the mare, gestation, maternal behavior, and so on).

Although field studies have revealed much about these subjects and about their causes and functions, that is, the how and why of the behaviors they give rise to, it would be premature to conclude that everything is now understood!

Consider social dynamics. On this subject, we now know more—for example, how and why young horses leave their natal group—than we do about some of the ways in which new groups are formed. Among horses, the departure of the young occurs through the drift of both sexes away from the original group; this is rare among mammals, but (as I have already mentioned) research has shown that this corresponds to a social mechanism for avoiding inbreeding (Monard and Duncan 1996). On the other hand, why is it fairly common for two males (and sometimes more) to band together to create a new harem? The answer to such a question is not really clear: A variety of hypotheses dealing with situations of cooperation or competition have been proposed (Feh 1999; Linklater and Cameron 2000; Feh 2001). Although each of these hypotheses is based on careful observation, the observations themselves have been carried out in different environments. Consequently, definitive conclusions will require further investigation.

That being the case, the obvious questions will not necessarily be answered in the field. Experimental work is helping to fill in some of the lacunae and to enrich field observations. The study I referred to earlier, carried out by Françoise Clément and a team of French scientists on equine behavior in France and elsewhere (Clément et al. 2002), offers an overview of the situation as of several years ago. The study classifies work into four broad thematic categories: well-being, temperament, handling, and sexual behavior. Salient findings, with some qualifications in the light of recent developments, include the following:

- Research on equine well-being (Heleski and Anthony 2012) is still embryonic, except as regards transport (Friend 2001) and stereotypies (Mills, Taylor, and Cooper 2005); other (different) perspectives, however, have emerged, for example, using adults to alleviate weaning stress in foals (Henry et al. 2012) or the significance of play among adult domestic horses as an indicator of their level of stress (Hausberger et al. 2012).
- At the international level, temperament, its causes, and its assessment, as well as equine-human relations, are similarly barely developed, except in France, where this topic is of substantial interest. I note, in particular, the body of work produced by the group of Martine Hausberger at the University of Rennes (Hausberger and Muller 2002; Henry et al. 2005; Henry, Richard-Yris, and Hausberger 2006; Hausberger et al. 2007; Hausberger et al. 2008; Hausberger, Muller, and Lunel 2011), as well as by Marie-France Bouissou and Léa Lansade (Lansade, Bouissou, and Boivin 2007; Lansade and Bouissou 2008; Lansade, Bouissou, and Erhard 2008a; Lansade, Bouissou, and Erhard 2008b; Lansade and Simon 2010).
- Some work has been done on the usefulness and effects of early handling of foals. The ideas proposed by the American veterinarian Robert Miller (1991), who first

described what is known as "behavioral imprinting of foals," have been strongly called into question by the experimental results of various studies, in particular by French researchers (Lansade, Bertrand, and Bouissou 2005). Moreover, this question comes up again in a broader context (interspecific social skills, lateralization of handling, and so forth) (Henry et al. 2009; Søndergaard and Jago 2010; de Boyer des Roches et al. 2011).

- Research on sexual behavior, especially among feral horses, is well documented. In terms of domesticated horses, however, studies still focus largely on aspects related to physiology.

As Clément concludes: "An immense amount of work remains to be done on well-being, temperature, human relations, and breeding conditions" (110), although some progress has been made since then.

Finally, the four thematic categories leave out one significant line of inquiry: cognitive ethology.

The Emergence of a New Field of Research: The Cognitive Ethology of the Horse

Cognitive ethology straddles the perspectives of animal or comparative psychology and ethology. In the words of Jacques Vauclair and Michel Kreutzer (2004): "The question is to understand how, based on its mental states and representations, an individual develops its behavior" (11). Consequently, "cognitive ethology focuses on the mental states associated with animals' relationships with their physical or social environment" (ibid.). The practical importance of such an approach is immediately apparent if the goal is to establish human-horse relationships based on real understanding of what we might call "the mind of the horse." What is more often the case is that such relationships are approached with preconceived ideas, projections of human ways of perceiving and reasoning about the world, and misplaced application to human-horse interactions of ethological findings on the social life of horses, not to mention manifold human fantasies.

And yet, curiously, the study of equine cognition remains somewhat patchy, even though the field has enjoyed a string of advances mainly over the last decade.

Although important revelations have been added to what was already known about the perceptual capacities of the horse, as Carol Saslow pointed out in 2002, the knowledge is not evenly distributed. Her assessment is still essentially valid.

As will be obvious in the chapters devoted to equine perception, the best-known perceptual modalities are clearly those that continue to capture the interest of researchers and, consequently, have been explored more carefully. Examples include visual perception (for example, including the difficult question of color perception) and, to a lesser degree, auditory perception.

In contrast, the literature on olfaction—whose role is key in the social life of the horse, whether close up or from a distance—is scant. The same is true, by and large, for tactile perception, though it is crucial to the practice of horsemanship and in the school of aids, whose finesse is evident, for example, in academic dressage in the quest to perfect the slightest touch of the legs (*souffle de la botte*).

An additional, recent new field of investigation worthy of mention, insofar as it affects the horse, is that bearing on hemispheric brain lateralization and the influence of specialization on perceptual processes in particular (McGreevy and Rogers 2005; Larose et al. 2006; Austin and Rogers 2007; de Boyer des Roches et al. 2008; Basile et al. 2009; Farmer, Krueger, and Byrne 2010; de Boyer des Roches et al. 2011; Sankey et al. 2011b). In France, this field is a speciality of the team headed by Martine Hausberger. We will come back to it in the chapter on the horse brain, as well as the chapters devoted to different perceptual modalities.

Progress over the last ten to fifteen years in research relating explicitly to learning in the horse (Fiske Godfrey 1979; McCall

1990; Nicol 2002; Nicol 2005; Hanggi 2005; Murphy and Arkins 2007; McGreevy and McLean 2010) is evident, but may to some degree be misleading. For the term *learning* covers a diversity of issues. Naturally, these include learning as such, for example, the relative effectiveness of different types of reinforcements (McCall 2007; Warren-Smith and McGreevy 2007; Heleski, Bauson, and Bello 2008; Innes and McBride 2008; Sankey et al. 2010a; Sankey et al. 2010b; Sankey et al. 2010c) and the potential advantage of learning spread over time rather than concentrated in a short period (Rubin, Oppegard, and Hintz, 1980; Dougherty and Lewis 1992; McCall, Salters, and Simpson 1993; Kusunose and Yamanobe 2002). But under the rubric one also finds work for which learning is only a means of exploring different cognitive processes and their associated representations. Thus, such studies encompass a wide field that may include perception (in discriminating colors, for example), memory (long or short term), the ability to categorize (consisting of grouping objects within the same class), representing space, and so on.

A propos of memory itself, however, aside from the many anecdotes about the proverbial long-term memory of the horse, to date very little research has been done that contributes in any major way to our fundamental store of knowledge. Indeed, except for two preliminary studies on fairly short-term memorization associated with learning (Marinier and Alexander 1994; Wolff and Hausberger 1996), the most telling work in this area is that of Evelyn Hanggi and Jerry Ingersoll (2009a). We will come back to it in chapter 3 during discussion of representations in the horse. Hanggi and Ingersoll showed that horses exhibit long-term memory that allows them to successfully carry out nearly all discrimination tasks based on categorization criteria learned six to ten years previously, with no other training required and no other prior exposure to the stimulus presented since the initial experiments.

Assessment of short-term memory has given rise to different experiments based on how long it takes a horse to remember the location of a container where food was placed in its presence. The results of these experiments are varied, and sometimes contradictory. Bernhard Grzimek (1949, cited by Waring), who tested two horses, found a retention time (beyond which the subject responded randomly) to be 6 seconds for one horse and one minute for the other. More recently, after Andrew McLean (2004) failed to observe a retention time above 10 seconds (twelve horses), Jack Murphy (Murphy, Hall, and Arkins 2009) compiled significantly positive results for retention times of 3, 6, 9, and 12 seconds (eight horses). Evelyn Hanggi (2010a) reports that she obtained equally positive results of 5, 10, 15, 20, and 30 seconds (four horses). Further comparisons of her own experiments with those of McLean and Murphy led her to conclude the importance of taking into account attentional factors and experimental conditions more or less pertinent to ecology to explain certain discrepancies in the results. Finally, Paolo Baragli and colleagues (2011c), who also obtained positive results of 10 and 30 seconds with two groups of different horses, note that using different groups avoids the possible risk of a learning effect on the length of the longest retention time. It does now appear that there is agreement on conditions for exploring longer retention times.

Cognition of the Physical World

Although it does appear that the horse is endowed with a certain aptitude for forming categories by building object classes based on perceptual cues (Hanggi 1999a)—shapes, in particular—to date the animal shows only a limited capacity to extract abstract ideas or concepts from such exercises (Sappington and Goldman 1994; Flannery 1997; Hanggi 2003), albeit with impressive results for concepts of identity and relative size, and existing studies clearly are merely initial forays in a large field of research.

A question such as object permanence and how it relates to working memory obviously has practical implications for learning. Fairly comprehensively studied in other domesticated species like the dog (Fiset, Beaulieu, and Landry 2003) and the cat (Fiset and Doré 2006), this issue does not appear to have been addressed in the horse aside from a few rare efforts that basically center on evaluation of short-term spatial memory, as mentioned above.

The same holds true for the representation of space, although a recent study (Hothersall et al. 2010) suggests that, in a discriminative learning task, very young horses (foals several months old) favored spatial cues over shape or color, thus revealing a specific bias toward spatial information. How does the horse structure its home range? How does it learn to form a cognitive map (Tolman 1948) of its space that eventually enables it to travel up to 25 kilometers to drink (Stoffel-Willame and Stoffel-Willame 1999), to move from one "good pasture" to another one, as is generally the case for ungulate herbivores (Howery et al. 1999; Brooks and Harris 2008), or even to undertake major seasonal migrations in response to ecological conditions (Olsen 1996)?

As Christine Nicol (2005) has pointed out, studies on the formation of abstract concepts, particularly with regard to the potential capacity of the horse to represent time, would be welcome.

The interaction between horses' perception of space and time is critically important with respect to their feeding strategies (Leblanc and Duncan 2007). One of the greatest problems animals living in the wild must confront is the spatial distribution of food resources in different environments. The capacity of horses (and of other animals) to return to lush locations in the prairies is a critical determinant of their success in foraging. It appears that horses are capable of choosing the richest sites when the information they have is recent; in contrast, when they have no recent information, they proceed by "dynamic averaging," choosing their feeding sites based on their long-term average richness (Devenport et al. 2005). It would be interesting to test the generalizability of the conclusion that horses use dynamic averaging, and moreover that they take into account information received at different times to optimize their feeding choices.

Social Cognition
Similarly to the cognition of the physical world, the field of equine social cognition is still relatively untrammeled, particularly with respect to the fundamental social nature of the horse.

Furthermore, one of the findings of the studies carried out to date may appear paradoxical in this regard, for no one has yet managed to show that the horse is capable of imitation or, more precisely, learning to accomplish an action by observation (Baer et al. 1983; Baker and Crawford 1986; Clark et al. 1996; Lindberg, Kelland, and Nicol 1999; Krueger and Flauger 2007; Ahrendt, Christensen, and Ladewig 2012). This fact throws into question the fairly widespread belief that stereotypies are acquired in stables through imitation.

Work on the human-horse relationship ongoing for several years now, especially in the Hausberger group at Rennes, does suggest a cognitive dimension (Baragli et al. 2009; Fureix et al. 2009; Sankey et al. 2010a; Sankey et al. 2010b; Baragli et al. 2011a; Sankey et al. 2011a), and some studies have, for example, focused on the degree to which horses understand referential communication signals given by humans, such as pointing and gaze, as well as posture (McKinley and Sambrook 2000; Maros, Gácsi, and Miklósi 2008; Verrill and McDonnell 2008; Proops and McComb 2010; Proops, Walton, and McComb 2010; Birke et al. 2011; Krueger et al. 2011). Interestingly, it would seem in this respect (Lesimple 2012) that the more attentive horses are to humans, the less likely they are to succeed by themselves at a learning task, even one that is appealing (in this case opening a chest containing food), which would seem in part to support the hypothesis that domestication may have favored the development of certain cognitive skills in terms of animal-human relationships and may have lowered cognitive skills related to individual task solving. It remains striking that, up until very recently, almost nothing existed apart from anecdote on recognition of individuals, either between horses or with regard to recognizing members of other species, including humans.

Indeed, the question of intraspecies recognition of individuals was the subject of only three preliminary, already outdated studies (Ödberg 1974; Wolski, Houpt, and Aronson 1980; Leblanc and Bouissou 1981). However, such recognition is strongly suggested by different ethological considerations, such as the existence of hierarchies in family groups (Tyler 1972) or those of bachelor males (Feist and McCullough 1976), as well as among stallions of different harems

(Miller and Denniston 1979), the existence of interindividual pref-erential relations within groups (Feist and McCullough 1976), the selectivity of mares vis-à-vis their foals (Tyler 1972), and so on. Only in the last few years have several experimental papers appeared, deal-ing either with social recognition in horses based on vocal cues (Basile et al. 2009; Lemasson et al. 2009) or olfaction (Hothersall et al. 2010; Krueger and Flauger 2011), or more specifically with the intermo-dal character of interindividual recognition within species (Proops, McComb, and Reby 2009), which among other things brings into play the horse's sense of hearing.

Work on equine recognition of individual humans was also thin up until recently, aside from the study of Grzimek (1944), which was inconclusive. Writing in *Année Psychologique,* Paul Guillaume (1942) summarized the field thus: "It is very unlikely that many animals possess an innate schema corresponding to the human form. What is observed, thus, reflects knowledge acquired from a global percep-tion whose main details, a priori, may be whatever; it is consequently not surprising that, in the recognition of a familiar person, clothes, for example, may play a greater role than the person's face or voice. Grzimek has conducted experiments with horses that would only submit to being saddled or bridled by a person they were used to. In general, the experiments show that a change of clothes sufficed to prevent recognition; other studies show that, by wearing a familiar person's clothes or approaching a horse that has just been bridled by a familiar, a stranger may be taken for that person." Although newer studies, in contrast, have suggested that horses may be equally capa-ble of discriminating between humans, in particular using facial cues (Koba et al. 2004), as appears to be widespread among mammals (Leopold and Rhodes 2010), and moreover that they may be able to transfer this power of discrimination to photographic representations of these humans (Tanida et al. 2005), these studies have not yet been published. It is also only very recently that three experimental stud-ies have built on the evidence for recognition of individual humans by horses. The first of these investigations (Stone 2010) deals with discriminative learning between photographed pairs of human faces, and on transferring this learning to the persons photographed. The two other studies (Lampe and Andre 2012; Proops and McComb 2012) used an experimental paradigm of intermodal visual and auditory recognition (Spelke 1985).

In any event, both social and individual recognition assume an initial phase during which the "recognizing" individual learns the features (categorical or individually distinct) of the "recognized" individual.

Can such learning be more thoroughly validated? If yes, the practical implications would be undeniable. How long can learning be maintained? Although sheep have been shown to be able to recognize fifty different sheep faces after more than two years (Kendrick et al. 2001), field observations of horses are less conclusive. Claudia Feh (1999, 2005) reports stallions that, "separated" from their mares during a year, subsequently recognized them within clusters of family groups. However, because some of them lived in semiliberty in a circumscribed space of 350 hectares, frequent encounters were not inconceivable. The observations of Joël Berger (1986), in contrast, suggest that stallions cannot recognize young males from their family group beyond a separation of eighteen months. However, mothers could recognize their foal following a separation of fifteen to nineteen months (McDonald and Warren-Smith 2010).

As I stated in the introduction, the picture is one of contrasts, with some parts now well defined and others still in shadow. These will be our focus as we explore the landscape that constitutes what we know about the nature of the horse, its behavior, how it adapts to its environment (which determines its behavior), and in particular its "mental faculties."

Before going more deeply into the topic of animal cognition, and more precisely equine cognition, let us skip briefly back in time to examine bygone notions of "equine intelligence."

2

EQUINE INTELLIGENCE

Are Horses Smart? One Question, Several Answers

The first question that arises in considering the mental capacities of horses is, "Are they intelligent?" This question has provoked a variety of responses, some contradictory. By way of evidence, let us look to the literature, sampling authors of different backgrounds from the nineteenth and twentieth centuries, including renowned horsemen.

Ernest Menault (1869) provides a good starting point. He is an intellectual descendant, so to speak, of French naturalist Georges-Louis Leclerc Comte de Buffon (Buffon, 1781).[1] Menault wrote a book titled *The Intelligence of Animals* that was initially published in 1868 in the *Bibliothèque des Merveilles* (Library of Wonders) to great acclaim. It had no fewer than twelve French editions and four English ones. In fact, Menault's appreciative and enthusiastic commentary sounds like Buffon: "The noblest conquest that man has ever made is, without any doubt, that of the horse. Everything in

---

1. "The noblest conquest ever made by man over the brute creation, is the reduction of this spirited and haughty animal . . . , which shares with him the fatigues of war, and the glory of victory. Equally intrepid as his master, the horse sees the danger, and encounters death with bravery; inspired at the clash of arms, he loves it, and pursues the enemy with the same ardour and resolution" (93–94).

this animal breathes out vivacity and energy. That need of continual movement, that impatience during repose, that nervous movement of the lips, that stamping of the feet, all indicate a pressing need of activity. The fullness of the skull, and the expansion of his forehead, show intelligence. The usual marks of the intelligent horse—one easily understanding his master's orders—are a well-developed head, eyes full and deep, jaws short, broad forehead, ears erect and diverging one from the other, and both eyes and ears very sensitive . . . Not only is his brain developed and provided with circumvolutions, but he also possesses exquisite senses" (Menault, 272). Although I previously noted (Leblanc 1984) that this argument in favor of equine intelligence was "articulated with more poetry than rigor," it is nonetheless true that it is in keeping with certain scientific notions of the times. After all, it was in the decade spanning 1810–1819 that Franz Joseph Gall and his student, Johann Caspar Spurzheim, published their multivolume work that would influence generations to come: *The Physiognomical System of Drs. Gall and Spurzheim; Founded on an Anatomical and Physiological Examination of the Nervous System in General, and of the Brain in Particular; and Indicating the Dispositions and Manifestations of the Mind.*

A much more nuanced opinion than Menault's was that of his contemporary, the psychologist George John Romanes (1884), a disciple and friend of Darwin who invented comparative psychology and who we will come back to later: "The horse is not so intelligent an animal as any of the larger Carnivora, while among herbivorous quadrupeds his sagacity is greatly exceeded by that of the elephant, and in a lesser degree by that of his congener the ass. On the other hand, his intelligence is a grade or two above that of perhaps any ruminant or other herbivorous quadruped . . . The emotional life of this animal is remarkable, in that it appears to admit of undergoing a sudden transformation in the hands of the 'horse-tamer'" (328). Referring to the "celebrated results" of a then well known tamer, Romanes wrote: "The untamed and apparently untamable animal has its foreleg or legs strapped up, is cast on its side and allowed to struggle for a while. It is then subjected to various manipulations, which, without necessarily causing pain, make the animal feel its helplessness and the mastery of the operator. The extraordinary fact is that, after having once felt this, the spirit or emotional life of the animal undergoes a complete and sudden change, so that from having been 'wild' it becomes 'tame'" (328–29). Continuing in this vein, he notes: "Another curious emotional feature in the horse is the liability of all the other mental faculties of the animal to become abandoned to that of terror. For I think I am right in saying that the horse is the only animal which, under the influence of fear, loses the possession of every other sense in one mad and mastering desire to run. With its entire mental life thus overwhelmed by the flood of a single emotion, the horse not only loses, as other animals lose, 'presence of mind,' or a due balance among the distinctively intellectual faculties, but even the avenues of special sense become stopped, so that the wholly demented animal may run headlong and at terrific speed against a stone wall" (329). Note Romanes's emphasis on the importance of emotional factors in the exercise of mental faculties. More conventionally, most of his remaining commentary is given over to the "remarkably good" memory of the horse and to relating several anecdotes regarding the animal's "intelligence" and "sagacity."

Gustave Le Bon was a prolific, eclectic sociologist of the late nineteenth and early twentieth centuries. His most famous contribution was *The Crowd: A Study of the Popular Mind.* But he also published works on archeology, anthropology, and physics, as well as a comprehensive study of equitation titled *L'equitation actuelle et ses principes. Recherches experimentales.* The third volume, fully revised in 1895, was remarkable for its inclusion, aside from fifty-seven figures, of an atlas of 178 "cinematographic photographs" intended to show the different speeds and positions of a single horse in response to changes in balance imposed on the animal by the dressage.[2] This work also contained a chapter on the "mental constitution" of the horse, much

---

2. As a reference point, in 1878 Eadweard Muybridge, an English-born American photographer, deconstructed the gallop by means of twenty-four cameras placed along a track, thus proving the existence of "suspension time" and establishing himself as a motion picture pioneer. The Lumière brothers' first public cinema showing did not take place until 1895.

admired by Lieutenant Colonel Henri Blacque-Belaire, Écuyer en chef at the École de Cavalerie in Saumur, who wrote the preface to the fourth edition (Le Bon 1923): "Your chapters on the mental constitution of the horse and 'Bases psychologiques du dressage' [psychological foundations of dressage] are masterpieces, and those without access to them, and the advantage of the rules you prescribe, cannot aspire to equitation" (ix–x). Le Bon commented, in particular: "The essential feature of horse psychology is memory. Though not very intelligent, the horse appears to have a representational memory far superior to that of man" (106). Moreover, "the aptitudes and personality of the horse differ markedly from one animal to another. Their degree of intelligence is similarly variable. One finds in them examples of very high intelligence, especially among thoroughbreds, as well as very limited ones" (55).

Ironically, the theme of intraspecies, interindividual variability, which went to the heart of Darwinian theory, failed to excite interest among specialists for many years thereafter. In fact, it reemerged only in recent decades, having been more or less outside the scope of behaviorists, whose major concern was general theory of learning, whatever the species. Nor did variability prove much fodder for proponents of the objectivist school of ethology, whose research bore essentially on species-specific behavior. Studies that reflect the importance of variability in modern investigations include that of Sappington and Goldman (1994), which we will return to in the next chapter and which shows the diverse cognitive capacities of horses subjected to fairly complex discriminative learning tasks, as well as a number of studies focusing on the variability, degree, determinants, and consequences of temperament (or personality) in horses (Mills 1998; Morris, Gale, and Duffy 2002; Hausberger et al. 2004; Hausberger and Richard-Iris 2005; Lansade, Pichard, and Leconte 2008; Lansade and Simon 2010).

Anecdotal accounts include those of a cavalry officer, Lieutenant Colonel Eblé (1981), who claimed to "have been around horses since childhood, trained dozens, and seen hundreds trained." Eblé saw in their variability the evidence of equine intelligence, which he expressed in a typically incongruous fashion: "People like to say that horses are not intelligent. That does not mean that horses have no intelligence at all, but rather that we attribute relatively little to them compared with, say, monkeys or dogs. We would say exactly the same thing about a student who routinely fails his exams. But horses do possess intelligence, and the proof is that some of them are stupid" (48).

Several decades after Le Bon, a self-styled "physiologist and hippologist" named Jean de Goldfiem (1974) espoused similarly striking opinions, though for a different reason. In a volume of his treatise on hippology dedicated to horse psychology, Goldfiem stated: "My experiments with electromyogram recordings (electroencephalograms for the cerebral cortex, electromyograms for emotion) show that equids possess intelligence comparable to that of a normal child of three years" (28). He reiterated the claim further on: "As far as I am concerned, my observations and experiments with electrophysiological recordings of the brain and other internal organs enabled me to show the academies, learned societies, and conferences that the normal adult horse possesses an intelligence equal to that of a three-year-old human infant" (81)! The reader will be hard put to discover what the author means by intelligence of a three-year-old infant, but, of course, "comparison" with humans is not uncommon. For example, Paul Vialar (who does not pretend to be a scientist) wrote both with greater restraint and more ambitiously in *Cheval, mon bel ami* (1982): "Intelligence: Its level is comparable to that of many humans who are intelligent but no more" (67). This statement harkens back to a formerly widespread illusion of a universal level of intelligence, uniformly applicable to all the species of creation! I will return to this point.

Last, but not least, what do recognized *écuyers,* masters in the theory and practice of horsemanship, make of equine intelligence, based on their own experience? James Fillis and François Baucher provide two examples that show how radically opposed opinions on the subject can be.

According to Fillis (1902), the intelligence of the horse is quite limited: "The great difficulty in breaking is to make the horse understand what we want him to do, which is no easy matter, because a horse, contrary to what many think, has only a small supply of intelligence. His only well-developed mental quality is his memory, which is particularly acute, and should therefore be specially utilized . . . A horse is incapable of affection for man: he possesses only habits, which he often acquires far too easily, and frequently sticks to them with too much persistence, a fact we should always bear in mind" (5).

Baucher, on the other hand, finds horses to be perfectly intelligent, a point he makes repeatedly. Thus, in the *Dictionnaire raisonné d'équitation* (1843), he wrote: "I have always believed that horses are intelligent, and it is on this belief that I based my method and all the principles expressed in this work. Thanks to this method, in mastering the will of the horse, I have succeeded in demanding of him only that which had previously appealed to his intelligence . . . How, indeed, can one label as merely instinctive a creature that can distinguish good and evil, appreciate circumstances, and even judge the ability of the rider? . . . The horse perceives, just as he senses, compares, and remembers; ergo, he possesses judgment and memory; ergo, he is intelligent" (150–51). The same sorts of considerations motivated him to introduce his *Méthode d'équitation* (1850): "And if, now, we grant the horse the portion of intelligence to which he is entitled; if we recognize that this animal is capable of appreciation, of discernment; that he senses, remembers, and compares, we must necessarily conclude that he is subject to all the rules common to feeling, intelligent beings" (2).

Étienne Saurel (1964), for his part, was moved to remark: "It is rather paradoxical to claim that, in contrast to many *écuyers*—Fillis, for example—Baucher believed in equine intelligence—'One has only to have observed many horses, to have made a careful study of their nature, to discover that they are intelligent'—and, yet, all his equitation is aimed at transforming their 'instinctive' capacity into that 'transmitted' to turn them into a 'docile instrument, at the beck of any impulse of the will'!" (53). "In any event," Saurel concluded, "it is clear that most authors attribute little intelligence to the horse. They do, however, find its memory creditworthy" (55).

Before exploring exactly what intelligence means, I would be remiss in closing this brief review of hypotheses about equine intelligence without adding a few words on the investigations and controversies that animated the beginning of the last century with regard to "calculating," "clever," and "thinking" horses.

A Rash of Clever Horses

As Paul Heuzé[3] once wrote (1928), "Clowns and clever animals have always been with us" (26). Nonetheless, the most frequently cited case of a "clever horse" is surely that of Clever Hans,[4] which I will describe briefly and which I will come back to regarding perception of movement.

Hans was special. The talents of previous clever animals were generally acknowledged by their masters themselves as being the product of particular training and not evidence of an unusual mental capacity. For example, in October 1904, mention of the famous Clever Hans in a previous month's issue of the scientific journal *Nature* prompted Joseph Meehan to write a letter to the editor recounting the story of a horse named Mahomet that he had seen at an exhibition sideshow at the Royal Aquarium in London twelve or thirteen years earlier. The horse had performed additions and subtractions, told time, guessed people's ages, and so forth. The animal belonged to an American who made his living training horses, mostly for the circus. Having found in Mahomet an especially gifted subject, "a genius among his kind," the trainer had committed endless time to the horse, going so far as

---

3. Paul Heuzé, a journalist for the daily *Opinion,* published several critiques of paranormal phenomena, including *La plaisanterie des animaux calculateurs* (1928). It was at his initiative that Henri Piéron organized a famous series of experiments at his Laboratory of Physiological Psychology at the Sorbonne. The first experiment, with the participation of Louis Lapicque, Georges Dumas, and Henri Laugier (Lapicque et al. 1922), discredited the "ectoplasmic medium" Eva Carrière. The second, to which Paul Langevin, Étienne Rabaud, Henri Laugier, Andre Marcelin, and Ignace Meyerson contributed, unmasked medium Jean Guzik.

4. Originally Kluge Hans, the horse was best known by the English version of his name.

to sleep in the animal's stall at his head. Questioned by Meehan, the trainer replied that he had quite simply taught the horse to paw the ground when the trainer looked straight at him and to stop when he gazed at the floor. The trainer and Meehan also observed another act at the exhibition, a collie that responded to a different signal: a twitch of the gloves that the dog's master carried in his hand. Although Meehan quite naturally concluded that Hans, too, was responding to training, there is more to the story. Judge for yourself.

Had you been reading the *Militärwochenblatt* of June 28, 1902, you might have spied the following advertisement: "'I would like to sell my beautiful, gentle horse, aged 7 years, with whom I have conducted experiments demonstrating the mental capacities of horses. He can distinguish 10 colors, reads, knows the four basic arithmetical operations, and many other things.' The horse was named Hans . . . His master, Wilhelm von Osten . . . did not wish to sell the animal. He only wanted to draw attention to his horse so as to show that the animals are capable of reasoning" (van Rillaer 1980, 157).

In July 1904, an article appeared in another German magazine, *Weltspiegel,* signed by General Major E. Zobel and titled: "The horse that reads and calculates." The article profiled Hans and his owner, von Osten, a retired mathematics professor devoted to "educating" his horse, a Russian stallion he had acquired in 1900. By pawing the ground, Hans could spell words, compute (addition, fraction-to-decimal conversion, square roots), recognize musical notes, and indicate a person present based on his photograph. The renowned psychologist Édouard Claparède[5] (1912) subsequently described the horse thus: "Not only did Hans know how to calculate; he could read; he was musical, distinguishing harmonic chords from dissonant ones. He also had an extraordinary memory: He could give the date of every day of the current week. In short, he could do

---

5. Édouard Claparède was a leading psychologist of the early twentieth century. Having founded the *Archives de Psychologie* with his cousin Théodore Flournoy in 1901, from 1904 he directed the Laboratory of Psychology at the University of Geneva's Faculty of Sciences. In 1912, he founded the Jean-Jacques Rousseau Institute of Educational Sciences, where he was joined in 1921 by Jean Piaget.

all the operations that a good student of fourteen years is capable of" (264).

Hans became famous and found himself at the center of a passionate controversy among scientists who came from all over Germany and Europe to watch him work. The scientists were divided among those who felt the need to question the received wisdom about equine intelligence and those who believed Hans was a fraud. The German board of education in Berlin created a commission of experts that included a circus director, a veterinarian, a zoologist, Oskar Heinroth[6] (who would become a major figure in ethology), and also an eminent psychologist, Carl Stumpf,[7] director of the Institute of Psychology at the University of Berlin. In September 1904, the commission announced it was convinced that the case was not one of simple fraud and that there were grounds for investigating the question more in depth. Stumpf gave the task to Oskar Pfungst, one of his doctoral students in psychology.

After having determined that anybody could examine Hans successfully and thus having eliminated the question of fraud connected to von Osten, "Pfungst noted that Hans most frequently erred in his responses (1) when he could not see the questioner, (2) when the distance between the latter and the horse increased, and (3) when the questioner himself did not know the answer to the question posed . . . Hans relied on visual cues to answer questions . . . Crucially, Pfungst also showed that the questioners were not conscious of the cues they were transmitting to Hans . . . Pfungst's interpretations were ultimately confirmed experimentally, in the laboratory, where Pfungst himself played the role of Hans (tapping with his fist on a

6. In 1911, Oskar Heinroth, whom Konrad Lorenz considered a mentor, published an article on the ethology and psychology of anatides (the duck family) that revealed the existence of behavioral homologies within related species. It is him we have to thank for the term *imprinting* (*Prägung* in German), which was subsequently studied by Lorenz. The phenomenon of imprinting itself was observed as early as 1873 by Douglas Spalding, an early forerunner of ethology (Renck and Servais 2002; Reznikova 2007).

7. Carl Stumpf was a universally recognized personality. He was professor of philosophy and (later) psychology at Würzburg, Prague, Halle, Munich, and, finally, Berlin from 1894 onward. Under his direction, the Berlin laboratory became one of the leading centers of German psychology.

table) and in this way managed to guess, better than chance would have predicted, a number in the minds of experimental subjects, whose only instruction was to think of a number between 0 and 100. Just like von Osten, the subjects involuntarily produced various movements that Pfungst trained himself to detect" (Cleeremans 2000, 3–4). The cues to which Hans was responding were nearly imperceptible. They consisted of minuscule movements of the head or torso. Among the observations made by Pfungst (2000), one subsequently had important methodological consequences and explains why Hans has persisted in our collective memory: the fact that he frequently made mistakes when the questioner himself did not know the answer to the question. This last condition, which characterizes a so-called double-blind situation, eventually formed the basis of a classical process in experimental psychology (and also significantly in medicine, in drug trials) to prevent the experimenter from biasing the subject. An illustration of the reverberations of this affair was a colloquium organized by the New York Academy of Sciences in 1980 (Sebeok and Rosenthal 1981), devoted to various methodological and epistemological aspects of the "Clever Hans phenomenon." Heini Hediger (1981) points out another key aspect, which recalls statements made by Mahomet's trainer, mentioned earlier: "It is only on the basis of extraordinary familiarity between Clever Hans and his master, gained during the course of teaching, that the horse became able to interpret as decisive signs movements of the head of his master of even one-fifth of a millimeter deflection" (4).

Following publication of Pfungst's work, the weight of public opinion began to turn against von Osten, and he became increasingly bitter. He ceased his presentations and died shortly after in 1909, at the age of seventy-one.

Yet history does not stop there, for in 1908 von Osten passed Hans to a wealthy Elberfeld[8] jewelry merchant, Karl Krall, who himself become personally interested in educating the horse (see figure 2.1).

---

8. The town of Elberferd, which in 1929 merged with Barmen to become Wuppertal, is located in the district of Düsseldorf (in the state of North Rhine-Westphalia).

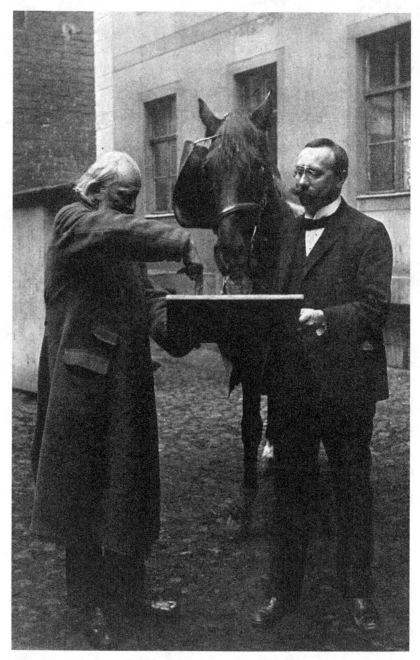

Figure 2.1 Wilhelm von Osten, Clever Hans, and Karl Krall. Source: Krall 1912.

Figure 2.2 Zarif at his reading lesson with Karl Krall. Source: Krall 1912.

Toward the end of that same year, Krall also acquired two Arabian stallions, Mohammed and Zarif,[9] with whom, according to Maurice Maeterlinck[10] (1925), he obtained results "incomparably swifter and more astounding than that of old Hans [see figure 2.2]. Within a fortnight of the first lesson, Mohammed did simple little addition and subtraction sums quite correctly. He had learnt to distinguish the tens from the units, striking the latter with his right foot and the former with his left. He knew the meaning of the symbols plus and minus. Four days later, he was beginning multiplication and division. In four months' time, he knew how to extract square and cubic roots; and, soon after, he learnt to spell and read by means of the conventional alphabet devised by Krall" (187). Although it is clear that, strictly speaking, Maeterlinck did no more than to report what someone else had told him, since he only met Krall several years later,

9. Krall subsequently acquired other horses, which he also exhibited, including Hänschen, a pony, and a blind stallion, Berto.
10. A Belgian writer of many talents, Maurice Maeterlinck received the Nobel Prize for Literature in 1911.

his account still reflects the ethos that permeated the demonstrations of the Elberfeld horses.

In 1912, Krall published a voluminous work on his research that provoked considerable uproar, including among scientists: "The publication, last March, of M. Karl Krall's *Denkende Tiere* is by far the most sensational event in the annals of animal psychology, and perhaps even in psychology in general," wrote Claparède (1912, 263) in his commentary on an article about the "clever horses of Elberfeld" submitted to the *Archives de Psychologie*. The article engendered a whirl of controversy, involving eminent figures (some of whom supported Krall) and challenging Pfungst's explanation of Hans's success. Among them were the great Swiss naturalist Paul Sarasin,[11] founder of the Swiss National Park, and Charles Richet, a respected French physiologist and member of the Academy of Medicine and the Academy of Sciences, who was awarded the Nobel Prize in Physiology or Medicine in 1913 for his discovery of anaphylaxis.

The acid test of the hypotheses—all more or less extraordinary—concerning the Elberfeld horses was the question of experimental control. Claparède (1913a) himself, however impressed he might have been at the time, having personally visited Elberfeld (see figure 2.3), acknowledged his own confusion in the case. During a meeting on the issue called by the French Philosophical Society on March 13, 1913, with such luminaries as psychologist Henri Piéron,

---

11. Sarasin made the following declaration during a meeting of the Society of Natural History of Basel (1915, 71): "Our eye has been almost dazzled by the results of the new science of animal pedagogy, and they must first get accustomed to the light that shines forth from the abundant observations for which we are indebted to the unflinching efforts of Karl Krall and Frau Paula Mökel, as well as to the important inferences for our conception of the world arising therefrom; but when we eventually succeed in doing so, however great the astonishment we feel at the Mökel results, however much we are taken aback or perplexed, we still are unable to admit that in one particular intellectual activity, that is, in working out difficult arithmetical problems in their heads, the animals mentioned, especially the horse, could be superior to us, that is, to the average man. Yet it is unquestionably established that Herr Krall's stallion 'Muhamed' in particular gives the right answers out of its head to sums involving square roots, which few of us could solve in a short time, while the right answer to the problem is always forthcoming with disconcerting speed" (Hediger 1955, 152). Sarasin was also the force behind the formation in Bern, in 1913, of an international commission for the worldwide protection of nature that comprised twenty member states and presaged the creation of the World Conservation Union in 1948.

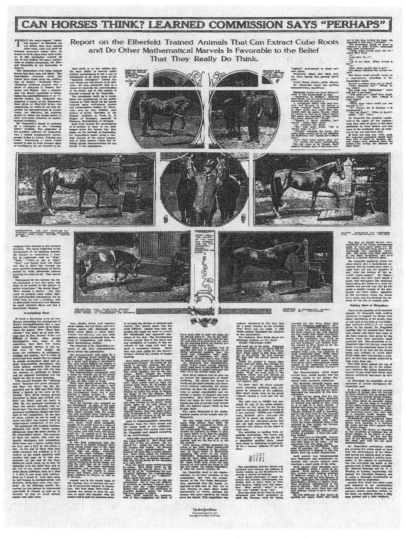

Figure 2.3 In its August 13, 1913, issue, the *New York Times* reported on Clever Hans, the Elberfeld horse, and Édouard Claparède's visit to Karl Krall.

philosopher André Lalande, mathematician Jacques Hadamard, and physiologist René Quinton in attendance, Claparède declared: "It is, however, to be regretted that the demonstrations were for the most part performed in conditions that did not exclude the possibility of Pfungst's distinctive cues . . . You will ask me why I did not do the crucial experiment, which would immediately have settled the question . . . If I have not done this experiment, it is partly owing to time, partly for extra-scientific, social reasons . . . To insist would have been to treat M. Krall as a liar" (118–19). Although Claparède's rationale is not particularly convincing, Maeterlinck's (1914) remarks on Krall's personality provide support: "For the rest, Herr Krall, though his faith is active, zealous and infectious, has nothing in common with the visionaries or illuminati. He is a man of about fifty, vigorous, alert and enthusiastic, but at the same time well-balanced; accessible to every idea and even to every dream, yet practical and methodical, with a ballast of the most invincible common-sense. He inspires from the outset that fine confidence, frank and unrestrained, which instantly disperses the instinctive doubt, the strange uneasiness and the veiled suspicion that generally separate two people who meet for the first time; and one welcomes in him, from the very depths of one's being, the honest man, the staunch friend whom one can trust and whom one is sorry not to have known earlier in life" (197–98).

In any event, during the same meeting, Claparède lamented: "Alas, I have no firm opinion on the nature of the abilities of the Elberfeld horses" (1913a, 117), which was reason enough to return to Elberfeld on March 26, only to find himself "back where he started," as he wrote later (1913b, 244). Thus, having recapitulated all the plausible explanations, all he could do was to wait and see: "My conclusion is that the existence of original thought in the Elberfeld horses is far from having been proved in any conclusive way, and that certain facts do not fit this hypothesis very well. On the other hand, to date no other explanatory system has managed to satisfactorily account for the totality of the events observed" (284).

The necessary control conditions were never satisfied. The next year brought the outbreak of the First World War, and the horses of Elberfeld perished in the ensuing tumult.

Over time, however, Claparède's skepticism grew. Paul Heuzé (1928) interviewed him twenty years later and recounted that, following a long conversation, Claparède reflected: "In my first report on Krall's horses, I was fairly positive, especially, I admit, in reaction to certain categorically negative views. My second report showed why a conclusion was overly hasty and imprudent. Today, I am more inclined to believe that the phenomena were purely and simply the result of training, whose method we never did manage to discern" (61–62).

Pierre Janet[12] (1936), whose *Intelligence avant le langage* tackles the topic of numbers and animals, provides his own amusing commentary: "But I found a different study extremely interesting; it was sent to me by its author, M. von Maday,[13] who published it in the Berlin *Journal de la cavalerie moderne* . . . He is surprised that Mr. Krall's horses only exhibit their intelligence in extracting cube roots, when in other aspects of their lives, they are so stupid. They do not know how to have fun or to show pride when they have solved a difficult problem. It is Mr. Krall who triumphs, whereas the horse is totally indifferent . . . If he really possessed a means of making horses intelligent enough to solve equations, he could exploit their intelligence in other ways and could begin by simplifying all the methods of dressage that everybody finds so difficult" (139–40).

It is also worth noting that, however incredible it seems in retrospect that such controversies could attract the interest of renowned scientists, the events are easier to appreciate in light of their historical context. Scientific psychology was still nascent, and physics itself was struggling with uncertainty and major paradigm shifts. Consider the pseudo N-ray affair, which nonetheless won physicist René Blondlot the Academy of Sciences prize in 1904. Or the discovery of radium by Pierre and Marie Curie in 1898, which completely upended conventional ideas about matter. Or the theory of general relativity developed by Albert Einstein between 1907 and 1915! No phenomenon,

---

12. Pierre Janet was one of the great figures of clinical psychology at the turn of the nineteenth and twentieth centuries. He founded the Psychological Society, which eventually became the French Psychological Society; he held the chair in experimental and comparative psychology at the Collège de France; and he was a member of the Institut de France and president of the Medical-Psychological Society.

13. Stefan von Maday was both a physiologist and a cavalry officer (Claparède 1913b). The reference to what appears to be this study, cited by Henri Piéron (1913), is S. de Maday (1913) *Gibt es "denkende" Pferde? Kavalleristische Monatshefte.*

however bizarre, could be ignored and consequently had to be investigated, whence the importance of both taking the extraordinary seriously and systematically testing it. Hoaxes and miracles mixed with science on the move, particularly in domains that involved the workings of the mind (Le Maléfan 2004; Marmin 2001). As Claparède recalled (1913b): "Each new observation to some extent contradicts established science up to that point . . . Certainly, the more an observation contravenes current theory, the more justified one is demanding that it be proved. The weight of the proof, M. Lournoy has correctly said, must be proportional to the strangeness of the observations" (268–69). That same year, in just such a spirit, Henri Piéron (1913) devoted an article to the "problem of thinking animals." Arguing that *L'Année Psychologique* cannot ignore the problem put to the public of the existence of complex abstract thinking in animals, especially with regard to the horses of Elberfeld," he concluded: "Unfortunately, to date Mr. Krall has opposed control tests on the grounds of inadmissibility, as he has opposed various similar efforts . . . The best thing, perhaps, would be to discourage his nonproductive discussions, and in the meanwhile to work on posing better questions. When control tests become possible, then we can talk about it" (225).

To be sure, such talents in horses have rarely been seen since. A few have made headlines, in particular in the United States, but only anecdotally, with the possible exception of a three-year-old mare named Lady Wonder.

The horse's owner, Claudia Fonda, had acquired her while she was still suckling, had nursed her with a bottle, and trained her from early on "to spell out small words and to make simple computations, having the colt touch the proper blocks with her nose" (Rhine and Rhine 1929a, 452). According to her owner, Lady Wonder "could make predictions, solve simple arithmetical problems, answer questions aptly and intelligently, and do all this without verbal command. All that was needed was that the question be written down and shown to Mrs. Fonda" (ibid.). During the winter of 1927–1928, Lady Wonder was investigated by the founder of parapsychology, Joseph Banks

Rhine, and his wife Louisa, both of whom at the time were young graduates in botany who had joined psychologist William McDougall at Duke University. The Rhines devised a series of precautions to avoid conscious or unconscious visual or auditory communication between Fonda and her mare. Although they did not employ the full complement of desirable precautions in any single experiment, they still concluded that what they had observed constituted telepathy. In response to criticism, they repeated the experiments a year later, but failed to produce the same results. Their revised conclusion was that Lady Wonder had lost her telepathic gifts, which made the initial results only more convincing (Rhine and Rhine 1929b). When, in 1956, the magician Milbourne Christopher (1970) went to see Lady Wonder, who was still active and at the time used a sort of large type-writer keyboard to produce her answers, he showed that she did react to subtle signals from her owner.

Today, it is tempting to agree with Thomas Sebeok (1981), who commented with reference to Henry Blake's *Thinking with Horses* (1977)[14]: "In the late 1970s, one is still distressed to read assertions, by a self-proclaimed 'centaur,' that horses communicate by means 'more subtle than man himself: extrasensory perception and telepathy'" (199).

Which brings us back to original question of equine intelligence. Would it not be useful to ask what is meant by "intelligence" in this context, and to consider the status of scientific investigation into the mental capacities of animals?

---

14. Extrasensory perception appears to be a recurring theme with Blake, as in *Talking with Horses* ("I talk to horses . . . They tell me") he devotes an entire chapter to telepathic communication between horses and people. Fortunately, other observers were more attentive, for example, Count Eugenio Martinengo Cesaresco (1906), who wrote at the beginning of the last century: "The intelligent horse is intent on observing the slightest movements of the rider he is bearing, and understands his intentions in this way. The rider, before guiding the horse to perform any particular action, gives him certain preparatory aids and likewise by force of habit and without noticing them, makes special movements with his body" (2).

ANIMAL INTELLIGENCE, COGNITION, AND REPRESENTATION

Intelligence and Cognition

Intelligence: The Eternal Question

Although it ranks high in the popular and professional discourse, and although we attach a significant amount of importance to it (even a particular kind of importance), the concept of intelligence is quite hard to define. And yet define it we must. I am mindful of the quip attributed to Alfred Binet, author (with Theodore Simon) of the first practical intelligence test at the beginning of the last century: "Intelligence is what my test measures."

A quick look at *The Merriam-Webster Dictionary* turns up the following: "The ability to learn or understand or to deal with new or trying situations . . . the skilled use of reason . . . the ability to apply knowledge to manipulate one's environment or to think abstractly as measured by objective criteria (as tests)." Here it is clear that the range of meaning covered by the word *intelligence* is fairly extensive, which may help in understanding, at least in part, the varying assessments vis-à-vis equine intelligence.

In *Vocabulaire technique et critique de la philosophie* (A technical and critical vocabulary of philosophy), André Lalande (2006) attempts to accommodate this variability, noting that the meaning of the term itself implies two adjectives equally, *intellectual* and *intelligent*. The former corresponds especially to the following definition: "The ensemble of functions whose objective is knowledge, in the broad sense of the word (feeling, association, memory, imagination, understanding, reasoning, consciousness). This term currently serves to designate one of the three major classes (or facets) of psychic phenomena, the two others being affective phenomena, and active or motor phenomena." In contrast, *intelligent* corresponds to definitions such as: "(Opposite of instinctive) voluntary action, thoughtful adaptation of means to an end . . . (Opposite of unintelligent) development of an average or above-average mind . . . (Opposite of fiction) the ability to easily understand the significance either of the facts or the ideas of others" (525–26). Lalande's observations echo, after a fashion, the difficulty scientists have in reaching a consensus on the definition of human intelligence.

In a report published some time ago by an ad hoc working group of the American Psychological Association (Neisser et al. 1996) on the subject of what is known and not known about intelligence, the authors remarked by way of introduction: "When two dozen prominent theorists were recently asked to define intelligence, they gave two dozen somewhat different definitions (Sternberg and Detterman 1986)." However, the authors were quick to explain, "[s]uch disagreements are not cause for dismay. Scientific research rarely begins with fully agreed definitions, though it may eventually lead to them" (77). They conclude: "Because there are many ways to be intelligent, there are also many conceptualizations of intelligence" (95).

Lingering for a moment in the field of human intelligence, though still mindful of the question "Are horses intelligent?," let us consider briefly what it means to be "pretty intelligent." Everyone knows that a given person may be very adept at conceptual speculation or dealing with logical abstractions, for example, yet be socially maladroit. Another person may have no abstract abilities whatsoever yet be extremely good at complex human relations or be a skillful negotiator. Which one is the more "intelligent"?

This question goes to the heart of a debate on intelligence that has endured for more than a century. In his earliest work on measuring intelligence, at the very beginning of the twentieth century, Alfred Binet—the "first to apply the notion of measurement to higher mental processes" (Zazzo 1993)—and his collaborator Théodore Simon used a series of fairly heterogeneous exercises involving memory, attention, imagination, and critical thinking to try and discern, empirically, the level of mental development of children. Far from reducing intelligence tests to a simple intellectual quotient, the IQ (which was only introduced in 1912, a year after Binet's death), these authors wrote (Binet and Simon 1905) about what they called their metric scale of intelligence: "Strictly speaking, this scale does not really measure intelligence, for intellectual attributes cannot be measured like dimensions; they are not superimposable" (194–95). On the use of intelligence tests in general, Binet (1911) underscored the need to take into account diverse aptitudes: "One might go so far as to say: 'Who cares about the tests, so long as there are lots of them?'" (201).

The use of factorial analysis by Cyril Burt and Charles Spearman sparked a major controversy regarding the possible existence of a general intelligence factor, called factor g, an idea that Louis Leon Thurstone (1940), among others, vigorously opposed: "Such a factor can always be found routinely for any set of positively correlated tests, and it means nothing more or less than the average of all the abilities called for by the battery as a whole. Consequently, it varies from one battery to another and has no fundamental psychological significance beyond the arbitrary collection of tests that anyone happens to put together for a factor analysis . . . As psychologists we cannot be interested in a general factor which is only the average of any random collection of tests" (208). More recently, for Howard Gardner (1985), for example, who introduced the theory of multiple intelligences, general intelligence does not exist. What does exist, in humans, are different specific forms of intelligence, each one corresponding to a distinct aptitude (linguistic, logico-mathematical, musical, spatial, bodily, interpersonal, intrapersonal, even relating to the natural environment, or existential). As for animals, Gardner has this to say: "Each animal species has a unique cognitive profile. Rats, for example, possess good spatial intelligence, and birds good musical intelligence. But man is the only species that has all the eight (or nine) forms of intelligence that I identified" (cited in Delacampagne 2000, 111). In any event, in each case of the possible existence of a general intelligence factor that to some extent takes into account intellectual performance, on the one hand, the correlations used to infer it are relatively weak, and on the other hand, the involvement of specific factors is not in doubt.

I do not intend to spend too much time on these debates. As Jacques Lautrey (2004) maintains, they are far from being resolved. Moreover, they basically concern human intelligence, which is beyond the scope of this book. Rather, I will leave it to Lautrey to recap the state of research on the subject: "The concept of intelligence is a fluid, very general one, whose definition may comprise many attributes. According to various authors and theories, definitions of intelligence emphasize this or that attribute, for example, the ability to adapt to new situations, to learn, to form abstract ideas, to maintain control, to solve problems, and so on. Despite these differences, it is still possible to extract the crux of the concept: Intelligence is generally defined as the capacity of an organism—or of an artificial system—to modify itself to adapt its behavior to the constraints of its environment . . . But this cognitive ability to adapt does not qualify as intelligence unless it is generalizable to a fairly high degree, that is, if it appears in different situations . . . The debates on extending the concept of intelligence and distinguishing between different forms of intelligence are age-old and no nearer to solution. Yet, the evolution of ideas regarding intelligence trends strongly to extension of the concept and to the proliferation of distinct forms of intelligence" (122–23).

If it is difficult to evoke degrees of intelligence in humans insofar as intelligence is not a "monolithic" entity (one person may thus be more intelligent than another in certain respects and less in others), it is even harder to establish a scale of intelligence among species, which may have developed cognitive competences of a completely different nature. This was the point Gardner made above concerning rats and birds. It is also the conclusion of a study by Brian Hare and Michael Tomasello (2005) showing the ability of dogs to spontaneously decipher human communication cues (which human babies also do from the beginning of their second year), such as pointing to an object, even while walking away from it, or to realize that a human may or may not be able to see. In contrast, chimpanzees, which exhibit remarkable talent in solving problems, especially involving tools, and which are considered particularly "intelligent" (as we are fond of saying, don't they share some 99 percent of their genetic heritage with us?) match the dogs' performance only in the best of cases, and then only with difficulty and after training.

From Intelligence to Cognition

As stated above, the concept of intelligence has a different meaning for scientists (Neisser et al. 1996, 95). In addition, the many ways in which the concept is used in common language, including the expressions cited earlier regarding equine intelligence, only hint at the complexity of the concept and further increase the uncertainty of the meaning attributed to it. Moreover, the term *cognition* is preferred because it avoids some of the confusion in the context of the emergence and development of the cognitive sciences.

Citing David McFarland (1989), for whom intelligence refers to the "assessment of performance judged by some functional criteria" (Vauclair 1996, 10), Jacques Vauclair (1995 1996) proposed to draw a distinction between intelligence and cognition (see table 3.1).

As soon as a species capable of adaptation and survival is considered intelligent, one can speak of adaptive intelligence based primarily on the implementation of preprogrammed routines in the central nervous system of the particular species in response to particular situations. Such is the case, for example, of the navigation system of the pigeon, which enables it to return to its nest, or the piping plover. "When a plover 'pretends' to have a broken wing to discourage a predator from its nest," writes Joëlle Proust (2000), "[i]t does what is required of the situation thanks to an innate motor mechanism triggered by the representation of the situation" (36).

Table 3.1 Distinguishing between the concepts of intelligence and cognition.

INTELLIGENCE

Area: Specific adaptations to specific problems.

Tools: Partially prewired processing systems→**evaluate performance** on the basis of a given functional criterion.

COGNITION

Area: Behavioral adjustment in response to changing environmental conditions.

Tools: Learning strategies and information processing→**implement means suited to reach a given goal.**

Distinct cognitive criteria: Flexibility, novelty, capacity for generalization.

Source: Vauclair 1996, 19.

"The concept of cognition is reserved for the manifestation of learning and information processing within a given individual. Appropriate criteria allowing the identification of cognition must be used to distinguish these processes from behavioral organizations, complex or not, that result from hard-wired rules in the nervous system. In general terms, cognition is seen to allow an individual to adapt to unpredictable changing conditions in its environment. Thus, behaviors that would aid adaptation would reflect several characteristics, such as *flexibility, novelty,* and *generalization.* Flexibility of a behavior designates the possibility of constructing an adapted response to unusual external conditions. The response must also be novel in the sense that it does not express the existence of a prewired program. Finally, the novel behavior, established to solve a novel problem, must be susceptible to generalization to situations that differ partially or totally from those in which they were initially acquired" (Vauclair 1996, 10–11).

Naturally, the semantic distinction between intelligence and cognition makes it possible to precisely separate anything related to the cognitive domain and facilitates a differentiated analysis of apparently similar behaviors. However, it does not follow that making intelligence nonsynonymous with cognition should limit the common use of the word. Vauclair himself obviously found the distinction no impediment because it figures in one of his works entitled *L'intelligence de l'animal* (1995), even though the topic was more specifically animal cognition. Another example is Zhanna Reznikova's (2007) *Animal Intelligence,* subtitled *From Individual to Social Cognition.*

In chapter 1, I introduced a number of topics relating to animal cognition and pointed out various lacunae in research on the cognitive ethology of the horse. A succinct characterization of the overall area covered by the study of cognition can be found in the work of Roy Lachman and colleagues (Lachman, Lachman, and Butterfield 1979) on cognitive psychology: "It is about how people take in information, how they recode and remember it, how they make decisions, how they transform their internal knowledge states, and how they translate these states into behavioral inputs" (99).

In other words, reapplying this description, the field of animal cognition deals with questions relating to perception, learning, categorization, memory, representation of space and use of resources, communication and social transmission of knowledge, social and individual recognition—all questions that are asked in a specific way for each animal species.

In short, wondering about the intelligence of the horse is meaningless if the goal is to determine whether the animal is more or less intelligent than a representative of another species (especially humans). Rather, the aim should be global comprehension of equine cognition and of all the horse's unique mental capacities (that is, specific to the species) as shaped by evolution.

Animal Behavior, Cognition, and Representation

Breaking with the traditional concept of the animal-machine bequeathed by René Descartes in the seventeenth century,[1] Charles

---

1. However, Edmond Perrier, who wrote the preface to the French translation of George Romanes's (1887) book on animal intelligence, noted: "Réaumur, Leibniz, and Voltaire believe that animals are intelligent. At the end of the eighteenth century, Erasmus Darwin endeavored to prove the fundamental sameness of psychic phenomena in all living beings" (xviii).

Darwin (1871) vividly formulated an evolutionary continuity between animals and humans that was not limited to physical characters but applied equally to mental faculties: "Nevertheless, the difference in mind between man and the higher animals, great as it is, certainly is one of degree and not of kind" (105).

This proposition was not indifferent to the quest for a better understanding of the human mind, whose origins could now conceivably be elucidated through study of animal mental function. Darwin's idea is also at the core of what would become comparative psychology, known additionally as animal psychology in French.

The Beginnings of Comparative Psychology

The origin of the discipline goes back to George Romanes (1884), a friend and close disciple of Darwin, whom I cited in the previous chapter as having initiated research on animal intelligence.

At the time, the approach generally taken by psychologists in studying mental capacity relied on introspection. To the extent that one can never have direct access to the mental states of animals, nor to ask them to take us into account, Romanes undertook to imagine, based on the human model, how horses felt and thought, by practicing what he called subjective inference. It consisted in attributing our own mental states to the animals based on observation and description of their behavior in an extensive collection of anecdotes. This somewhat anthropomorphic approach aroused serious objections on two levels.

On the epistemological level, Conwy Lloyd Morgan (1894) proposed a principle of parsimony, known as Morgan's canon, according to which "in no case may we interpret an action as the outcome of a higher psychical faculty, if it can be interpreted as the outcome of the exercise of one that stands lower in the psychological scale" (53). Morgan's canon fit naturally within the framework of Ockham's razor, put forward at the start of the fourteenth century by the philosopher William of Ockham: "Plurality should not be posited without necessity." In the second edition of his work (1903), Morgan responded to various attempts at a reductionist interpretation of his canon, in the following way: "To this, however, it should be added, lest the range of the principle be misunderstood, that the canon by no means excludes the interpretation of a particular activity in terms of the higher processes if we already have independent evidence of the occurrences of these higher processes in the animal under observation" (59).

On the methodological level, Edward Lee Thorndike (1898), who fully embraced Morgan's canon, also objected to an approach based on anecdote and promoted adoption of the experimental method, pledging his support to its development. It was Thorndike who described learning by trial and error, according to which, among an organism's responses to a situation, those that it finds most satisfying will be most closely linked to the situation and most readily invoked when the situation comes up again. This is a primitive form of the concept of reinforcement, later introduced by Burrhus Frederick (B. F.) Skinner, the great behavioral theorist.

For almost all of the first half of the twentieth century, comparative psychology was dominated by the behaviorist school, founded by James Broadus Watson, for whom "the study of animal behavior should be limited to relations between environmental stimulations and reactions of organisms to those stimulations" (Vauclair 1996, 4). Completely rejecting mentalism—"The time seems to have come when psychology must discard all reference to consciousness; when it need no longer delude itself into thinking that it is making mental states the object of observation" (Watson 1913, 164)—and limiting itself only to observable phenomena, behaviorism was founded on the principle that the animal brain was a blank slate at birth and that it is only learning, natural or induced, to which animals are exposed by life circumstances that shapes their behavior.

At core, pushing Darwin's idea of evolutionary continuity to its extreme, Watson's work relied on the idea that the same laws of learning could be applied in the same way to all organisms, from the simplest to the most complex, no matter the species, including humans: "The behaviorist, in his efforts to get a unitary scheme of animal response, recognizes no dividing line between man and brute. The behavior of man, with all of its refinement and complexity, forms only a part of the behaviorist's total scheme of investigation" (ibid., 158).

In this way, during the first half of the twentieth century, behavioral psychologists developed what is known as white rat psychology: Most of the studies on the acquisition and modification of behavior were basically related to situational learning in the laboratory and involving only a few species (including, in particular, rats and pigeons). At the time, it was thought that, with regard to learning, no real differences existed between species and consequently that a single species could efficiently suffice to demonstrate everything one might want to know about learning.

Recall that these learning experiments consisted of nonassociative learning (habituation, related to the complementary phenomenon of sensitization); associative learning, namely, classical or Pavlovian conditioning, practiced by the Russian physiologist and Nobel laureate Ivan Pavlov; and operant or instrumental conditioning, conceptualized by Skinner. In classical conditioning, through repeated association of an initially neutral stimulus with an unconditioned stimulus that naturally produces a marked response called unconditioned response, the neutral stimulus acquires the ability to trigger the response on its own, the unconditioned stimulus having served as a reinforcement for the acquisition. We say that the neutral stimulus has become a conditional stimulus and the response a conditioned response. Unlike classical conditioning, where the subject relies on the stimulus being produced locally, and over which he has no control, in operant conditioning, the subject's own behavior is the source of the appearance or suppression of an agreeable or disagreeable outcome. In other words, the subject's behavior in some way serves as a tool to obtain an effect. Depending on the nature of the effect, a distinction is made between reinforcement, which increases the probability of reappearance of a behavioral response and thus tends to reinforce it (depending on whether the outcome is a reward—food, for example—or suppression of discomfort—releasing the reins or leg pressure, for example, in the context of equitation—we speak of positive or negative reinforcement), or punishment, which translates as a decreased probability that the behavioral response will reappear, and consequently tends to cause it to disappear.

Moreover, a number of laws and properties govern the establishment, maintenance, and restoration of learning, notably with respect to the temporal contiguity of events to be connected, sensitivity to their correlation, and phenomena of generalization and discrimination. In practice, these laws and properties can also be combined.

Thus, for behavioral psychologists, all behavior results from learning that consists of acquiring a simple nervous system association between a stimulus and a response, during the course of individual development (ontogenesis).

Despite the early dominance of behaviorism, two trends appeared in the first decades of the 1900s that would prove a serious challenge to it: The first, biological trend basically contested the abandonment of interspecies comparisons and led to the development and recognition of ethology; the other, psychological trend, rejected the behaviorist reductionism that ignored mental activity, thus contributing to the birth of cognitive psychology. Both trends were instrumental in the emergence of cognitive ethology.

The Development of Ethology
Around the mid-1900s, ethologists' doubts concerning behaviorism reached a crescendo. In the wake most notably of the work of Oscar Heinroth, the research of Konrad Lorenz and Nikolaas Tinbergen on the behavior of animal species in their natural habitat simultaneously showed the existence of instinctive behavior specific to each species and the importance of evolutionary history (phylogenesis) for currently existing species. Ethology stressed the predispositional variety of diverse species as a function of the conditions in which they evolved and the adaptations—both behavioral and morphological—that enabled them to survive to the present day. Although controversy flared over these findings for a time, in fact they did not question the importance of the general laws of learning estab-

lished by the behaviorist school, which ultimately suited the ethologists very well. As Lorenz (1981) later wrote: "What we reproach behaviorists for is certainly not what they do; they do what they do in the most excellent manner. Our criticism refers only to the belief that there is nothing else to investigate" (70). Earlier in the same work, Lorenz elaborated his critique more explicitly: "At that time the corrective criticism made by the behaviorists concerning the opinions held by the purposive psychologists was salutary in every way. But, unobserved, a ruinous logical error crept into behaviorist thinking: because only learning processes could be examined experimentally and since all behavior must be examined experimentally, then, concluded those of the behavioral school, all behavior must be learned—which, naturally, is not only logically false but also, factually, complete nonsense" (2). Thus, the two approaches should be more complementary than they were: "Phylogenetic adaptation and individual adaptative modification of behavior are the only two possible ways extant for the acquisition and storage of information" (81), which obviously accords with the ethological "program" espoused by Tinbergen (1963): "Huxley likes to speak of 'the three major problems of Biology': that of causation, that of survival value, and that of evolution—to which I should like to add a fourth, that of ontogeny. There is, of course, overlap between the fields covered by theses questions, yet I believe with Huxley that is useful both to distinguish between them and to insist that a comprehensive, coherent science of Ethology has to give equal attention to each of them and to their integration" (411). Tinbergen's position is frequently summarized by evoking his four questions: What are the proximate (external or internal) causes of behavior? What is its function and its impact on survival? How does it arise over the course of development of the individual (ontogenesis)? How does it arise over the course of evolution of the species (phylogenesis)?

Although it was possible to demonstrate general laws of learning, the specific capacities and means of modifying or acquiring behavior were simultaneously dependent on and constrained by predispositions that varied according to the species.

Even hardline behaviorists such as Keller and Marian Breland (Breland and Breland 1961) were brought to the same conclusion. Their story is worth a digression given that they are the originators of "clicker training." This method of learning by operant conditioning is founded on a system of positive reinforcement accompanied by a secondary or conditional reinforcing agent: ultrasonic whistling in initial experiments with dolphins and later (and more famously) a clicker with dogs and horses. Popularized by Karen Pryor (2008), who coined the expression clicker training, this method went on to enjoy great notoriety, including in dressage for horses (McGreevy 2004). Pryor herself was responsible for training dolphins at Sea Life Park in Hawaii, where she worked on applying operant conditioning to encourage new behaviors (Pryor, Haag, and O'Reilly 1969).

Keller and Marian Breland were both students of Skinner at the University of Minnesota, Marian from 1938 and Keller beginning in 1940. They later worked with him in 1942 and 1943, during the Second World War, on what Skinner himself called a "crackpot" idea: Project Pigeon (Skinner 1960). The goal was to use pigeons as an "organic control" for missiles. The head of each missile was to be equipped with a camera that would send back images to tactile screens situated in compartments that contained pigeons whose purpose was to guide the missile to its target. As long as the pigeons pecked the center of the image, the missile stayed on course; when the pigeons erred, the trajectory veered off. Each missile was intended to carry three pigeons. The "decisions" made by the guidance system were based on a majority rule depending on the signals emitted by the pigeons. The pigeons were successfully trained to distinguish ship silhouettes in images and to peck at them. The project was ultimately never implemented by the U.S. Navy (for reasons, apparently, having to do both with the peculiarity of the project and technological problems related to missile guidance). Nonetheless, mindful of German achievements in guided missiles, beginning in 1948, the year in which Skinner joined Harvard University, the U.S. Navy financed an extension of the project called Orcon (for Organic Control). But that is another story.

In 1943, the Brelands purchased a small farm and began carrying out operant conditioning experiments on a variety of species (including dogs, cats, chickens, parrots, turkeys, pigs, ducks, and hamsters). Their results led them, despite Skinner's reservations, to abandon their studies and to create an animal training company to serve the needs of advertising and show producers (Bailey and Gillaspy 2005).

In 1951, they published a glowing article about their control experiments in animal behavior (Breland and Breland 1951). Around the same time, they invested in training marine mammals, for which, necessarily working remotely, they used an ultrasonic whistle as secondary reinforcement, thus presaging clicker training.

Ten years later, they published a new article (Breland and Breland 1961) that caused a stir. After having trained more than six thousand animals from thirty-eight different species, including "such unlikely subjects as reindeer, cockatoos, raccoons, porpoises, and whales" (681), they recognized the utility of Lorenz and Tinbergen's cautions and confirmed that, to understand and predict the behavior of organisms, it is essential to completely familiarize oneself with the instinctive behavioral structures of the species that one wishes to study. They supplied various illustrations of what they called instinctive drift, in which "learned behavior drifts toward instinctive behavior" (684). For example, they tried to train a raccoon (a species that has a good appetite, is easily trained, and whose "hands" resemble those of primates) to put tokens into a metal box. After having trained their subject—a perfectly tamable and agreeable animal—to grasp a token, they introduced the box and encountered their first difficulty in getting the raccoon to put the token inside it. The animal would not let the token go: It rubbed it against the inside of the box, then took it out and held it firmly for several seconds. Finally, it released the token and received a portion of food as a reward. When the Brelands gave the raccoon two tokens, they had a real problem: Not only did the raccoon refuse to let go of the tokens, but it spent whole minutes rubbing one against the other like "a miser" and dipping them in the box. The behavior went from bad to worse, despite nonreinforcement, such that the initial idea of "a display featuring a raccoon putting money in a piggy bank" (682) had to be jettisoned. Actually, the raccoon was exhibiting typical "washing behavior," which it employs, for example, in deshelling a crayfish, behavior that is an integral part of its repertoire for obtaining food. After describing several similar incidents, the Brelands stressed that demonstrating instinctive drift was not intended to "to disparage the use of conditioning techniques" but revealed the weakness of the philosophy underpinning these techniques and the need to revise three of its important assumptions, to wit, "that the animal comes to the laboratory as a virtual *tabula rasa,* that species differences are insignificant, and that all responses are about equally conditionable to all stimuli . . . After 14 years of continuous conditioning and observation of thousands of animals, it is our reluctant conclusion that the behavior of any species cannot be adequately understood, predicted, or controlled without knowledge of its instinctive patterns, evolutionary history, and ecological niche" (684).

The Origins of the Cognitive Revolution
Despite Watson's (1913) categorical rejection of mentalism, taking mental states into account—both experimentally and conceptually—in the behavior of animals proceeded along with the rise of behaviorism. Indeed, it is credited with introducing the concepts of insight, latent learning, and cognitive maps, which no longer refer to associative learning but to what might be called cognitive learning (Lassalle 2004).

In *The Mentality of Apes,* which appeared in the original German in 1917 and was translated into English in 1925 and then French by the psychologist Paul Guillaume in 1927, Wolfgang Köhler, a founder of Gestalt theory (Guillaume 1979), showed the phenomenon of insight, or sudden comprehension ("Aha!"), by giving chimpanzees problems to solve. The phenomenon would occur after a phase of trials, followed by a phase of reflection (or "reorganization of the perceptual field" according to the tenets of Gestalt psychology). For example, a chimpanzee named Sultan was given two hollow

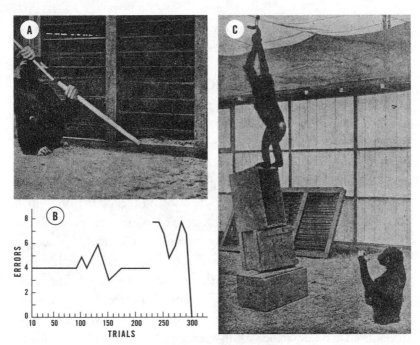

Figure 3.1 (a) Sultan the chimpanzee building with a two-part stick. Source: Köhler 1925, 132. (b) A Julius learning curve. No learning was observed until the 290th trial. Then, suddenly, the task was accomplished perfectly (Yerkes 1916a, 27). Note: At the 220th attempt, the situation was modified to increase motivation, but the outcome was even worse. (c) Stacking of boxes to access fruit placed beyond reach. Source: Köhler 1927, 105.

reeds, similar to ones he had already used to obtain fruit. One of the reeds was substantially narrower than the other, and the animal fit it into the larger reed, creating a double stick. He then immediately began trying to reach fruit previously beyond his grasp and in which he had lost interest after several abortive attempts (see figure 3.1): "After this there is no doubt that he makes use of the double-stick technique intelligently, and the accident seems merely to have acted as an aid—fairly strong it is true—which led at once to 'insight'"[2] (Köhler 1925, 167).

---

2. The English translator of the original German version notes that she was led to translate the German term *Einsicht* either as *intelligence* or *insight*, due in no small part to the lack of an English adjective for *insight*. Paul Guillaume never uses the English term *insight*, choosing instead to convey its meaning by alluding to intelligence or, as here, understanding.

Although Köhler's description of insight remains authoritative, he was not alone in accounting for mental states in solving problems, seeing as how he had been a student of the eminent behaviorists Watson and Thorndike. Robert Yerkes also followed this line of investigation and reached similar conclusions. In fact, one year before Köhler, he published an experimental study on the mental life of monkeys and apes (Yerkes 1916a) in which an orangutan named Julius was given relational-type problem-solving tasks (for example, to the left of, to the right of, in between). The method developed by Yerkes, called multiple choice (Yerkes 1916b), which consisted of finding the right solution among several possibilities, enabled him to produce learning curves, unlike Köhler's setups, which only allowed for qualitative descriptions. The learning curves show a sharp drop-off in the number of errors Julius made after several days: "The curve of learning plotted from the daily wrong choices had it been obtained with a human subject, would undoubtedly be described as an ideational, and possibly even as a rational curve; for its sudden drop from near the maximum to the base line strongly suggests, if it does not actually prove, insight. Never before has a curve of learning like this been obtained from an infrahuman animal" (Yerkes 1916a, 68), which led Yves Delage (1916), director of *L'Année biologique,* to comment: "Study of ideational reactions. The two (two 'small monkeys' and an orangutan) appear to make use of representational processes, which are quite sophisticated in the orangutan" (378), thereby anticipating representation, an idea that would form a cornerstone of the cognitive sciences.

Yerkes's article included a number of tasks similar to those of Köhler, such as stacking boxes to reach fruit placed high out of reach (see figure 3.1c). Although both Yerkes and Köhler briefly alluded to each other's work, the citations served primarily to underscore their meager mutual awareness due to the First World War.

According to Edward Tolman (1948), who introduced so-called neobehaviorism as opposed to the radical behaviorism of Watson, Hugh Blodgett (1929) was responsible for formulating the concept of latent learning and for conducting the first experiments relating to it as part of his doctoral work, under Tolman's supervision.

Blodgett used three groups of rats, one a control group and two experimental groups, which he taught to negotiate a labyrinth at a rate of one trial per day. The control group (group 1) followed classical procedures: The rats consistently found food at the exit, situated at the far end of the labyrinth. For the first experimental group (group 2), food appeared only on day seven, and consistently thereafter, whereas for the second experimental group (group 3), food appeared from day three. The learning curves show a sudden drop in errors from the day following the introduction of food for the experimental groups (2 and 3), which brought their performance on level with that of the control group.

Tolman and Charles Honzik (1930) repeated this experiment with several modifications. These included more animals and a more complicated labyrinth (fourteen corridors instead of six). Moreover, the animals were divided into two control groups—one (C1) for whom the exit was consistently supplied with food from the outset and another (C2) for whom there was never any food at the exit—and an experimental group (E) that only received food after the eleventh day. The researchers obtained results similar to those of Blodgett, with a classical learning curve for the first control group, a curve that showed practically no learning for the second control group and the first eleven days of the experimental group, followed by a sudden jump in learning from day twelve in terms of fewer errors and a faster time to completion (see figure 3.2). In other words, in both experiments, the sudden improvement in performance of the rats in the experimental groups shows that, even with no reinforcement in an early phase, they still managed to learn the arrangement of the labyrinth in a latent fashion and were only motivated to use their knowledge—called a cognitive map (Tolman 1948)—after reinforcement (food) was introduced: "We believe that in the course of learning something like a field map of the environment gets established in the rat's brain . . . a tentative, cognitive-like map of the environment. And it is this tentative map, indicating

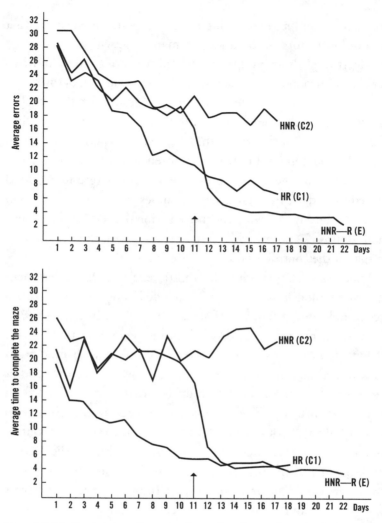

Figure 3.2 Whether in terms of lessening the number of errors (entering dead-ends) or improving the time required to get through the labyrinth, introducing a reinforcement on the eleventh day results in signs of latent learning for the experimental group (HNR-R: hungry nonrewarded-rewarded). Source: Tolman and Honzik 1930.

routes and paths and environmental relationships, which finally determines what responses, if any, the animal will finally release . . . Once, however, they knew they were to get food, they demonstrated that during these preceding non-rewarded trials they had learned where many of the blinds were. They had been building up a 'map,' and could utilize the latter as soon as they were motivated to do so" (192 195).

This latent learning, which goes hand in hand with exploratory behavior, is clearly shown by animals placed in an unfamiliar environment. In particular, it enables them to locate themselves in their habitat. Horses, to take just one example, are capable of orienting themselves and heading to their home range, even though it may extend to several dozen square kilometers. They do not wander haphazardly in search of a watering hole that may be very far away from where they are feeding. Similarly, horses have not been observed to move randomly between grazing sites and their home range, but are able to localize sites that are more likely to offer the best food.

Although very few studies have been done on the way in which horses actually structure their space and how they orient themselves, it is nonetheless possible to formulate a realistic hypothesis based on what is known. First of all, it seems quite probable that, in following its mother through displacements of the family group throughout its youth, a foal or a colt accumulates information about its environment, and thus by latent learning acquires knowledge of different reference points or the "physiognomy" of various places. At the same time, it will have learned way finding through borrowing. Consequently, the young horse progressively forms a sort of mental cartographic representation of its home range—a cognitive map—by memorizing places, their relative positioning, and routes, in short, by familiarizing itself with its environment by means of visual and especially olfactory cues. Moreover, as is now known, the mammalian brain contains place cells, particularly in the hippocampus (O'Keefe and Nadel 1978), and head direction cells found in areas anatomically and functionally related to the hippocampus (Taube, Muller, and Ranck 1990), which provide neuronal support for such a representation.

This explanatory model also takes into account the capacity of domesticated horses to return to the stable after unseating their rider in familiar environments, aside from the olfactory cues supplied by the stable itself when the winds are favorable.

But it remains very general and leaves open many questions, for example, on the respective role of proximity cues and perhaps more global cues contributing to orientation to the distance of the location to be reached, or in the way the horse remembers (for greater or lesser amounts of time) and uses the state of the vegetation in various parts of the pasture. Although the concept of cognitive map retains its relevance, after waning a bit prior to the 1970s, it cannot in and of itself fully explain the complexity of real-world situations.

Coming briefly back to Tolman, following this digression regarding our favorite animal, let me add that his contribution was significant in at least two respects: His theoretical contribution to learning constitutes a sort of prologue to an approach to animal cognition. In particular, he introduced not only the concept of intermediate variables intervening in the relationship between stimulus and response, but also those of the cognitive map and of hypothetical and goal-oriented attention, corresponding to the idea that animal behavior is purposeful, which called the associative stimulus-response model into question. We also have Tolman to thank for showing the importance of taking into account mental states in the context of a rigorous experimental approach, without having to resort to introspection or to treating animals as humans.

Cognitive Sciences and Animal Representation

What is called the cognitive revolution did not really take off until the 1950s, most notably with the emergence of information theory and the development of computers and formal languages and more generally with the advent of the cognitive sciences.

The revolution unfurled mainly between 1950 and 1970, through a gradual movement of cross-fertilization between disciplines that were just emerging or were being reshaped. Alan Turing's article presenting a theoretical model for a universal programmable computer, for which he invented the concepts of (symbolic) program and programming, appeared in 1936. Claude Shannon's publications introducing information theory, and Norbert Wiener's laying out the principles of cybernetics (especially the concept of feedback) date from 1948.

During the 1950s, simultaneously with rapid technological progress that led to the arrival of real "universal" computers, important,
mutually interacting conceptual advances began to be felt. Marvin
Minsky and John McCarthy launched artificial intelligence. Allen
Newell and Herbert Simon set about simulating cognitive processes
using computers. And in linguistics, Noam Chomsky published his
work challenging the behaviorist approach of Skinner on language
and how it is learned.

Also during the 1950s, the influence of behaviorism diminished,
due not only to the advent of ethology but also the development of
cognitive psychology. In truth, as George Miller (2003) has remarked,
behaviorist hegemony in psychology was basically focused on North
American experimental psychology (although the mind had never
disappeared from social or clinical psychology). The reintroduc-
tion of the mind as central to experimental psychology during this
period translated into a reopening of exchanges with eminent foreign
psychologists who had never denied the mind: "In Cambridge, UK,
Sir Frederic Bartlett's work on memory and thinking had remained
unaffected by behaviorism. In Geneva, Jean Piaget's insights into the
minds of children had inspired a small army of followers. And in
Moscow, A. R. Luria was one of the first to see the brain and mind as
a whole" (142). Jacques Vauclair and Patrick Perret (2003) observed,
moreover, that the contribution of European cognitive psychology
was totally original (especially for taking into account the development perspective), substantially more important and more audacious
than Miller lets on, and that its influence is still apparent today. In
France, recognition of cognitive psychology as a subfield of psychology crystallized with the appearance in 1988 of the first special issue
of the *Bulletin de Psychologie*, titled "Cognitive Psychology" (Plas
2004) and in 1989 of the first French textbook bearing the same title
(Bonnet, Ghiglione, and Richard 1989).

As Miller explained: "These fields represented, and still represent,
an institutionally convenient but intellectually awkward division.
Each, by historical accident, had inherited a particular way of looking at cognition and each had progressed far enough to recognize that

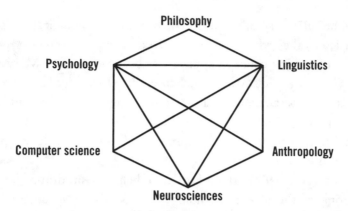

Figure 3.3 Cognitive sciences in 1978. The lines connect the disciplines that were already engaged in interdisciplinary research in 1978. Source: Miller 2003, 143.

the solution to some of its problems depended crucially on the solution of problems traditionally allocated to other disciplines" (Miller 2003, 143). The following years were devoted in part to reinforcing existing links and to forging missing ones (see figure 3.3).

The concepts of representation and information processing are central to the cognitive sciences (Lachman, Lachman, and Butterfield 1979, 99). Analogously to the computer, the brain is conceived as an organ that extracts, stores, and combines information incoming from the environment by making inferences. As Michael Kreutzer and Jacques Vauclair (2004) state with reference to animal cognition: "The mental apparatus is assimilated into a system of information processing . . . The cognitive approach proposes a general model of animal cognition . . . This model relies on the concept of representation, defined as a correspondence between one or several environmental aspects and processes (behavioral and neural) that enable adaptation of the animal's behavior to the environment" (14). Cognitive ethology "favors research on mental states that accompany the relational life that animals have with the physical and social milieu" (13). In other words, the animal constructs and uses representations of the physical and social environment to adapt its behavior.

Representations in the Horse

Observation of natural animal behavior suggests the existence of representations in the horse. Examples include, as already noted, the way in which horses structure their space by constructing cognitive maps that enable them to adapt to the character of their environment (for example, protective or grazing space and shade).

But it is also possible to show the existence of animal representations experimentally. For instance, recent work reveals the capacity of the horse to compare two small quantities.

Claudia Uller and Jennifer Lewis (2009) have reported that horses shown two apples being put into a container whose contents are not visible and three apples being put into a second container will spontaneously—without prior learning—head for the second container. They also maintain in working memory a representation of the number of apples deposited in the containers that they compare in deciding. This capacity for representation does appear to have its limits. Although the horse is able to successfully distinguish one apple from two apples, or two apples from three apples, it has trouble comparing four apples with six apples. The animal appears to experience limits in numerical representation similar to those shown not only by adult rhesus monkeys (Hauser, Carey, and Hauser 2000) but also newborn babies (Antell and Keating 1983) and freshly hatched chicks (Rugani et al. 2009), which is consistent with Elizabeth Spelke's theory of core knowledge (Spelke 2000; Spelke and Kinzler 2007). This theory holds that basic mental representations in different areas (for example, object representation and representation of intentional agents) exist from birth, including numerosity, core knowledge that is also phylogenetically ancient and shared among vertebrates. Note, however, that Uller and Lewis also point out that the nature of this representation has not been experimentally confirmed; it might well be related to the overall volume of the apples as opposed to their number, which additional experimentation would hopefully determine. In any event, clearly in this case the horses acted in accordance with mental representations that they had retained in working memory.

Other examples with horses further illustrate the theme of categorization and formation of concepts involving long-term memory. To categorize basically consists of grouping stimuli—objects or events—within the same class, thus constructing equivalent classes, which assumes being able to remember them. The capacity to categorize is obviously adaptive to the extent that it allows generalization, potentially extending to unfamiliar stimuli the knowledge acquired by the animal with known stimuli. According to the level of abstraction of the cognitive processes involved, a global distinction is made between perceptual and conceptual categorization (Doré and Mercier 1992; Fagot, Wasserman, and Young 2004).

The former is carried out based on concrete or absolute properties such as shape (for example, everything that is round), color (for example, everything that is blue), orientation (for example, all vertical things), texture (for example, smooth things), and so forth. It assumes representation in memory of simple or combined characteristics, or perceptual configurations.

The latter refers to abstract relationships. The properties are not perceptible as such in the stimulus. They must be extracted from it and constructed in the representational system, based, for example, on relations of similarity or of relative sizes, or animate/inanimate. These relationships may also be of a functional nature, as is the case, for example, in categories of tools, means of transport, edible objects, or shady places.

As will be evident in the chapters devoted to the study of equine perception, many experimental protocols rely on the capacity of horses for perceptual discrimination. This is the case, for example, of their visual acuity (Timney and Keil 1992), which is based on their capacity to discriminate uniform surfaces from surfaces marked by parallel lines, which will be discussed in more depth in chapter 7, which deals with shape perception. Continuing in the area of visual perception, Evelyn Hanggi (1999a) showed the ability of horses to discriminate between two-dimensional filled black shapes and open-center shapes—whether the same or different—at the end of discriminative learning by operant conditioning. During the learning, the choices relating to open-center shapes were treated as the correct ones, and thus rewarded, in contrast to the filled shapes, which were not. Thus, after initial learning, the horses were able to choose correctly, even when the orientation of the stimulus was varied or when the stimulus consisted of a mixed pair (for example, an open-center triangle with a filled hexagon). The horses had clearly managed to categorize "open-center shapes" versus "filled shapes" based on the discriminative learning to which they had been exposed.

Barbara Flannery (1997) also found that horses are able to solve more complex discrimination problems involving the acquisition of a relational concept of identity. Flannery used a matching-to-sample paradigm that makes it possible to determine whether subjects establish an identity relationship between different stimuli presented to them.

Hanggi (2003) showed further that horses are capable of abstracting comparative size relationships ("larger than" and "smaller than") based on discrimination tasks between stimulus pairs of the same shape, but with different sizes or colors (which were varied for a pair of stimuli in a ratio of about 1:2, for four degrees of size), whether the stimulus is two or three dimensional (see figure 3.4).

Categorizing assumes mental representation, as noted above. But the notion of representation is difficult to disassociate from that of memory. According to Vauclair (1996), an animal possesses a representation if "it is able to activate and use information that is not available in its present environment. A representational activity thus implies an ability to form a trace of a stimulus or, in other terms, to encode a stimulus in memory and then to use this trace to react appropriately to current environmental conditions . . . Functional differences between *memory* and *representation* are hard to make, and in fact both terms are often used interchangeably" (9). Experiments by Hanggi and Jerry Ingersoll (2009a) on long-term memory for categories and concepts in horses is particularly relevant to our thesis. Essentially repeating with the same horses the experiments described above (aside from an earlier experiment on image recognition—Hanggi 2001—to which I will return later in chapter 7), the

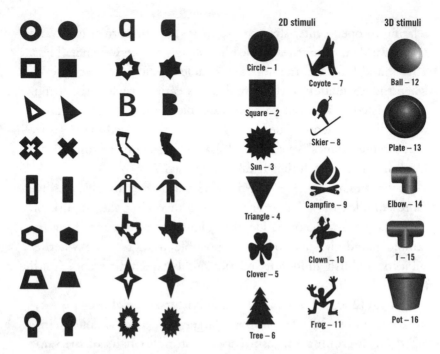

Figure 3.4 Left: stimulus used to discriminate between open and filled shapes. Right: forms used to discriminate based on conceptual categories (larger or smaller, 2D or 3D). Sources: (left) Hanggi 1999a, 246; (right) Hanggi 2003, 205.

researchers showed that, with no other previous learning and with no interim exposure to the stimulus presented since the initial experiments, the horses succeeded at nearly all the tasks of discrimination based on the categorization criteria acquired six to ten years before.

Remember, however, that there is also variability among individuals within a species. That is, all the individuals of a single species are not all equal. This caveat applies not only to their physical potential but also their mental capacities, and one cannot extrapolate to the species based on individual performance. Indeed, in a study by Sappington and Goldman (1994), four horses (AK, AZ, NK, and AL) were subjected to six types of discriminative learning by operant conditioning, four of them simple and the two others involving formation of a concept of triangularity. For each type, the horses had to choose the correct stimulus within a presented pair. For types

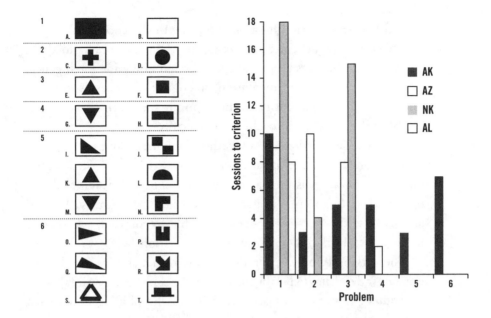

Figure 3.5 Left: stimulus used for each of six types of learning to reach the established success criterion. Right: the number of sessions required for each of the horses. The absence of a result of a horse for a given type of learning indicates that the horse was not able to satisfy the success criterion. Source: Sappington and Goldman 1994, 3082–83.

five and six, each of the correct stimuli (in this case, the triangle) was successively paired, in random order, with each of the other three stimuli. Learning was considered complete when the horse had achieved a success score of 85 percent of correct responses during two consecutive sessions of thirty trials each (see figure 3.5).

The findings suggest in particular that one horse did not progress past the initial type of learning, and a second did not succeed past the third type. Only one horse achieved the success criteria for the sixth type (characterized by three new triangles and nontriangular shapes that the horses had not seen before), an outcome that clearly suggests substantially different cognitive performance in different individuals.

Finally, individual equine capacity for learning, in addition to depending on what is to be learned, appears also to be linked to characteristics not only of age, sex, and breed but also of temperament (Lansade and Simon 2010).

A Brief Backtrack to the Question of Equine Intelligence
Before ending this chapter and recapping the major points made in it, let me say that it is clear that the question of equine intelligence, understood as the implementation of cognitive processes and not only preprogrammed routines, is not a purely speculative exercise but a subject that can be approached scientifically.

In short, the question is not whether the horse belongs to a species that is more or less intelligent than others, which makes little sense, but in what ways it is intelligent. Which cognitive capacities developed within the species throughout its evolutionary history? Looked at this way, the question suggests testable hypotheses, whether in terms of fieldwork or laboratory experimentation.

The variability between individuals within a species should also not be underestimated. Within a given species, each animal is an individual. It stands to reason that if, on the physical level, each individual has unique characteristics within the framework of the global organization that in turn is unique to the species, the same must be true (though less immediately obvious) on the mental level. I will come back to this point later, in discussing, for example, work in which certain individuals had to be omitted from the experimental protocol not least for lack of motivation or, for those participating in the experiments, learning curves or the final results differed substantially among individuals. Moreover, two given individuals may be gifted in different ways, one with one talent and one with another.

Finally, an observation that is valid for the group cannot automatically be attributed to the individuals, which unfortunately is a commonly made error. While it may be true that men are taller than women, that says very little about the relative height of this or that particular man or woman.

Representation Is Not Awareness
Note that acknowledging the existence of representations in animals does not imply that animals think. Representations are a necessary but not sufficient condition for awareness (Latto 1988).

And it is not self-evident that animals share with humans the same mental experiences, nor do their capacities for anticipation amount to conscious intention. It is striking that the very term *awareness* is essentially absent from an edited work specifically dedicated to animal representation (Gervais, Livet, and Tête 1992). As Kreutzer and Vauclair (2004) explain: "The use of concepts such as representation or memory by cognitive psychologists does not imply awareness or subjective mental experiences. As suggested by Terrace (1984), one of the advocates of animal cognitive psychology, the demonstration of representations, aided by the objective procedures of experimental psychology, did not lead to hypotheses on the conscious nature (or not) of the 'thinking' animal. Rather, based on solid data, the approach infers the nature of cognitive processes, as do experimental psychologists working with humans. In this, comparative psychologists distinguish themselves fairly clearly from the positions adopted by theorists of cognitive ethology such as Griffin (1984)" (14–15).

While Donald Griffin, whose contributions in particular on the mechanisms of echolocation in bats had already proved significant, is indeed credited with originating the expression *cognitive ethology*, his efforts to identify conscious phenomena in animals are not without problems. The very concept of awareness is ambiguous; one cannot compare the simple fact of being conscious, in the sense of being awake, with not being conscious, or with being aware of something, or a fortiori with being able to demonstrate conscious reflection, of which self-awareness is an extreme case (Jennings 1998; Lehman 1998; Piggins and Phillips 1998). As Vauclair (1992) stresses: "First, one might ask what is meant by consciousness. For his part, Griffin is noncommittal and uses the concepts of consciousness, mental states, mind, and thought interchangeably. Second, Griffin proposes no systematic method of testing the observations called 'ambassadors of thought.' Finally, and especially, the approach that he suggests relies on extremely dubious hypotheses according to which the animals studied share the same conscious experiences as humans" (139).

I cede the final analysis to Kreutzer and Vauclair (2004): "Current research in cognitive ethology is typified, on the one hand, by the fear of slotting the behaviors studied into an adaptive framework, and on the other, of relying on behaviors implemented by animals with their nervous substrate" (16).

No wonder, then, that the next chapter is devoted to the equine brain.

4

THE EQUINE BRAIN

The brain is simultaneously both an integral part of the nervous system and an orchestra conductor. As Robert Barone and Ruggero Bortolami (2004) write: "The nervous system governs an organism's relationships with its environment as well as the functional coordination of the rest of the apparatus. To this end, it perceives every change in the external or internal milieu and responds with appropriate action. In higher mammals, this function of adaptation reaches its most perfect form with the development of the psyche; one might say that intellectual functions are its highest expression" (1). For this reason, it is well worth briefly recalling the general organization of the mammalian nervous system, before moving on to the specific structures of the equine brain and their function.

Nervous Tissue and the General Organization of the Mammalian Nervous System

Nervous Tissue
The nervous system is made up of two types of cells: neurons and glial cells.

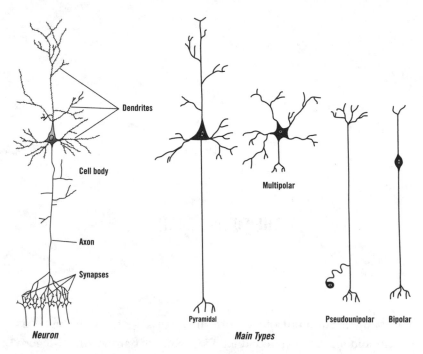

Figure 4.1 Left: schematic diagram of a neuron. Right: main types of neurons.

The *neuron* (or neurocyte) is the functional unit of the nervous system (see figure 4.1). Its speciality is the generation and conduction of nerve impulses in the form of a succession of electrical pulses called action potentials. Each neuron comprises several components:

- Dendrites. A few microns in diameter (by way of comparison, the thickness of a human hair is around 50 to 100 microns), these branches bear structures in the form of buds called dendritic spines, which enable synaptic connections. Dendrites conduct nerve impulses (excitatory or inhibitory) received from the synapses.
- Cell body. The soma (varying in size from 5 microns to 1 millimeter) serves to integrate excitatory or inhibitory signals sent from the dendrites.
- Axon. Measuring 5 to 10 microns in diameter, the axon moves the nerve impulses toward its branched extremity,

the terminal tree, at the end of which synaptic boutons form synapses with other cells, either at the level of dendrites or directly at the level of the cell bodies. In vertebrates, the axons of all motor neurons (or moto-neurons) are coated with a myelin sheath that speeds the propagation of the electrical impulses. The axons of sensory neurons, in contrast, may or may not possess myelin sheaths. Myelinated axons are also called white fibers. Axon length varies widely, from a few microns up to more than 3 meters for the recurrent laryngeal nerve that supplies the equine larynx (Hahn 2004).

A *synapse* is a junction between a synaptic bouton on the terminal branch of a neuron and another cell (neuron, muscle cell, sensory receptor, glandular cell). Very generally, it consists of a presynaptic membrane, the synaptic bouton, a space or synaptic cleft (very narrow: 0.03 to 0.05 microns), and a postsynaptic membrane (see figure 4.2). Transmission of a nerve impulse between two cells at the synaptic level occurs chemically, through release into the synaptic space of neurotransmitters contained in the presynaptic membrane (here, the

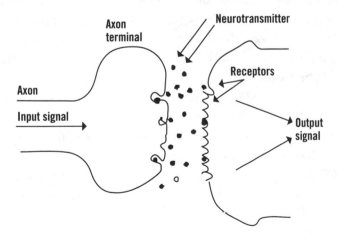

Figure 4.2 Schematic illustration of a synapse.

mechanism is electrochemical conduction; the cellular membrane of neurons is excitable, that is, it generates electrical activity in response to stimulation, be it electrical, chemical, or mechanical). Synaptic action is very brief (0.3 to 0.5 milliseconds). More rarely there are electrical synapses comparable to simple electrical conduction, in the case, say, of heart muscle or smooth muscles, for which the transmission of nerve impulses requires only a few microseconds.

A neuron may possess several thousand terminal boutons, which explains the frequently cited claim that the human brain contains a hundred billion neurons that, on average, each connect to another ten thousand neurons.

Neurotransmittors may either be excitatory (acetylcholine, adrenaline, dopamine, and glutamate) or inhibitory (in particular, gamma-aminobutyric acid, or GABA). Neurons thus integrate excitatory and inhibitory inputs. When the sum of these inputs exceeds a certain threshold, a nerve impulse is triggered and propagates along the axon.

More globally, afferent neurons carry sensory information from the periphery to the center of the nervous system, and efferent neurons are those that control the actions of the muscles and glands. So-called interneurons are confined in the central nervous system.

Glial cells, which make up the glia, occupy the vacant space between the neurons. Ten times more numerous than neurons, they do not carry nerve impulses but play a determining role in the functioning of the nervous system. They isolate the neurons of the central nervous system from other tissues and ensure their proper operation particularly by supplying needed energy. There are different categories of glial cells, fulfilling complementary functions, in particular oligodendrocytes, which form the myelin sheath that covers the axons, and astrocytes, whose importance is becoming increasingly apparent. By supplying mechanical support to the neurons, as well as nutrients (they are the only storehouse of glucose in the nervous system), and ensuring a balanced extracellular environment, these cells also play a central role in brain plasticity (Chneiweiss 2002). This has as much to do with their major role in the formation (synaptogenesis) and development of the synapses as with their activity in controlling neurogenesis (that is, in manufacturing new neurons) in the hippocampus, a brain structure that is essential to memory and learning.

This leads us to a brief discussion of *brain plasticity* in mammals, a question that has engendered an explosion of knowledge over the last twenty years, although much remains to be discovered. Such a question is obviously of interest, since it bears on the capacity of the horse to reorganize its neural network based on external stimuli or individual experience or to adapt its functioning in response to trauma or illness. Broadly speaking, plasticity can be divided into two categories, synaptic and neuronal.

At the beginning of the 1970s, a form of synaptic plasticity was demonstrated in the hippocampus and considered to be one of the main physiological mechanisms responsible for learning and memory: long-term potentiation, which refers to a sustained increase in synaptic transmission resulting from activation of neurons by high-frequency stimulation (Bliss and Lomo 1973). (A complementary phenomenon known as long-term depression leads to inhibition of synaptic transmission.) This phenomenon was subsequently shown in many areas of the cortex.

Around the same time, a theory of learning by selective stabilization of synapses over the course of development was proposed (Changeux, Courrège, and Danchin 1973). Accordingly, until the 1980s, in the words of Alain Privat (1988), synaptogenesis was "a relatively simple event, limited in time and space, but strongly dependent on prior experience" (24). As asserted by Jean-Pierre Changeux (1985): "Synapses . . . proliferate in successive waves from birth to puberty in man.[1] . . . To learn is to stabilize preestablished synaptic combinations, and to *eliminate* the surplus" (248–49). In the last decade, however, researchers have found that there is much more to synaptic plasticity. Long-term potentiation is accompanied throughout life by structural modification of synapses (in shape and size),

---

1. In humans, it is estimated that only 10 percent of synapses are present at birth.

as well as formation of new dendritic spines and the creation of new synapses (Muller 2004).

Moreover, since the discovery of neurons at the end of the nineteenth century, the reigning dogma held that a person was born with a supply of neurons that could only diminish through age or illness.[2] It would be many years before a convergence of observations called the paradigm into question. The first seeds of doubt were sown in the early 1960s, with the report of the birth of new nerve cells in the hippocampus and olfactory bulb of adult rats (Altman 1962). Yet uncertainty persisted regarding whether the cells were neurons or glial cells. During the 1980s, the question resurged after canaries were shown to lose neurons in autumn and to replace them in spring, which enables them to learn new songs each year (Nottebohm 1981).

But only in the next decade did several different efforts, particularly those of Elizabeth Gould (Gould and Gross 2002), clearly show the existence of neurogenesis in certain areas of the brain—the hippocampus and the olfactory bulb—and apparently even within the neocortex. Certain features of these discoveries are particularly relevant here. For example, the environment appears to influence the formation of neurons. Thus, stress inhibits neurogenesis, as does an impoverished environment, whereas an enriched one favors it. Similarly, learning promotes growth of new functional neurons. Indeed, experimental blocking of neurogenesis in animals has been shown to cause learning deficits.

Finally, knowledge relating to these questions is more than likely to evolve in the coming years in view of the link established between neurogenesis and the existence of multipotent adult stem cells (capable of differentiating into neurons or glial cells) in different areas of the central nervous system, which were discovered quite recently (Lemasson and Lledo 2003; Lledo and Gheusi 2006).

---

2. Santiago Ramón y Cajal, who received the Nobel Prize in Physiology or Medicine in 1906 for his work on the structure of the nervous system, wrote in 1902: "We are born with a certain number of brain cells that decrease with age. Everything must die in the brain or spinal cord—nothing can regenerate."

General Organization of the Mammalian Nervous System

The nervous system has three main functions: sensory, integrative (including mental function), and motor. It comprises both the central nervous system and the peripheral nervous system.

The *central nervous system* (or neuroaxis) consists of the brain, situated in the cranial cavity (the skull), and the spinal cord, located in the vertebral column. The right and left sides of the brain and the spinal cord have bilateral symmetry; thus, with a few exceptions, most of the structures of the central nervous system are paired. Bathed in cerebrospinal fluid (or cephalorachidian liquid), which ensures the stability of its environment and transports hormones to the cerebral ventricles, the central nervous system contains all the brain's emotional centers, as well as some of those belonging to the vegetative nervous system.

The central nervous system can further, albeit roughly, be classified into gray and white matter. Gray matter consists of parts of neurons devoid of myelin, that is, cellular bodies. In the brain, gray matter makes up the cerebral cortex (the outer portion of the brain) as well as the basal ganglia (subcortical centers) within. Inside the spinal cord, gray matter is surrounded by white matter. White matter consists of myelinated axons. Within the brain, it is located at the base between the cerebellum and the brainstem.

The spinal cord carries various nerve pathways toward or coming from the brain. It also provides an integrative pathway that coordinates much subconscious—reflex—neural activity (pain, scratching, and postural reflexes). In addition, it appears that the spinal cord in and of itself serves as a complex system of information processing capable of learning (Grau 2002). Because it supports major cognitive functions, the brain constitutes the main area for integrating thought, emotions, and decision making, as we will see below.

The *peripheral nervous system* is made up of somatic and visceral components. The somatic peripheral nervous system links the central nervous system to receptors (sensory) and effectors (motor). It consists of nerves (axons covered in protective tissue that connect the central nervous system to various parts of the body) coming from the neural

Pair	Name	Root	S/M	Main function
Table 4.1 The twelve pairs of cranial nerves.				
I	Olfactory	Telencephalon	S	Smell
II	Optic	Diencephalon	S	Vision
III	Oculomotor	Mesencephalon	M	Eye movements, pupillary constriction and accommodation; eyelid muscles
IV	Trochlear (or pathetic)	Mesencephalon	M	Eye movements
V	Trigeminal	Metencephalon	S + M	Sensitivity of the face, mouth, and cornea; masticatory muscles
VI	External oculomotor (or abducens)	Myesencephalon	M	Eye movements
VII	Facial	Myesencephalon	S + M	Mobility of the face, gustatory sensitivity, salivary and tear glands
VIII	Vestibulocochlear	Myesencephalon	S	Hearing, sense of balance
IX	Glossopharyngeal	Myesencephalon	S + M	Gustatory sensitivity; pharyngeal sensitivity and mobility
X	Vagus (pneumogastric)	Myesencephalon	S + M	Vegetative functions of the digestive column; sensitivity of the pharynx and muscles of the vocal cords, swallowing
XI	Accessory	Myesencephalon	M	Muscles of the neck and shoulder
XII	Hypoglossal	Myesencephalon	M	Tongue movements
S: Sensory. M: Motor.				

axis: that is, the twelve pairs of cranial nerves, which, except for the olfactory and optic nerve pairs, emerge from the brainstem and innervate the head (see table 4.1), and the rachidian nerves, which originate in the spinal cord and innervate the neck, the trunk, and the limbs. The nerves may be sensory (like the optic and auditory nerves), composed of afferent fibers that direct sensory information from the receptors to the brain, in general through the intermediary of the spinal cord. Or they may be motor (such as the ocular motor nerves), which possess efferent fibers that transmit motor signals emanating from the central nervous system to the periphery (in particular, the skeletal muscles). The nerves may also be mixed, composed of both sensory and motor fibers (for example, the facial or pneumogastric nerves).

The peripheral nervous system, also known as the vegetative or autonomic nervous system, controls homeostasis, that is, the balance of the internal environment of the organism necessary to life, and innervates the viscera (organs contained in the cranial, thoracic, and abdominal cavities: brain, heart, liver, intestines, lungs, spleen, kidneys, and uterus). It functions automatically, reflexively, and comprises the sympathetic and parasympathetic systems, which act antagonistically vis-à-vis the same organs.

The sympathetic system prepares the organism for action and, in response to stress, mobilizes energy (by increasing cardiac and respiratory rhythm, blood pressure, pupil dilation) and readies reactions of fight or flight. This system is called adrenergic because it involves the neurotransmitters noradrenalin and adrenalin.

The parasympathetic system maintains basal activities. On activation, it slows the functions of the organism to save energy. It speeds activity only in the case of digestive and sexual functions. This system is called cholinergic by association with the neurotransmitter acetylcholine.

General Organization of the Equine Brain

Before getting to the crux of our subject, it is worth touching on a terminological ambiguity stemming from the different meanings we give to the word *brain*. As Georges Chapouthier (2001) notes: "The word 'brain' does not in fact have the same meaning for scientific specialists as for others. For everyone, the brain is what is contained in the head, which the scientist refers to as the encephalon. The encephalon, which thus corresponds to what we usually call the 'brain,' has a certain number of components (brainstem, cerebellum, cerebral hemispheres), all contained in the skull. Roughly speaking, the scientist reserves the term 'brain' uniquely for the two cerebral hemispheres, which is to say, only part of what is contained in the skull" (22–23). Thus, depending on context, the word *brain* corresponds to the entire encephalon (that is, all the nervous structures within the skull) or only one of its parts, for example, the forebrain, which comprises the telencephalon and the diencephalon, or even the telencephalon by itself.[3]

However, let us be guided by a second observation of Chapouthier (2001) on this subject: "That said, as the essence of the reflections contained in this work bear on the highest structures of the brain, those situated in the cerebral hemispheres, they relate to the brain in the scientific sense as well as the popular sense" (23).

I will begin by briefly reviewing the major subdivisions of the brain (and their various names), whose principal neural structures are presented in table 4.2 as well as in illustrations from the atlas by Barone and Bortolami (2004) (see figures 4.3 to 4.6). These major subdivisions correspond to the successive stages of differentiation of the neural tube over the course of embryogenesis. Following an initial stage where its anterior portion presents a bulge corresponding to a single vesicle (the posterior part gives rise to the spinal cord), the tube goes through a stage of three vesicles (prosencephalon, mesencephalon, rhombencephalon) before arriving at a five-vesicle stage (telencephalon, diencephalon, mesencephalon, metencephalon, myelencephalon). Note that the

Table 4.2   Major subdivisions of the brain.

Brain							
Brain				Brainstem			
Forebrain or prosencephalon				Midbrain	Hindbrain or rhombencephalon		
Telencephalon or cerebrum		Diencephalon		Mesen-cephalon	Metencephalon		Myelencephalon
Cerebral cortex	Basal ganglia	Thalamus	Hypo-thalamus	Cephalic isthmus/ cerebral peduncles	Pons or (annular) protru-berance	Cerebellum	Brainstem or (rachidian) bulb

---

3. This last choice is the one made by Barone and Bortolami (2004) in their work, consistent with the Latin term *cerebrum,* also used to designate the telencephalon.

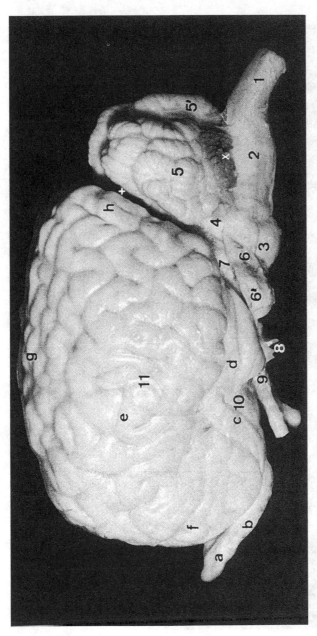

Figure 4.3 Equine brain. Left lateral aspect. 1. Spinal cord. 2. Myelencephalon: Medulla oblongata. 3–5. Metencephalon with: 3. Pons; 4. Middle cerebellar peduncle; and 5. Cerebellum (right hemisphere) and 5′. Vermis. 6. Trigeminal nerve with 6′: its ganglion. 7. Mesencephalon. 8 and 9. The only visible parts of the diencephalon, with 8. Infundibulum (the hypophysis has been removed) and 9. Optic chiasm extended by the optic nerves. 10 and 11. Telencephalon: Brain (left hemisphere) with: 10. Rhinencephalon: (a) olfactory bulb; (b) olfactory peduncle; (c) piriform lobe, rostral portion; and (d) piriform lobe, posterior portion and 11. Neopallium with: (e) convex surface; (f) rostral or frontal pole; (g) dorsal edge; (h) posterior or occipital pole. x: Choroid plexus of the fourth ventricle. + Transverse fissure. Source: Barone and Bortolami 2004, 2.

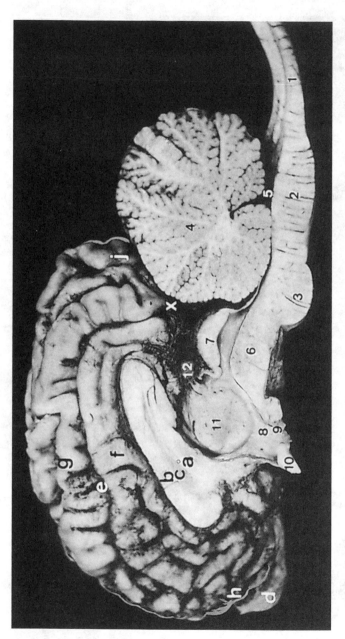

Figure 4.4 Median section of a horse brain. Right half, medial aspect. 1. Spinal cord (note the central channel). 2. Myelencephalon. Medulla oblongata. 3 and 4. Metencephalon with 3. Pons and 4. Cerebellum, whose white matter forms a tree of life. 5. Fourth ventricle. 6 and 7. Mesencephalon with 6. Cerebral peduncle and 7. Mesencephalic tectum; the mesencephalic aqueduct is situated between them. 8–12. Diencephalon with: 8. Third ventricle; 9. Infundibulum; 10. Optic chiasm; 11. Interthalamic adhesion; 12. Pineal gland. (a–j) Telencephalon, showing: (a) fornix body; (b) corpus callosum; (c) septum pellucidum; (d) olfactory bulb; (e–j): right cerebral hemisphere, which forms a wall along the longitudinal fissure of the brain; (e) splenial sulcus; (f) cingulate gyrus; (g) marginal or saggital gyrus; (h) rostral or frontal pole, and (j) posterior or occipital pole of the brain. x: Transverse fissure. Source: Barone and Bortolami 2004, 4.

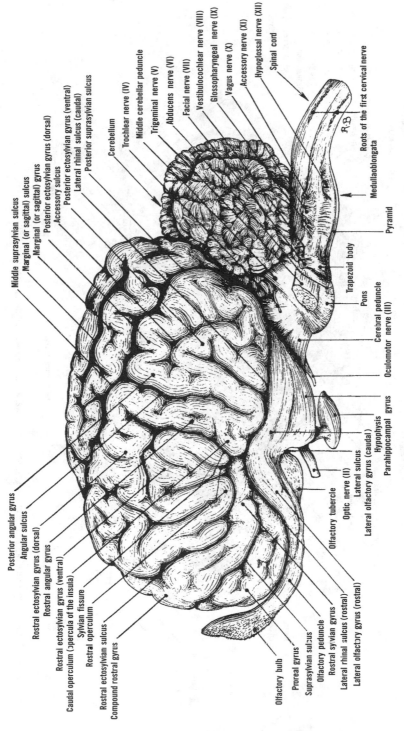

Figure 4.5 Equine brain. Left lateral aspect. Source: Barone and Bortolami 2004, 90.

Posterior angular gyrus

Angular sulcus

Rostral ectosylvian gyrus (dorsal)

Rostral angular gyrus

Rostral ectosylvian gyrus (ventral)

Caudal operculum (opercula of the insula)

Sylvian fissure

Rostral operculum

Rostral ectosylvian sulcus

Compound rostral gyrus

Middle suprasylvian sulcus

Marginal (or sagittal) sulcus

Marginal (or sagittal) gyrus

Posterior ectosylvian gyrus (dorsal)

Accessory sulcus

Posterior ectosylvian gyrus (ventral)

Lateral rhinal sulcus (caudal)

Posterior suprasylvian sulcus

Cerebellum

Trochlear nerve (IV)

Middle cerebellar peduncle

Trigeminal nerve (V)

Abducens nerve (VI)

Facial nerve (VII)

Vestibulocochlear nerve (VIII)

Glossopharyngeal nerve (IX)

Vagus nerve (X)

Accessory nerve (XI)

Hypoglossal nerve (XII)

Spinal cord

Roots of the first cervical nerve

Medulla oblongata

Pyramid

Trapezoid body

Pons

Cerebral peduncle

Oculomotor nerve (III)

Hypophysis

Parahippocampal gyrus

Lateral olfactory gyrus (caudal)

Lateral sulcus

Optic nerve (II)

Olfactory tubercle

Rostral sylvian gyrus

Olfactory peduncle

Suprasylvian sulcus

Proreal gyrus

Olfactory bulb

Lateral rhinal sulcus (rostral)

Lateral olfactory gyrus (rostral)

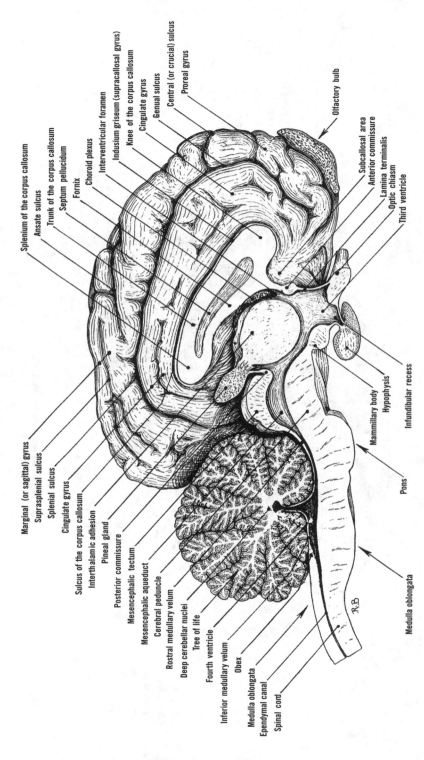

Figure 4.6 Median section of an equine brain. Left portion, medial aspect. *Source: Barone and Bortolami 2004, 42.*

Marginal (or sagittal) gyrus
Suprasplenial sulcus
Splenial sulcus
Cingulate gyrus
Sulcus of the corpus callosum
Interthalamic adhesion
Pineal gland
Posterior commissure
Mesencephalic tectum
Mesencephalic aqueduct
Cerebral peduncle
Rostral medullary velum
Deep cerebellar nuclei
Tree of life
Fourth ventricle
Inferior medullary velum
Obex
Medulla oblongata
Ependymal canal
Spinal cord

Splenium of the corpus callosum
Ansate sulcus
Trunk of the corpus callosum
Septum pellucidum
Fornix
Choroid plexus
Interventricular foramen
Indusium griseum (supracallosal gyrus)
Knee of the corpus callosum
Cingulate gyrus
Genual sulcus
Central (or crucial) sulcus
Proreal gyrus

Olfactory bulb
Subcallosal area
Anterior commissure
Lamina terminalis
Optic chiasm
Third ventricle

Mammillary body
Hypophysis
Infundibular recess

Pons

Medulla oblongata

brain also contains four internal cavities that communicate with each other and within which the cerebrospinal liquid circulates: the lateral ventricles (in the cerebral hemispheres), the third ventricle (between the two thalami), and the fourth ventricle (in the brainstem).

"Viewed externally," write Barone and Bortolami (2004), "the brain is egg-shaped with a large caudal pole, roughly one-and-a-half times as long as it is wide, and exhibiting many narrow and tight gyri. In the horse, each cerebral hemisphere is, on average, 125 mm long (and ranging from 105 to 150 mm). Its largest tranversal dimension is on the order of 5 mm and its height 80 mm . . . The total weight of the equine brain typically comes to 530 g. The most extreme weights we have recorded were 410 and 655 g . . . The weight of a single isolated hemisphere was 210 g on average . . . At equal body weight, the brain of a mare is slightly less heavy than that of the stallion" (429).

According to Antonio Barasa (1960, cited by Barone and Bortolami 2004), the gray matter of the horse contains some 11,500 neurons per cubic millimeter, and the total number of neurons in its cerebral cortex is estimated to be 1.2 billion (Roth and Dick 2005).

The Forebrain (Prosencephalon)
Comprising the telencephalon and diencephalon, the forebrain features the following main functional and structural characteristics.

The Telencephalon
"In all vertebrates, the telencephalon is divided into two large areas, the pallium[4] and the subpallium. The first, which forms the cortex of mammals, is further divided into four parts (hippocampus, neocortex, rhinencephalon, and claustrum), and the subpallium in three parts (striatum, amygdala, and septum)" (Vincent 2004, 2).

In mammals, write Barone and Bortolami (2004), the telencephalon "is by far the most voluminous and evolved part of the nervous system. It is in this class of vertebrates that it attains its full develop-

---

4. In Latin, *pallium* means "coat."

ment and maximal expansion. It is the terminus of all sensory signals, and its cortex is the departure hall for all voluntary motor responses. It controls all the organism's functions; it is the seat of the cognitive faculties, of memory, and of intelligence in whatever form" (393).

The telencephalon is divided into two cerebral hemispheres overlaid by the cerebral cortex. In higher mammals, the development of the brain within the skull over the course of evolution can be read as folds on the cortical surface: These are gyrencephalons, that is, their cortex comprises convolutions (called gyri), folds of gray matter separated by furrows (sulci). In contrast, the wall of the telencephalon remains smooth in lower mammals (rodents, marsupials, monotremes [egg layers], and insectivores), hence the term *lissencephalic*.

The ventral portion of the telencephalon contains primitive structures that form the *rhinencephalon*,[5] with the paleocortex, the oldest cortex, positioned laterally, where one finds the bulbs and the olfactory tubercle. The archicortex, a more recent structure, occupies the medial position and is the locus of the hippocampus, a formation with an essential role in memory. Finally, the mesocortex, a transitional structure between the archicortex and the neocortex, has links to the parahippocampal cortex and the cingulate gyrus, which surrounds the corpus callosum.

The dorsal telencephalon is the most evolved and contains the *neocortex* (or isocortex), which is found for the most part only in mammals. The neocortex consists of six layers of neurons several millimeters thick and makes up a large portion of the *gray matter*. As with other higher mammals, the horse is a gyrencenphalon, that is, its cerebral cortex contains convolutions. However, Barone and

---

5. The term *rhinencephalon* (from the Greek *rhino*, "nose") designates a large band of gray matter situated on the internal surface of the cerebral hemispheres around the corpus callosum and is part of the limbic system. Strictly speaking, the term is used variably, from its most restrictive sense (as essentially limited to the structures underlying olfaction) to its broadest sense (synonymous with the entire limbic system). According to Barone and Bortolami (2004): "The term rhinencephalon is used for convenience and has no meaning other than topographic. It suggests the main site occupied by the olfactory formations in lower vertebrates. In mammals, this area is minor, and most of the rhinencephalic structures are devoted to important functions of control of the hypothalamus and coordination between the neopallium and the diencephalon" (397).

Bortolami (2004) note that from a structural vantage, for the horse as for other domesticated mammals, "the central and lateral sulci,[6] which plow so deeply through the human brain, are shallow, even absent . . . At most, one can speak of frontal, parietal, temporal, or occipital regions, but their borders are imprecise, and no lobe is recognizable" (521). Yet, the same authors observe: "The distribution of types of structures presents a precise topography, roughly similar in all species. Consequently, one can recognize in the neocortex a certain number of cortical areas or domains where it is possible to establish the topography, which is constant in each species" (513). Functionally speaking, "electrophysiology and the study of disturbances linked to various kinds of lesions have shown that the activity of the neocortex is spread among specialized areas whose topography is fixed in each species and is broadly comparable in all mammals" (515).

Worth recalling here is that the visual areas are located in the occipital region, the auditory areas in the temporal region, and somesthetic inputs (involving sensitivity to touch, heat, and pain) and motor outputs terminate in and originate from the back and front, respectively, of the central sulcus of the hemispheres in the parietal and frontal areas of the limbic system, about which more later. Generally speaking, the cortex can be divided into primary, secondary, and associative areas of information processing. For example, the primary sensory areas consist of termination zones for projections from subcortical sensory pathways: They perceive basic sensations and are organized according to a precise "somatotopy." To wit, each contiguous area of the body—an entire body schema— is represented, including the primary sensory cortex as well as the

---

6. The central (or crucial) sulcus of the hemispheres corresponds to what in humans used to be called Rolando's fissure, which clearly separates the frontal lobe from the parietal lobe. Barone and Bortolami (2004) specify in addition (37) that it is "much shorter" but "quite recognizable" in carnivores, and that while also present in ungulates, in them it is nonetheless "minimal and less evident." Its localization, indicated in figure 4.6, shows the weak development of the frontal regions in the horse compared with humans, where these regions are especially associated with "executive functions" corresponding to high-level cognitive processes (selective attention, planning action, learning rules, abstract thinking, and so forth).

primary motor cortex, at the level of adjacent cortical regions. This somatotopy exists at all levels of the central nervous system, which allows nerve impulses to circulate "in an orderly fashion" between their point of origin and their final destination. The primary areas are surrounded by secondary sensory areas that elaborate these elementary sensory signals (not only noise, for example, but whether the sound is melodious or discordant, sonorous, or shrill) by filtering them and, if needed, correcting them. (In the next chapter, we will see that the horse is sensitive to certain optical illusions, such as the Ponzo illusion.) Finally, the associative areas assemble and integrate the information from diverse areas of the cortex, thus giving rise to different sensory modalities.

As Barone and Bortolami (2004) write: "Each sensory, motor, and (apparently) associative function is governed not by a single area but by the cooperation of several areas. In particular, the primary area is assisted by one or two secondary areas situated nearby . . . Each of the groupings formed by the primary and secondary areas overlaps more or less with the borders of the pre- or postcentral gyrus that it characterizes. But it is also assisted by complementary associative areas that stretch over a large adjacent region . . . The sensory and motor areas occupy only a small part of the neocortex. Most of the cortex consists of association areas. For some of these, it is nearly impossible to determine the meaning of the organization particular to each species" (515, 517, 518). Note also that the somesthetic inputs are crossed, and thus at the level of the left hemispheric cortex they project sensory information coming from the right side of the body and vice versa. According to whether these inputs are fine mechanoreceptive sensory signals (specifying the location and intensity of touch, detecting light movements on the surface of the skin) or more global ones (pain, heat, cold, rough touch), they borrow fast pathways (the posterior strands of the white matter in the spinal cord), which crisscross at the level of the bulb, or slow pathways (the anterolateral fibers in the spinal cord), which cross in the spinal cord. The motor inputs of the pyramidal pathways, which control voluntary movements, cross at the level of the bulb just before its junction with the spinal cord. Thus, the left hemisphere controls the right side of the body, and the right hemisphere the left side (this crisscrossing is called decussation). In contrast, the extrapyramidal pathways that govern involuntary motricity do not cross.

The cerebral hemispheres contain the nuclei of the gray matter—the *basal ganglia* (sometimes called the central gray nuclei)—which are located in the middle of the white matter composed of the myelinated cortical axons. The ganglia are interconnected subcortical centers (primarily the caudal nucleus, the putamen, and the pallidum, which form the striatum, and especially the claustrum and the amygdala). Linked to the motor cortex, they are principally responsible for autonomic motricity, accompanying the execution of voluntary movement, by controlling the balanced tonicity of the agonist and antagonist muscles. They are also involved in learning, memory, and emotions.

The innumerable nerve fibers that make up the white matter serve as projection pathways that link the cortex with other parts of the central nervous system: association pathways, which connect different cortical zones of the same hemisphere, and commissural pathways, which link the homologous parts of the two hemispheric cortexes (see figure 4.7). The *corpus callosum* (which is only present in placental and eutherian mammals) occupies a substantial

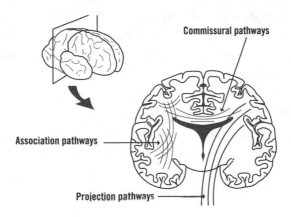

Figure 4.7 Nature of the nerve fibers that make up the white matter of the brain.

portion of the cortex and represents a significant network of nerve fibers. It is situated under the longitudinal fissure that separates the two hemispheres for about a third of its length. The corpus callosum transfers information from one hemisphere to the other and coordinates their activities. "When the brain is separated from the diencephalon, the two hemispheres are connected only by their commissures, the largest of which is the corpus callosum" (Barone and Bortolami 2004, 403).

The Diencephalon

Situated between the posterior portions of the two cerebral hemi-spheres, and in front of the mesencephalon, the *diencephalon* links the former to the latter. It comprises two major gray matter nuclei, the thalamus and the hypothalamus, as well as the epithalamus and the subthalamus. The optic pathways, which originate in the retina, partially cross in the optic chiasm, at the level of the inferior surface of the hypothalamus, near the hypophysis (pituitary gland), the crossed and uncrossed fibers joining in the thalamus. "The percentage of crossed fibers that pass into the opposite optic tract . . . is on the order of 81 percent in the horse[7] . . . The fibers coming from the temporal region of the retina are those that remain homolateral, whereas those coming from the nasal region cross in the raphe[8] with their counter-parts from the opposite eye" (Barone and Bortolami 2004, 367–68).

The thalamus, a paired, symmetrical structure surrounding the third ventricle, is essentially a switchboard for relaying sensory infor-mation (eye, ear, skin). Containing numerous nuclei (around thirty), it serves a sorting function by immediately distributing all the infor-mation to the appropriate areas of the cerebral cortex and into the deep regions of the telencephelon. It also plays a role in the motor system; subcortical information (cerebellum, basal ganglia) on its way to the motor cortex synapses in the thalamus. In addition to its

---

7. For purposes of comparison, this relationship is on the order of 53 percent in humans, 65 to 70 percent in cats, and around 75 percent in dogs.
8. In anatomy, the raphe is a ridge or seam marking the union of the halves of two symmetri-cal parts of a body organ.

relay function, the thalamus acts as a modulator, in particular during activation of the cortex in situations of alarm. It is indispensable to the functioning of the cortex. The destruction of a thalamic nucleus entails a loss of function of the cortical region to which it is linked.

The hypothalamus consists of a small number of gray matter nuclei and has immense functional importance. It is connected to nearly all parts of the central nervous system. It also controls essentially all of the hormonal secretions from the hypophysis (itself considered to be the "orchestra conductor" for the other endocrine glands in the organism). On the functional level, "as its many connections suggest, the hypothalamus is involved with the entire organism, both through the 'autonomic' nervous system and through the hypophysis and the endocrine system" (Barone and Bortolami 2004, 385). It also comprises the centers of heat regulation, hunger, thirst, reproductive behavior, biological rhythms, and emotional reactions (fight or flight). It is one of the most important structures in maintaining homeostasis.

Among the myriad connections of the epithalamus, observe Barone and Bortolami (2004), "those linking it to the rhinencephalon are the most remarkable; it is involved in olfactory pathways and especially in activities related to the limbic system . . . The pineal gland[9] is attached to the epithalamus by virtue of its topography but not directly linked to its nuclei" (347). The subthalamus helps to control the extrapyramidal motor pathways.

The *limbic system* is not, strictly speaking, a structure. Rather, it corresponds to a complex forebrain assembly that groups together different structures surrounding the brainstem (cingulate gyrus, hippocampal formation, parahippocampal gyrus, subcallosal gyrus), which are all connected with each other as well as with the main subcortical structures (amygdala, thalamus, hypothalamus, basal ganglia). "Limbic lobe structures are also phylogenetically older than the surrounding neocortex . . . The limbic system participates

---

9. At night, the epiphysis, or pineal gland, secretes melatonin, a hormone that plays a central role in regulating chronobiological (circadian or annual) rhythms. Incidentally, Descartes considered the pineal gland to be the seat of the soul.

in emotional processing, learning, and memory. With each passing year we discover new functions of this system" (Gazzaniga, Ivry, and Mangun 1998, 56), which underscores the link between emotions and learning during development of the individual.

The Brainstem

An important crossroads between the forebrain, the cerebellum, and the spinal cord, the brainstem comprises the mesencephalon, the metencephalon (the dorsal part of which includes the cerebellum), and the myelencephalon (see figure 4.8). It is a complex structure that

- contains all major ascending (sensory and cerebellar) and descending (motor) pathways, transporting information destined for the cortex and transmitting messages coming from it;
- constitutes a pathway for sensory and motor cranial nerves (pairs III to XII), whose nuclei are distributed all along the length of the brainstem, on both sides of the medial line; and
- shelters the reticular formation, an important area consisting of a group of nerve cells, the reticular nuclei, disseminated through the white matter and arranged in a narrow network within the brainstem, along its length.

In contrast to the other neurons of the central nervous system, the neurons of the reticular formation share numerous axonal connections and are also linked to the neurons of the thalamus, the hypothalamus, the cerebellum, and the spinal cord. Situated at the interface of the sensory, motor, and autonomic systems, the reticular formation represents a notable hub for afferent inputs from all sorts of receptors situated throughout the body. It integrates the intensity of the neural messages that converge there, whatever their origin. It coordinates the activity of the nuclei of the cranial nerves, plays a central role in regulating vigilance, and controls tonus (posture and

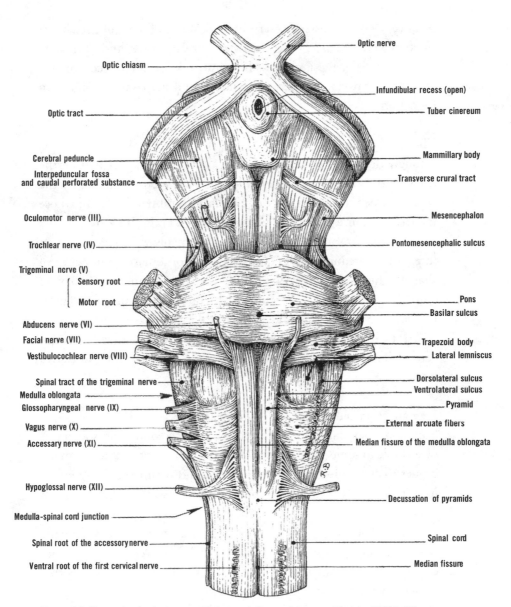

Optic nerve

Optic chiasm

Infundibular recess (open)

Optic tract

Tuber cinereum

Cerebral peduncle

Mammillary body

Interpeduncular fossa
and caudal perforated substance

Transverse crural tract

Oculomotor nerve (III)

Mesencephalon

Trochlear nerve (IV)

Pontomesencephalic sulcus

Trigeminal nerve (V)

Sensory root

Motor root

Pons

Basilar sulcus

Abducens nerve (VI)

Facial nerve (VII)

Trapezoid body

Vestibulocochlear nerve (VIII)

Lateral lemniscus

Spinal tract of the trigeminal nerve

Dorsolateral sulcus

Medulla oblongata

Ventrolateral sulcus

Glossopharyngeal nerve (IX)

Pyramid

Vagus nerve (X)

External arcuate fibers

Accessary nerve (XI)

Median fissure of the medulla oblongata

Hypoglossal nerve (XII)

Decussation of pyramids

Medulla-spinal cord junction

Spinal root of the accessory nerve

Spinal cord

Ventral root of the first cervical nerve

Median fissure

Figure 4.8 The equine brainstem, ventral aspect. Source: Barone and Bortolami 2004, 170.

balance). The brainstem determines the level of general activity of the nervous system, whether emotional or vegetative, and ultimately of the organism, and it controls how information such as commands is treated. It oversees selective attention, and prepares the organism for action.

The Cerebellum
Strongly developed in mammals and abounding in gray matter, the cerebellum is linked to the dorsal part of the metencephalon by the cerebellar peduncles. It is an unpaired and symmetrical organ whose internal structure is comparable to that of the brain. It receives inputs from all the sensory pathways, but is not involved in perception. It integrates these inputs in a complex system of somatic motor control that it achieves in close cooperation with the brain and the spinal cord. In particular, the cerebellum helps in maintaining balance, regulating muscle tonus, and coordinating movements (for example, it is critical to realizing rapid muscular activities such as running). As I will show later, it is also involved in memory for associative learning.

Lateralization and Left-Brain, Right-Brain Specialization
Functional specialization of each of the cerebral hemispheres is observable to some extent in many species of vertebrates, across many classes (fish, amphibians, reptiles, birds, and mammals). In general, it manifests as a contralateral bias due to crossing (decussation) of nerve pathways.

Recall that laterality, as commonly understood, is a systematic preference for the use of one side of the body over another (hand, eye, or foot). In other words, it manifests both on the perceptual as well as the motor level (and we will see that it involves various perceptual modalities in the horse and other species). Moreover, it is also expressed at the individual or at the population level. For a given function, individuals of this or that population will show a marked right or left preference that varies from each to each. Within a population, the majority will express the same net laterality preference. Note, however (and I will return to this point), that in the latter case the bias manifests in a majority, not all members of the population, generally 65 to 90 percent (Vallortigara and Rogers 2005). For example, based on 9,362 observations of racehorses (thoroughbreds, Arabians, and American quarterhorses), Dan Williams and Brian Norris (2007) showed a 90 percent preference among them for galloping on the right foot and 10 percent for galloping on the left foot. However, it is important to note that, in practice, it may be necessary to precisely specify the criterion by which laterality is defined. For example, in a population of 106 English thoroughbreds, Paul McGreevy and Lesley Rogers (2005) observed a population bias where the horses generally extended their left foreleg while grazing. However, Alexandra Wells and Dominique Blache (2008) subsequently showed that, in fact, the preference for one or the other foreleg that manifests when the animal is galloping is independent of what occurs during grazing.

Laterality bias is also subject to variations within the population based on factors such as sex or task (Corballis 2009). Thus, although 90 percent of humans are right-handed, a study of throwing involving a million subjects showed the incidence to be 92.4 percent for women and 89.9 percent for men (Gilbert and Wysocki 1992). In chimpanzees, 65 percent of which are estimated to be right-handed, a similar throwing test involving 92 monkeys revealed only 53 (57 percent) to be right-handed, 16 (18 percent) to be ambidextrous, and 23 (25 percent) left-handed (Hopkins et al. 2005). Moreover, the same phenomenon appears to be characteristic of horses. In an experiment testing different motor tasks carried out on 20 males and 20 females, Jack Murphy, Alastair Sutherland, and Sean Arkins (2005) found a left lateral preference among males and a right preference among females. More recently, in an experiment investigating how 10 mares negotiated an obstacle—symmetrically or asymmetrically, right or left—to attain a bucket of food, Paolo Baragli and colleagues (2011b) observed different strategies depending on the individual: Some chose based on laterality and others according to the shortest path (which in the latter may relate to more highly developed spatial reasoning abilities). Variations in laterality relating to age or the influence of training cannot be excluded, as suggested in a study by Paul McGreevy and Peter Thomson (2006) on motor laterality in a population of 80 horses.

Work on the hemispheric lateralization of emotions has produced two main, and competing, models. According to the first model, emotions are basically governed by the right hemisphere. The second holds that hemispheric dominance depends on the emotional valence of a stimulus: The right hemisphere mostly processes negative emotions (such as fear or aggression), whereas the left hemisphere processes positive emotions (such as a reward of food). A recent review on the issue relating to nonhuman vertebrates (Leliveld, Langbein, and Buppe 2013) suggests that, generally speaking (except perhaps for fish), the second model may be the most useful.

In a review devoted to the evolution and genetics of brain asymmetry, Michael Corballis (2009) refers a variant of the second model proposed by Richard Davidson (1995), based initially on his research on humans, according to which the left hemisphere tends to be specialized for approach and the right hemisphere for avoidance. Moreover, he provides a rather amusing animal example featuring man's best friend: Dogs have a tendency to wag their tail to the left in the presence of unfamiliar dominant dogs, which involves right hemispheric dominance, whereas they wag their tail to the right in the company of their owner, under the control of the left hemisphere (Quaranta, Siniscalchi, and Vallortigara 2007).

More generally, and especially in mammals and birds, there are many examples of a dominant role of neural structures localized in the right hemisphere (accompanied by greater reactivity to stimuli situated to the left) for detecting novelty, possibly in connection with neophobic reactions (in particular, in the visuospatial area), the expression of intense emotions, and the execution of rapid responses to emotional content (such as flight or flight behaviors) (Rogers, Zucca, and Vallortigara 2004). The right hemisphere appears to be more directly involved in agonistic reactions of attack or avoidance in many species (Hook 2004; Corballis 2009). Reactions to predators are generally stronger vis-à-vis those appearing on the left (Ghirlanda and Vallortigara 2004). Thus, for example, many different species appear to be more reactive to the appearance of predators in their left visual hemifield than in the right. Although there seems to be a marked preference for conspecifics to be situated to the left, interactions—including aggressive ones—are often more intense (Hook 2004). In addition, the right hemisphere appears to be more specialized for vocal communication signals of modulated frequency, which transmit information concerning the identity of the emitter, the existence of a threat, and what type, as well as the contextual intensity of the threat (Hook 2004). The left hemisphere appears more directly involved in categorizing stimuli, control of responses that require choices, and inhibition of responses until decisions have been made (Vallortigara 2000; Rogers 2002; Hook 2004). In particular, it plays a preferential role in feeding choices: recognizing prey, say, or selecting foods (Corballis 2009). Although it is mostly concerned with producing vocal communication signals, in many species it is also preferentially involved in recognizing species-specific vocalizations (Ehret 1987; Hauser and Anderson 1994; Siniscalchi, Quaranta, and Rogers 2008).

In the horse, it has also very recently been observed that cognitive activity linked to intermodal individual recognition is associated with a laterality bias reflecting left hemispheric dominance (Proops and McComb 2012). More generally, a variety of research efforts (to which we will return in reviewing perceptual modalities) have shown lateral biases in the horse in perceptual areas concerned with vision (Larose et al. 2006; Austin and Rogers 2007; de Boyer des Roches et al. 2008; Farmer, Krueger, and Byrne 2010; Sankey et al. 2011b), hearing (Basile et al. 2009), and smell (McGreevy and Rogers 2005), with special emphasis on emotional and motor components (McGreevy and Rogers 2005; Murphy, Sutherland, and Arkins 2005; McGreevy and Thomson 2006; Williams and Norris 2007; Wells and Blache 2008; Austin and Rogers 2012).

The existence of functional brain specializations and the lateralization phenomena into which they are translated raise at least two questions: first, how hemispheric specialization and the corpus callosum coexist in placental mammals; and second, why functional brain asymmetries developed within fundamentally symmetrical organisms. (The few organic asymmetries that exist, such as those associated with unpaired organs like the heart or the liver, are not just predominant but characteristic of the entire population of the species in question, aside from several malformations, and can be explained by optimal placement in the available body space.)

As indicated above, the corpus callosum transmits information from one hemisphere to the other and coordinates their activities. James Ringo and colleagues (Ringo et al. 1994) showed that the difference between the time required for this transmission, taking into account the length and thickness of the corpus callosum fibers, and the relatively short time needed by the intrahemispheric circuits to process the information becomes prohibitive in tasks for which speed is essential. In other words, even if interhemispheric information transfer does take place and establishes the "cognitive unity" of the individual, this not only does not preclude hemispheric specialization but may even contribute to its development in a complementary fashion.

Apart from that, the omnipresence of lateralization phenomena within different classes of many species of vertebrates does call into question their evolutionary advantage. According to Stefano Ghirlanda and Giorgio Vallortigara (2004): "The traditional explanation of brain lateralization is that it avoids costly duplication of neural circuitry with the same function, as well as decreasing the interference between different functions" (853). They note, however, that although this argument may explain lateralization at the individual level, it does not explain why, at the population level, a given lateralization is more frequent than another nor why most vertebrates show lateral biases at the population level.

Considering the advantages and disadvantages of hemispheric specialization at the population level (Ghirlanda and Vallortigara 2004; Rogers, Zucca, and Vallortigara 2004; Vallortigara and Rogers 2005), the fact that most individuals of a given species show a laterality bias in the same direction appears a priori an inconvenience, to the extent that the behavior of this majority becomes more predictable for the other species that it confronts. Thus, better detection of predators on the left could lead them to attack their prey from the right! Consequently, a lateralization that appears to be an advantage on the individual level could become a disadvantage at the population level, . . . until a selective pressure over time favors the emergence of a population-level lateralization. It is this kind of selective pressure, for example, that exists in social species, where an individual's chances of survival depend on the behavior of other individuals in the group and where it is in the individual's interest to coordinate the various behaviors. To take the example of predation again, if prey are being attacked on their "bad" side, and practically all of them flee in the same direction, each will see its chances of survival enhanced by the effect of dilution in the group (Treisman 1975). Of course, minority prey lateralized opposite to the majority will have detected the predator earlier and been able to assure their survival. But it is possible to model the evolutionary dynamics of the situation on the basis of game theory (Ghirlanda and Vallortigara 2004) and to show that the survival of the minority is not appreciably enhanced. A stable evolutionary strategy develops, ensuring the continued existence of lateralization at the population level. In this way, it has been noted (Vallortigara and Bisazza 2002, cited by Ghirlanda and Vallortigara 2004) that in testing the direction of flight when approached by a simulated predator, out of 20 species of fish (10 solitary and 10 living in schools), 6 of the solitary species show lateralization only at the individual level, whereas the 10 species living in schools were all lateralized at the population level.

Brain and Mind in the Light of Evolution

As Theodosius Dobzhansky (1973) famously put it: "Nothing makes sense in biology except in the light of evolution." Accordingly, in addition to the functional anatomical elements of the central nervous system described thus far, it might now be useful to elaborate on what we know about the way in which evolution has shaped the brain and mind of vertebrates in general and the horse in particular.

During the 1960s, Paul MacLean proposed a synthetic theory relating to the evolution of the structure of the human brain that he called the triune brain. The idea is that the human brain is the result of a nested hierarchy of three brains appearing in succession over the course of evolution: a primitive, reptilian brain comprising the brainstem and striatum; a limbic, paleomammalian brain characteristic of the first mammals and consisting primarily of the hippocampus, the amygdala, and the hypothalamus (in other words, the basis of the neuroanatomical circuit identified by James Papez in 1937 that he believed to be involved in the formation of emotions); and, finally, a neocortex or neomammalian brain, which developed in later mammals, reaching its apogee in "higher primates." For MacLean, the reptilian brain was the seat of genetically programmed behavior such as the basic impulses of eating, defense, flight, and reproduction. It also controlled the organism's vital functions (respiration, cardiac rhythm, body temperature). The limbic brain was responsible for emotional reactions, and the neocortex for "intellectual life" and higher mental functions. These included singular capacities for learning and culminated in humanity with abstract thought, language, and consciousness.

Although MacLean's theory of the triune brain had the virtue, at a certain era, of drawing attention to the importance of the evolutionary past in the construction of the brain, it is nonetheless incompatible with what is known today. As Steven Pinker (1997) points out: "One problem for the triune theory is that the forces of evolution do not just heap layers on an unchanged foundation. Natural selection has to work with what is already around, but it can *modify* what it finds. Most parts of the human body came from ancient mammals and before them ancient reptiles, but the parts were heavily modified to fit features of the human lifestyle, such as upright posture . . . The circuitry for the emotions was not left untouched, either . . . Emotional repertoires vary wildly among animals depending on their species, sex, and age. Within the mammals, we find the lion and the lamb" (371). Moreover, writes Marc Jeannerod (2002): "MacLean's triune theory is evidence of the enduring persistence of a model of evolution, advancing in successive layers, where increasing complexity is treated as the result of the hierarchy of species, constructed over time . . . Yet we know that this vision of the evolution of species in the form of continuous progress hardly corresponds to current thinking, which is based on mutation and selection" (60–61). Finally, remarks Pierre-Yves Risold (2008): "The mammalian cortex is no more recent than that of current reptiles: both have evolved, each in its own way, as a function of the constraints to which they were subject . . . In fact, MacLean's entire edifice is suspect . . . It is wrong to assert that the Papez circuit is specialized in the formation of emotions . . . Today, it appears that MacLean's triune theory rests on false assumptions about the anatomy of the brain, how it works, and the way in which it evolved from one species to another over millions of years" (70).

Before briefly going into several aspects of current ideas about the evolution of the brain and mind, let me say that my reason for highlighting MacLean's dated notions is that they are still to be found in many pseudoscientific commentaries regarding the brain of the horse and its mental faculties. This apparently has something to do with the fact that MacLean himself compared the "limbic brain" to that of the horse (1989). As he wrote (1962): "Speaking allegorically of these three brains within a brain, we might imagine that when the psychiatrist bids the patient to lie on the couch, he is asking him to stretch out alongside a horse and a crocodile" (300). What MacLean meant only as metaphor—and which in fact was adopted by some authors (Koestler 1967; Carlier 2002; Jouvent 2009)—was taken literally by others who have, unfortunately, forgotten its figurative intent. Moreover, considering that in such a context, a horse is treated as an inferior mammal, like the lissencephalic rat or rabbit, the basis of many wrongheaded speculations about the neuroanatomy of the horse becomes obvious.

Without going into detail regarding evolutionary theory, recall that evolution is a fundamentally opportunistic process: Its basic mechanism, natural selection, creates nothing; rather, it plays on the variability associated with sexual reproduction by selecting among existing characteristics the ones that, at a given moment, are the most adapted to the immediate environment—in other words, those that best contribute to the reproductive success of individuals. To summarize, natural selection is based on

- interindividual variety (morphological, psychological, and behavioral);
- hereditary variations associated with the appearance of new hereditary characters, either by mutation (physical alteration of genes, which modifies their effects) or by genetic recombination (separation of groups of genes joined within the same chromosome, which produces new genetic combinations); and
- the difference in reproductive success associated with hereditary variations, arising from either the rate of reproduction (a function of sexual selection—intermale competition and choice of females—and fertility rate of females) or juvenile mortality.

To paraphrase a well-known aphorism, mutation or genetic recombination proposes, selection disposes; individuals are selected, the population evolves.

To arrive where they are today, each extant species followed its own path, according to the various selection pressures to which it was exposed over its history as a result of fluctuations in its physical and social environment (and in particular as a function of the other species sharing its ecological niche at any given moment in time) and to the specific characteristics it possesses by virtue of its own evolutionary history. To put it another way, to some extent, each species that exists today is the fruit of a "successful" evolution. With respect to the horse, for example, I described in the first chapter the successful evolution that it represents in terms of its remarkable adaptability to diverse ecological conditions (feral horses are found in regions with widely varying features); moreover, this adaptability is linked to an equally impressive behavioral adaptability.

In the context of evolution, one long-standing idea is that of a differentiated relationship between the weight or size of the brain and that of the body, depending on the species. Jean Gayon (2000) recalls that, as early as the 1890s, Eugène Dubois[10] and Louis Lapicque suggested that, in mammals, the relationship between the two obeyed a power law of the type $E = K \times P^{0.56}$ (where E is the weight of the brain and P that of the body), K being the encephalization quotient, thus anticipating the work of Julian Huxley and Georges Teissier between 1920 and 1930, who introduced the concept of allometry to designate the concept of relative growth. An idea related to the encephalization quotient was that a greater relative brain size (that is, a higher encephalization quotient) indicated a more developed degree of intelligence. As expressed by Paul Chauchard (1956): "Both from the static perspective of comparative anatomy and the dynamic view of paleontology, the progress of intelligence parallels the increasing complexity of the higher nerve centers and, in vertebrates, the brain which contains them. The desire to approach the problem quantitatively is longstanding, but the question is difficult; for aside from that fact that the precise measure of intelligence is fairly arbitrary, there is the elementary issue that the size of the brain depends not only on the degree of intelligence but also the dimensions of the body" (101).

And yet, having considered various mathematical formulas for calculating encephalization quotients based on all or part of the brain and purporting to take into account the level of intelligence achieved, Chauchard concludes that "any significant increase in the brain from the point of view of intelligence depends on the development of cortical connections" (104). Nonetheless, research on encephalization quotients continued up to the 1970s, in particular by Harry Jerison

10. The Dutch anatomist Eugène Dubois is a celebrated figure in the annals of paleoanthropology for having studied the "cradle of humanity" in Asia at the end of the nineteenth century and consequently having discovered, in 1891, Java Man, whom he named *Pithecanthropus erectus* in 1894.

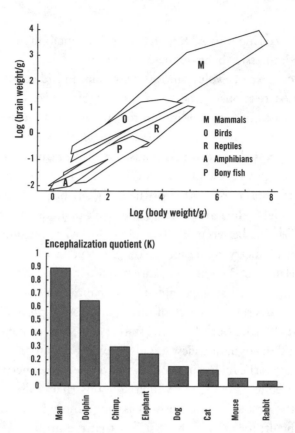

Figure 4.9 Top: relation between the weight of the brain and the weight of the body for different orders of vertebrates (Jerison 1973). Bottom: encephalization quotient, with K = E/P2/3. **Chimp.**: chimpanzee. Source: Vauclair 1995, 164.

(1973) (see figure 4.9). Although Jerison did show a variety of general relationships between the relative size of the brain and cognitive capacities, including the capacity for learning in different orders of vertebrates, anomalies were evident, in particular in mammals. As Jacques Vauclair points out (1995): "The application of this quotient does not necessarily reveal a regular and progressive increase in the size of the brain between the top and the bottom of the phylogenetic tree" (165).

In a review titled "Evolution of the Brain and Intelligence," Gerhard Roth and Ursula Dicke (2005) present several telling figures in this regard. The authors define intelligence in terms of the level of mental

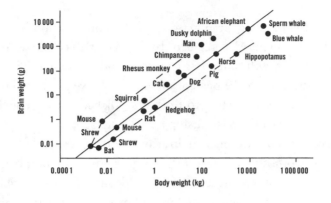

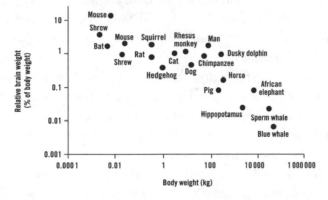

Figure 4.10 Top: weight of the brain as a function of body weight. Bottom: percentage of body weight represented by the brain. These charts include two different species of mouse, as well as shrews. Source: Roth and Dick 2005, 252.

or behavioral flexibility, which, faced with challenges, translates into the appearance of new solutions outside the normal repertoire of the animal. Accordingly, the authors gathered features particular to a set of mammals and constructed two logarithmic charts relating relative brain weight to body weight as a function of an allometric power law with an exponent of 0.6 to 0.8 and the percentage of body weight that the brain represents. They note especially that mammals with larger brains are assumed to be more intelligent and that, according to Jerison (1973), among others, with increasing body size the brain becomes absolutely larger but relatively smaller (see figure 4.10, top). For all that, among large mammals, the human brain has the greatest relative weight (2 percent of body weight), yet shrews (the smallest mammals with much less cognitive and behavioral flexibility) have brains of up to 10 percent of their body weight (see figure 4.10, bottom). Thus, according to Roth and Dicke, nothing conclusive can be said about the relationship between relative brain size and intelligence, nor about the encephalization quotient (see table 4.3).

Moreover, Roth and Dicke (2005) suggest that the number of neurons combined with the conduction velocity of cortical fibers (which can be estimated by the thickness of myelinated fibers) probably correlates best with intelligence, which they nonetheless recognize to be fairly speculative given the lack of data on both the thickness of myelinated fibers and on meaningfully comparative features with respect to animal intelligence.

I might add, more generally, that although evolution is based on successive adaptations to environmental, physical, and social constraints, which become modified over time, such adaptations—both mental and physical—are realized by "adjustments" to these constraints. Evolution is a dynamic process that continually builds new structures over old ones. It is thus hardly surprising to find that animal species cannot be classified according to a strict hierarchy of mental or physical capacities. By way of illustration, consider the chimpanzee, our closest relative, which, as we are constantly reminded, shares with humans 98 percent of its genome and whose cognitive capacities are acknowledged to be remarkable. Taking a hierarchical view of evolution, who would seriously maintain that the dog, which humans consider close on the basis of affection, though not necessarily of intelligence, is somehow superior to the chimpanzee with respect to mental faculties? And yet! Recent research (Hare and Tomasello 2005) in fact shows that, although chimpanzees are irrefutably better than dogs at solving problems, dogs have the advantage in certain aspects of social communication, in particular in interpreting human cues. "When the wise man points at the moon, the fool looks at his finger" goes the proverb. It is precisely in responding to a finger pointed at an object in the distance that dogs trump fools and

Table 4.3   Brain weight, encephalization quotient (relative to that of the cat, conventionally set at 1), and number of cortical neurons in selected mammals.

Animal taxa	Brain weight (in grams)	Encephalization quotient	Number of cortical neurons (in millions)
Whales	2,600–9,000	1.8	
False killer whale	3,650		10,500
African elephant	4,200	1.3	11,000
Man	1,250–1,450	7.4–7.8	11,500
Bottlenose dolphin	1,350	5.3	5,800
Walrus	1,130	1.2	
Camel	762	1.2	
Ox	490	0.5	
Horse	510	0.9	1,200
Gorilla	430–570	1.5–1.8	4,300
Chimpanzee	330–430	2.2–2.5	6,200
Lion	260	0.6	
Sheep	140	0.8	
Old World monkeys	41–122	1.7–2.7	
Rhesus monkey	88	2.1	480
Gibbon	88–105	1.9–2.7	
Capuchin monkeys	26–80	2.4–4.8	
White-fronted capuchin	57	4.8	610
Dog	64	1.2	160
Fox	53	1.6	
Cat	25	1.0	300
Squirrel monkey	23	2.3	480
Rabbit	11	0.4	
Marmoset	7	1.7	
Opossum	7.6	0.2	27
Squirrel	7	1.1	
Hedgehog	3.3	0.3	24
Rat	2	0.4	15
Mouse	0.3	0.5	4

Source: Roth and Dick 2005, 251.

chimpanzees fail, aside from a few rare adult individuals raised in a human environment. Similarly, from early youth onward, dogs show an amazing capacity for interpreting the direction of the human gaze, comparable to that of infants less than a year old and much better than that of chimpanzees. Such abilities appear to go hand in hand with selection pressures linked to the domestication of the dog, dating back some fifteen thousand years (Leonard 2002).

Returning to my main thesis, the combination of specific adaptive adjustments acting selectively on this or that characteristic suggests varied evolutionary pathways depending on the pressure that some observers—Chapouthier (2001), for example—call mosaic and that translate into specialized nervous structures or particular psychic modules.[11]

The way in which the brain recognizes faces serves to illustrate this point. Although it has long been known that human recognition of faces depends on specialized processing (Gazzaniga, Ivry, and Mangun 1998) whose localization is now well known (Schwarzlose, Baker, and Kanwisher 2005), it has been shown that such specialized processing exists as well not only in primates but also, and especially, in an ungulate: sheep (Kendrick, Leigh, and Peirce 2001; Tate et al. 2006). In this case, a specialized cognitive module does correspond to a particular brain structure. Although similar studies have not been done for the horse, there is reason to think that the same holds true for them.

However, here the term *structure* must be understood in a broad sense: It applies to networks of interconnected neurons as much as to a well-delineated anatomical entity. This goes back to ideas of brain localization, as they were conceived in the past, "having not much to do with current localized modifications of cerebral activity observed using brain imaging techniques. What these techniques reveal is the

---

11. "Between the barely acquired juxtaposition and perfect integration within a higher-order structure (assuming such perfect integration is in fact achievable) are long laps of time where integration is occurring, but where the basic structures maintain substantial hierarchical autonomy. The term *mosaic* appears to best describe this state" (Chapouthier 2001, 70–71).

existence of networks of activity connecting diverse brain regions. These networks are activated during execution of a given task and then give way to others. The same region (the same localization) can belong to several different functional networks, which does not mean, however, that it can fulfill several functions: It would be more correct to say that different functions require the participation of the same region" (Jeannerod 2002, 24–25).

Various studies have dealt with the influence of certain categories of selection pressure on cognitive capacities, for example, constraints associated with the mode of localization and acquisition of food resources (McLean 2001), which would be related to higher-level mental faculties in carnivores than in herbivores, or the importance of the size of the group (Croney and Newberry 2007) on particular capacities related to social cognition.

Let me touch briefly on a few other research efforts that focus more particularly on the evolution of the brain itself. Among them are investigations of possible links between functionally specialized brain systems and the evolution of the brain (Barton and Harvey 2000; Brown 2001; de Winter and Oxnard 2001). These studies tend to support the idea of the role of mosaic-type evolution in the brain. As Barton and Harvey (2000) put it: "The most striking feature is the combination of highly significant relationships within functional systems and the general absence of such relationships between systems" (1057). Other studies use comparisons between species to examine relationships between behavioral characteristics and the development of brain structures, honing in, for example, on the importance of social factors within different mammalian orders (Pérez-Barberia, Schultz, and Dunbar 2007), as well as ecological factors such as habitat and the feeding regime of ungulates (Pérez-Barberia and Gordon 2005; Shultz and Dunbar 2006) on the size of part or all of the brain.[12] More specifically, note that there is a general tendency toward reduction of brain size linked to domestication (Kruska 1988). In any event, a review (Healy and Rowe 2007) of more than fifty studies comparing the size of the brain carried out over the last decade inspires caution in drawing any conclusions, given the importance, complexity, and controversial nature of the problems raised by the investigations, in particular the methods and even the criteria used. By way of illustrating these difficulties: Is the brain measured in the same way from study to study, for example, directly by weight, or indirectly by volume? Is the brain sample used to measure the average value for a given species adequately representative?[13] Have variations in induced weight, specific or seasonal differences in cerebral water volume, or the effect of rehydration of frozen brain after dessication been taken into account?[14] What exactly is meant by "social complexity"? Are different studies even talking about the same thing? Is correlation being too easily confused with causality?

In concluding this chapter, let me add that, although research on the brain, its structure, and its function is a matter of intense interest at present and promises much in the way of new findings in the relatively near future, what we know today about mammals, and especially the horse, already enables us to better understand how several mental processes work, such as memory and learning. But current knowledge also lays bare the complexity of the phenomena at work, persistent lacunae, and the necessity of resisting the temptation toward overly simplistic interpretations that await us at every turn.

---

12. Expressed by Hahn (2004) as "the relatively small size of the forebrain of equids compared with that of social carnivores and primates."

13. Nonnegligible variations in volume and weight occur, for example, with sex and age (shrinking due to aging). Apart from interindividual variability (within our own species, Anatole France's brain weighed in at 1,017 grams and Ivan Turgenev's at 2,012 grams).

14. Water is the main constituent of the mammalian body. In the adult human, for example, the total water content of the body is around 65 percent; in the brain, it is around 75 percent.

5

THE NATURE OF EQUINE PERCEPTION

Perception: A Dynamic Process That Constructs the World

To perceive the world, we use our sense organs. However, we know that in the animal world, all species do not share the same senses. Thus, some animals have senses that we do not and that enable them, for example, to detect electrical signals—like some fish and even insects (Clarke et al. 2013)—to navigate and forage by echolocation (like bats,[1] dolphins, and whales), and to orient themselves with the aid of magnetic fields (like migratory birds, especially garden warblers, which appear to do it using magnetic particles located on either side of their beak) (Heyers et al. 2007).

Other animals, in contrast, lack some of our senses. For example, although snakes have mechanoreceptors in their skin that allow them to feel ground vibrations (McFarland 1985), they do not "hear" sounds (in fact, it is the movements of the flute, and not the melody being played, that earns a snake charmer his living).

Moreover, the same senses do not function in the same way in all species. Dogs, for instance, perceive ultrasound frequencies to which

---

1. Donald Griffin, who was a pioneer in the field of cognitive ethology, and Robert Galambos demonstrated the features of echolocation experimentally in the bat in 1941.

111

we are insensitive (which is what makes it possible to train them using whistles that are totally inaudible to humans). Elephants, in contrast, communicate by means of infrasound frequencies (Payne, Langbauer, and Thomas 1986). Bees see ultraviolet light and are sensitive to light polarization (von Frisch 1955[2]). The antennae of the male silkworm moth *Bombyx mori* can detect the odor of a female 10 kilometers away, and a single molecule of the pheromone emitted by the female is enough to trigger male response (Kaissling and Priesner 1970). Finally, frogs possess a retina containing photoreceptors, called feature detectors, that react selectively to small, dark, moving objects (Lettvin et al. 1959).

In short, at the level of the sense organs, reality is not the same for everyone. Each animal species is more or less endowed with sensory filters unique to it that let through only some of the perceptual information from the environment, which we call sensations. As Arlette Stréri (2003) puts it: "The information extracted by the sensory systems gives rise to a sensation. They are specific to the system that detects them, and for this reason remain incomplete, fragmentary. At this level, the processing is automatic, prewired, largely unconscious, and thus modular" (329). These raw data are then taken into account by the brain to give rise to a perception: "From a cognitive perspective, perception has an interpretive function vis-à-vis sensory data and acts as an information processor" (ibid.).

By way of illustration, take the example of what are called perceptual constants. If I look at a round plate set flat before me, the image formed on my retina is oval, and yet I "see" the plate as round. If, now, I watch a door opening, the shape of the door does not change for me, although the image it projects on my retina is gradually deforming. Similarly, if I look at two identical objects that are at different distances away from me, the image on my retina of the closer one is larger than the one that is further away. Yet I continue to "see" the same size. Finally, if I see a car in the distance, and I focus

2. Karl von Frisch, one of the founding fathers of ethology, shared the Nobel Prize in 1973 with Konrad Lorenz and Nikolaas Tinbergen.

Figure 5.1 The human mental representation of the animal makes it possible to distinguish the dog despite the obvious bad quality of the image.

close up on a little model of it, even if the image of the model on my retina is larger than that of the real car, I will still perceive the real car as larger.

Returning to Stréri (2003): "Before interpreting a stimulus, grouping and segregation operate on the sensory flow based on the knowledge of the perceiving subject. This knowledge guides perceptual structuring and enables identification of the object" (329–30). In other words, existing mental representations contribute in equal measure to the perceptual process. Moreover, this is particularly obvious in our ability to process very poorly resolved stimuli, such as paintings or drawings degraded by time (see figure 5.1). As paleontologists in particular observe in the context of their fieldwork: You only find what you already know. The intervention of such preexisting mental representations also occurs in object recognition, which I will discuss in chapter 7, on equine visual perception of shapes and movement.

Consequently, far from being a copy of reality, the human brain perceives the world by transforming the data supplied by the senses: "Perception can work like inference by searching and obtaining information that directly validates a causal interpretation that a more detailed study of inferences could produce, and inference can work like perception by filling in gaps in information when perception is deficient or inadequate" (Massad, Hubbard, and Newtson 1979, 531). Another notable example of this is also found in the area of visual perception. Whereas cones, which enable color perception, become increasingly less dense just a few degrees away from the point of fixation (the fovea), the scenes are still perceived as being completely and uniformly colored, and not just at their center. What is at play here is an adaptive process that enables humans to restore to the environment the stability and coherence needed to act and survive in it.

In *The Brain's Sense of Movement,* a work devoted to what he calls the sixth sense, neurophysiologist Alain Berthoz (2000) writes: "The species that passed the test of natural selection are those that figured out how to save a few milliseconds in capturing prey and anticipating the actions of predators, those whose brains were able to simulate the elements of the environment and choose the best way home, those able to memorize great quantities of information from past experience and use them in the heat of action. Relationships between perception and action are the model of choice for studying the functions of the nervous system" (3).

Jacob von Uexküll (1957) observed early on that each animal species had its own universe (its *Umwelt*), noting at the same time that the *Umwelt* is intrinsically bound up with a perceptual world (*Merkwelt*) and an effector world (*Wirkwelt*), which Frederik Buytendijk (1965) describes in the following way: "Each animal species has its own environment. Within this environment, von Uexküll distinguishes the perceptual world and the effector world. Through its sensory organs, each animal is sensitive only to certain stimuli—signals capable of triggering action. But this action is determined by moving organs, limbs or mouthparts, by means of which the animal may influence some of its surroundings (the effector world). Animal and world adapt to each other, says von Uexküll, like a lock and key. The makeup of the animal and its corresponding environment are like the parts of an organism, in their mutually dependent functional inter-actions" (54–55). Von Uexküll (1957) provided an additional, pithy formulation: "Perceptual and effector worlds together form a closed unit, the *Umwelt*" (6).

Such remarks emphasize the adaptive function of perception; perception is not neutral. To perceive is to imbue something with meaning combined with action: "Figuratively speaking, every animal grasps its objects with two arms of a forceps, receptor and effector. With the one it invests the object with a receptor cue or perceptual meaning, with the other, an effector cue or operational meaning" (von Uexküll 1957, 10). Moreover, the animal's particular universe is constructed within the combined constraints of the characteristics of its sensory equipment, its mental faculties, and the possibilities enabled by its motricity.

And yet, this adaptive process of interpreting reality on which perception is based may also be tricked and give rise to perceptual illusions: "Over and above the complexity of its functions, the most striking attribute of gestalt perception is its specific weakness—the ease with which it can be misled to fallacious 'conclusions'" (Lorenz 1981, 41). At issue is what used to be called optical illusions insofar as most of the investigations and research carried out in the area relate to visual illusions, which are generally easier to demonstrate. In fact, perceptual illusions involve many other senses (illusions linked to contrast effects, in particular, apply to all the senses), and they not infrequently engage several sensory modalities simultaneously, as I will now show.

The illusion created by Franz Müller-Lyer in 1889 (see figure 5.2, top) is probably one of the most familiar. It consists of inserting arrowheads at the ends of line segments of the same length: When the arrowheads point out, the line segment (not including the arrow-heads) appears shorter; when the arrowheads point inward, the line appears longer. Similarly, in the T illusion, or inverted T, the vertical

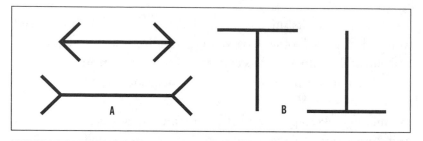

Figure 5.2 Top: (A) The Müller-Myer illusion. Despite appearances, the two lines are the same length. (B) The T illusion (right side up; upside down). Bottom: the Gateway Arch, in St. Louis, Missouri, is in fact as wide as it is high (630 feet). Source: © Fotalia.com/Gino Santa Maria.

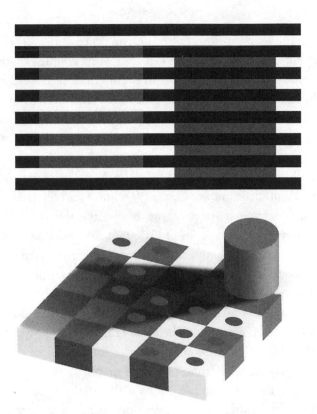

Figure 5.3 Top: The white illusion. The gray vertical stripes are the same shade. Bottom: shadow over an Adelson checkerboard. Similarly, there is no difference in shading between rounds, no matter the color (black or white) of the square they are in.[3]

appears longer than the horizontal, whereas they are exactly the same length. An architectural illustration of this illusion is supplied by the Gateway Arch in St. Louis, Missouri, whose height appears greater than its width, although both measure 630 feet. Other visual illusions play on effects of color or hue, such as the White (1979) illusion, or the spectacular effect of shade on a checkerboard, devised by Edward Adelson (2000) (see figure 5.3).

3. If you are not convinced by the example on the right side of the figure, take a sheet of white paper and make two holes in it such that they reveal only squares A and B when you lay the paper over the checkerboard. The contrast effects will disappear, and you will be able to see that the two squares have exactly the same shade.

As an example of illusions engaging the other senses, in the area of tactile perception, it is easy to do the following simple experiment: Simultaneously dip your left hand in a container of hot water and your right hand in a container of cold water and keep your hands there for a few seconds. Then dip both hands into a container of lukewarm water. The water will feel colder to your left hand and warmer to your right hand.

Tactile illusions are not limited to contrast effects. An example is the so-called cutaneous rabbit illusion (Geldard and Sherrick 1972). When clearly separated parts of the body, like the wrist and the elbow, are stimulated by a series of little taps in rapid succession, the perception is one of a "phantom" succession of regularly spaced taps between the two stimulated points (whence the sensation of a rabbit hopping between the fist and the elbow). This illusion has a counterpart in the auditory domain (Bremer et al. 1977), where there is no shortage of perceptual illusions, the most familiar being linked to the use of stereophonic channels that give the impression of orchestral instruments playing at different points in space.

In real life, then, perception in basically multimodal. Auditory information can modify visual perception. When a subject hears a rapid series of beeps at the same time he or she is exposed to a single light flash, the subject perceives a series of flashes (Shams, Kamitani, and Shimojo 2000). Interactions between visual and auditory senses can also be illustrated by means of auditory perception. A classic illusion, which concerns the perception of speech, involves the McGurk effect (McGurk and MacDonald 1976). When a subject observes a video sequence showing the head of a person whose lips are repeating the sound *ga* while the soundtrack is simultaneously playing a series of *ba* sounds, the subject perceives an intermediate *da*.

During this short review of perceptual illusions, I have so far stressed examples relating to human perception because they are varied and easy to explain. Indeed, to the extent that we broadly share a common perceptual world, we also largely share a similar experience of perceptual illusions and thus generally have an intuitive understanding of them. Moreover, the field of investigation is larger for humans than for animals.[4] It is certainly easier to ask people to say what they perceive and to describe, or even to measure, discrepancies between objective and perceived reality than to "ask" animals. Consequently, it is easier to find examples referring to different senses for humans than for animals. And yet, the existence of perceptual illusions that reveal the involvement of complex mental processes in perception has also been observed in the animal kingdom, although the number of such studies in this case is fewer.

In any event, using discriminative learning techniques (that is, training during which an organism learns to respond differently to different stimuli), researchers have been able to show (Nieder 2002) that insects, birds, and mammals can perceive illusory shapes. This is especially true of cats (Bravo, Blake, and Morrison 1988) when exposed to the Kanizsa triangle. It is also true of bees (van Hateren, Srinivasan, and Wait 1990) shown a Kanizsa square, as well as barn owls (Nieder and Wagner 1999) "asked" to tell a square from a triangle, which were first shown a real contour then, randomly, an illusory one (see figure 5.4).

In another example, researchers showed that different species, including pigeons (Fujita, Blough, and Blough 1991), rhesus monkeys, chimpanzees (Fujita 1997), and horses[5] (Timney and Keil 1996) are sensitive to the Ponzo illusion, which makes two identical line segments appear to have different lengths when they are placed inside two converging lines, which reveals a certain capacity for evaluating depth based on monocular clues. Along the same lines, baboons (Barbet and Fagot 2002) are sensitive to the corridor illusion (see figure 5.5).

These last illusions bring us back to the relationship between perception and action, given the importance of mechanisms of depth and distance perception to survival, if only in the context of competition between prey and predator. It is thus not so surprising

---

4. It goes without saying that here I am using the term *animals* to mean nonhuman animals and thus do not suggest that humans should be excluded from the animal kingdom.
5. I will come back to this point more specifically in chapter 7, in exploring the behavioral aspects of equine visual perception.

Figure 5.4 Top: Kanizsa triangle and square. Bottom: square and triangle with real (left and middle) and illusory (right) contour.

to discover that perception is not only multimodal (that is, hearing participates in evaluating the localization of partners, competitors, or adversaries) but that the same modality comes into play in different complementary ways. Accordingly, visual perception of depth and distance is based both on binocular vision and monocular clues, as well as on clues supplied by an individual's own movements.

Regarding the link between perception and action, I must add that selecting information based on the action to be taken is also a key element of perception. In other words, it is subject to the functional constraints of attentional processes that may be hardly conscious. Everyone knows that mothers can sleep through the rumble of a passing truck but wake at the least murmur of their baby slumbering in another room.

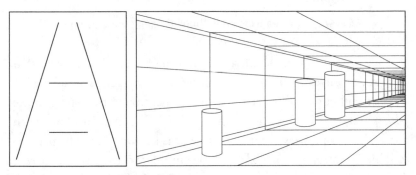

Figure 5.5 Left: The Ponzo illusion. The two horizontal lines are the same length. Right: corridor illusion. The three cylinders are the same height.

This selection of information occurs everywhere. It is particularly striking, even spectacular, in the so-called cocktail party effect (Cherry 1953). This effect refers to the ability of humans to focus on a conversation amid a noisy crowd. It involves both selective listening and a sort of "neutralization" of the surrounding sound. The ability is also present in animal species whose ecological conditions expose them to similar perturbations. A particular example is the royal penguin (Aubin and Jouventin 1998), which nests within colonies of several thousand individuals. When parents return from fishing, infants need to be able to recognize them by their calls.

"The brain is not a reactive machine," writes Alain Berthoz (2000). "It is a proactive machine that investigates the world" (1).

Let us look more closely at the horse and its version of the five senses usually attributed to humans. In the horse, these senses function somewhat differently. Moreover, because the equine brain was not constructed in the same way as that of humans over the course of evolution, I will briefly touch on questions posed by the study of animal perception in general and that of the horse in particular, before summarizing perception linked to the five sensory modalities and their chief characteristics.

A Few Issues Regarding the Study of Equine Perception

Without repeating in detail the points made thus far, to wit, that each species lives in its own world, which it perceives in a specific way, a number of additional issues merit mention with respect to the perceptual experience of nonhuman animals in general and horses in particular. In a review synthesizing what is known about the perceptual life of horses, Carol Saslow (2002) highlights the gaps in current understanding of how horses perceive.

Take, for example, the relationship between the emission of a given physical stimulus (visual, auditory, olfactory) and its reception by the sensory organ. Consideration of this aspect assumes taking into account the nature of the physical characteristics of the stimulus in question as well as those of the sensory organ. Accordingly, it requires, on the one hand, being able to carry out realistically controlled experiments in color vision: that the experiments be comparable; that colors be physically definable and their wavelength be known; that it be possible to determine how to obtain all the colors we recognize based on the primary colors red, yellow, and blue; and that perceptual parameters be specified, as well as a color's hue, purity (or saturation), and intensity (or luminosity,[6] brilliance, brightness, or sharpness). On the other hand, one has to be able to identify the kind of receptor cells involved (the cones), how many of these a given species has (three in humans and two in the horse), and the color characteristics of the photosensitive matter they contain (pigments). All this suggests that care needs to be taken with experimental protocols; if not, one risks not really knowing what one is measuring or how to interpret the results!

Without delving too deep into details, the complexity of investigations in the area of hearing is well illustrated by the need to experi-

---

6. The quantitative expression of luminosity is in fact given by luminance. Expressed in candelas per square meter (cd/m^2), it is the quotient of luminous intensity on a surface by the apparent area of this surface for a distant operator. Thus, it is a measure of reflected light perceived by the eye. The candela replaces the old notion of candle and thus corresponds approximately to the light provided by a candle.

ment with speakers or a soundproof room to differentiate between the stimulus produced by the experiment and that reaching the eardrums of the subject. This is a major concern, for example, in investigating the quantitative relationship between stimulus and sensation, that is, the branch of experimental psychology that, since its invention by Gustav Fechner (1860), has been known as psychophysics.

Saslow (2002) points out how tricky this subject can be and the amount of time required to adapt to nonspeaking subjects methods developed for humans by psychophysicists. Moreover, given differences between individuals, the number of animals tested must be sufficiently high to produce reliable data. Under such conditions, it is easy to imagine the problems posed by animals that are big, expensive, and hard to handle, like horses, especially inasmuch as the studies called for are not just behavioral but also neuroanatomical and neurophysiological.

Insofar as "perceptual mechanisms tend to be conservative in evolution" (Saslow 2002, 211), a comparative approach between species could prove useful. Nonetheless, mammals whose evolutionary histories and survival strategies are very different, as is notably the case with carnivores and herbivores, could very likely also present widely differing cognitive (and especially perceptual) characteristics. Writes Saslow (2002): "We must be particularly cautious when generalizing human perceptions to our horse companions" (212).

Along these same lines, although it seems reasonable to ask exactly what type of sensory information contributes to the adaptation of a given species in its natural milieu—and here, ethology is in a position to contribute significantly—observations too frequently emphasize vision because of its importance to humans and underestimate other perceptual modalities: "In the scant research literature on horse perception, the bulk of studies have been on vision while senses probably of more crucial importance to the horse Umwelt have been neglected" (ibid.).

While I do not fully share this point of view since, in any event, vision is obviously vital in detecting possible predators, it is true that, at present, vision and hearing are the only equine senses for which significant amounts of data are available (Timney and Macuda 2001). This point will become clearer in the next chapters on exteroceptive perceptual modalities, that is, corresponding to the traditional five senses, those fundamentally involved in interactions with the environment, thus excluding the kinesthetic modalities (proprioception and the vestibular system).

6

THE ANATOMICAL AND PHYSIOLOGICAL BASIS OF EQUINE VISUAL PERCEPTION

Visual perception is enormously complex. Consequently, I will begin this chapter by reviewing the main anatomical and physiological aspects relevant to visual perception in the horse, with special emphasis on how the sensory organ—the eye—captures, filters, and codes visual information. I will also make several observations about how this information travels to the brain and is processed there, as well as about some basic elements of color vision. I will then examine how the equine brain interprets visual information, in other words, how the horse perceives its visual world. To this end, I will refer to behavioral work that has enabled researchers to "question" the horse and consequently both complementary and more specific information regarding its "experience." Accordingly, the next chapter will be devoted to how the horse perceives shapes and movement (visual acuity, visual integration, lateral bias, pattern and depth perception, image and object recognition, motion detection). Color perception, which is a tricky and controversial issue, will be the subject of its own chapter.

Here, I will focus on the size and arrangement of the eyes and visual field, the overall structure of the eye, the anatomy of the retina, retinal composition and image quality, optical pathways and their distribution in the brain, color theory, and the anatomical and physiological basis of color perception.

Size, Arrangement of the Eyes, and Visual Field

The most remarkable thing about a horse's eyes is their size and the way they are set in the head.

Horses have especially large eyes. The nearly spherical eyeball, similar to that of humans, has a horizontal diameter of 5 to 5.4 centimeters, a vertical diameter of 4.5 to 5 centimeters, and an anteroposterior diameter of 4.2 to 4.3 centimeters (Barnett et al. 2004). That is about double the anteroposterior diameter of the human eye (around 2.4 centimeters). A study carried out on thirty adult horses of varying breeds indicates a margin of variation in volume, independent of breed, of 42 to 62.5 cubic centimeters (Evans and McGreevy 2007). That makes the equine eye seven to nine times larger than the human eye, whose average volume is 6.4 cubic centimeters. It seems too that the horse has the largest eye of all land mammals (McGreevy 2004). The size of the horse's eye contributes to its visual acuity, which as we will see is relatively good and better than that of most other mammals (Timney and Keil 1992; Timney and Macuda 2001; Timney and Wright 2007).

Contrary to legend, horses do not perceive objects bigger than humans do, and it is not for this "reason" that they startle so easily! It is true that the images formed on the equine retina, at the back of the eye, are larger than those formed on the human retina. Nonetheless, large as these images are, equine perception is as much a function of the brain as it is of the eye. Indeed, it is the brain that interprets visual sensations from amid a wide range of information. It goes without saying that, strictly speaking, the size of an animal's eye is unrelated to how it assesses the scale of what it sees.

Set to the side and at the back (widest) portion of the skull, the eyes project to the front, are fairly protruberant, and can rotate in their orbit. The pupil is oval, and when it contracts, it forms a horizontal slit instead of small circular point, as in humans. These features, as well as the retinal characteristics that I will describe below, combine to give the horse a very large, nearly circular visual field,[1] though with two blind spots: one to the rear and the other extending forward at eye level and directly below the nose (Timney and Macuda 2001; Timney and Wright 2007).

Each eye has a field of view extending up to 200 degrees on average and may reach up to 215 degrees. A forward-facing horse can accurately discriminate objects appearing laterally up to an angle of around 138 degrees; in contrast, objects nearly behind it, at an angle of 162 degrees or more, are difficult to identify (Hanggi and Ingersoll 2012). The vertical extent of the visual field may approach about 180 degrees (Roberts 1992, cited by Timney and Macuda 2001). At rest, the optic axis of each eye diverges about 40 degrees from the longitudinal axis of the body and is oriented downward at about a 20-degree angle with respect to the horizontal (Hughes 1977, cited by Waring 2003). As indicated above, a small region, perpendicular to the horse's forehead (of fairly shallow depth at the level of the eyes) and under the nose (where it may extend forward up to 2 meters) is blind but can be compensated by slight head movements. Also at the front, extending down and above the blind spot, is a region of binocular[2] overlap of between 55 to 65 degrees (Timney and Macuda 2001). By comparison, the human binocular field is around 120 degrees, and that of the dog or cat around 85 degrees. This allows the horse to view the ground in front at a distance of around 2 meters when its head and neck are in the normal position of rest. When it lowers its head to graze, its binocular visual field enables it to see "what is on its plate." Simultaneously, the animal's monocular visual fields enable wide scanning for danger (Harman 1998). The horse is thus a priori

---

1. Generally speaking, the eyes of mammalian prey (such as cows, sheep, goats, and rabbits) are positioned to the side, which allows panoramic vision conducive to detecting predators. The eyes of predators, on the other hand (such as cats and canines), are mostly positioned at the front of the head, which gives them a larger forward stereoscopic visual field. In this way, they can better scan the distant horizon for their prey.

2. As is well known, binocular vision is genuinely stereoscopic. Nonetheless, monocular clues permit some distance and depth perception.

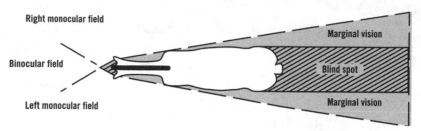

Figure 6.1 The equine visual field.

well equipped for simultaneously seeing the vegetation in front of it—no small thing given that it spends two-thirds of its time feeding (Duncan 1992)—scoping out the ground under its hooves, and detecting the arrival of possible enemies (see figure 6.1). Naturally, these attributes amount to a survival factor for a species with weak natural defenses: Horses owe their continued well-being to flight and thus to early detection of potential predators.

Moreover, as I will show below with respect to the composition of the retina, the downward orientation of the optic axis of each eye is linked to the horse's need to tilt its head upward to see into the distance, for example, when mounted and on the bit, with its forehead nearly vertical, and especially when it is in a hyperflexed position (Harman et al.,1999; McGreevy 2004; McGreevy et al. 2010).

Finally, I would like to lay to rest the erroneous notion that a horse shown the same object at different times, first with one eye and then another, will not recognize the object. At first blush, this idea would appear to explain, for example, why a horse might react, or even be frightened, by an object that previously caused no alarm. But anatomically and physiologically speaking, it is clearly contradicted by the existence of the corpus callosum, the very large network of nerve fibers that, as we saw in a previous chapter, links the horse's two cerebral hemispheres, as it does in all placental mammals. And in fact, as I will show in discussing behavioral approaches to visual perception, Evelyn Hanggi (1999) has experimentally shown this idea to be baseless.

Anatomical Structure of the Eye

Proceeding from the outside in, and keeping in mind the essential structures of the equine eye associated with specific ways of seeing, one feature of note is that the iris, contained in the ciliary body, contains at its center a "window," the pupil, which allows light rays to penetrate the inner eye by crossing through the lens in front of it (see figure 6.2). The pupil acts like the aperture of a camera, constricting or dilating depending on the light intensity, under the influence of the circular smooth muscles of the iris. As indicated above, the pupil is oval and forms a horizontal slit on constricting. The pupillary response to light is a subcortical response. In the horse, it is normally "somewhat lazy" (Barnett et al. 2004), which is consistent with adaptation to changes in light that is slower than that of humans[3] (Williams 1976; Wouters, De Moor, and Moens 1980). I will return to this point later in the chapter.

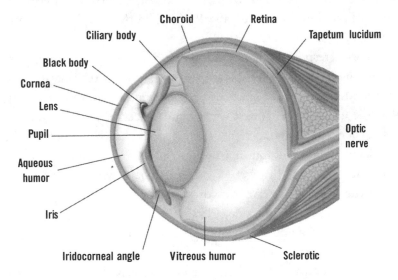

Figure 6.2 Schematic cross-section of the horse eye. Source: Leblanc et al. 2004, 80.

---

3. While the pupil adapts rapidly to moderate variations in light, the eyes themselves adapt to extreme light by changing the state of the visual pigments (Eckert et al. 1999).

The crystalline lens is a biconvex structure and, in the horse, fairly rigid. The primary function of the lens is to optimize the image on the retina by using the muscles of the ciliary body to accommodate—that is, increase or diminish—the lens's convexity based on the distance of the object. In the horse, the ciliary muscles (which in fact are bundles of smooth muscles that operate automatically) are relatively weak (McGreevy 2004). Along with the limited elasticity of the lens, the muscular apparatus apparently enables only weak, slow dynamic accommodation (as is generally the case with ungulates). This limitation does not pose a major problem a priori: In fact, in good conditions, the horse, which is rather emmetropic[4] (Sivak and Allen 1975; Harman et al. 1999), can generally see in focus from about 1 meter ahead to infinity. For objects right in front of its nose, it relies on skin sensitivity and its whiskers (McGreevy 2004).

There does, however, appear to be some interindividual variability in this regard. An anatomical study of thirty eyes in 15 horses showed 7 with myopia, 10 with emmetropia, and 13 with hypermetropia (Farrall and Handscombe 1990, cited by Murphy, Hall, and Arkins 2009).

A horse's responses, such as raising its head by tipping up its nose or arching its neck to see distant objects, or even turning its head to the side to see near objects, led to the conclusion, until recently, that the animal is endowed with a static accommodation mechanism, the so-called ramp retina (which does exist in some fish, like the skate or the shark, but not in mammals), in which the distance between the lens and the retina varies. Such an arrangement would allow the horse to reserve the lower portion of its field of vision for near objects (for example, tufts of grass right where it is grazing), whereas the upper portion would serve for distant objects (in particular, potential predators).

This concept, which had already been largely discredited (Sivak and Allen 1975), now appears to have been abandoned for good.

---

4. Emmetropia corresponds to normal vision: The relaxed eye focuses rays from infinity perfectly onto the retina.

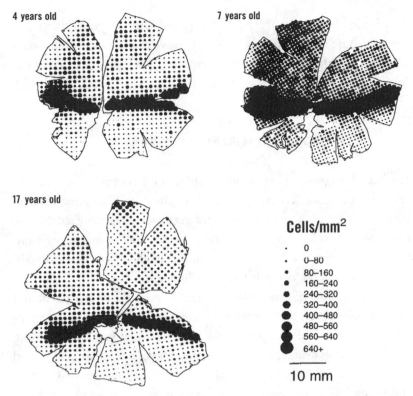

Figure 6.3 Visual band and area centralis of three horses (temporal periphery at left and nasal periphery at right represented by the density of the ganglion cells on retinas cut and laid flat). Source: Harman et al. 1999, 387.

Alison Harman and colleagues (1999) carried out a precise, detailed investigation of eyes removed from deceased horses, one to two hours after death. They found that different points along the retina were normally equidistant from the lens, except for the dorsal and ventral periphery, where they were closer. This finding contradicted the idea of the ramp retina, which proposed that the retina was farther from the lens in the lower part of the eye. In fact, what the researchers found was a narrow horizontal band (visual streak), dense with cells, situated a bit below the center of the retina that widens slightly at its temporal periphery to form an area centralis (see figure 6.3).

It is mostly within this streak that visual acuity is good, as we will see later. The retinal images below or above the streak produce only blurry vision that is, however, sensitive to slight changes in light and minimal displacements of moving stimuli (Ehrenhofer et al. 2002).

A Short Tour of the Anatomy of the Retina

It is worth digressing briefly to consider the structure of the retina, the neuronal tissue covering the back of the eye. The retina receives images (inverted, upside-down, and right/left) transmitted by the lens and transmits them to the brain along the optic nerve pathway. The characteristics of the cells that make up the retina determine in part how visual sensation is recorded (see figure 6.4).

The inversion of the retinal image is actually of no particular interest: We do not see our retinal images, which are only the departure point for nerve inputs processed by the visual system, including the brain. Moreover, seen up close, the images are so-so at best and involve none of the complex coding of information carried out by the retina itself, which in contrast does enable precise vision: "The fact is that one does not see the retinal image; one sees with the aid of the retinal image. The incoming pattern of light provides information that the nervous system is well adapted to pick up . . . The retinal images of specific objects are at the mercy of every irrelevant change of position; their size, shape, and location are hardly constant for a moment. Nevertheless, perception is usually accurate: real objects appear rigid and stable and appropriately located in three-dimensional space" (Neisser 1968, 204). In the same vein, Michel Imbert (2006) emphasizes the importance of the role of the brain in visual perception: "The ocular optic, composed of the cornea and the crystalline lens, creates a mediocre image of the outside world . . . Helmholtz remarked that if an optics manufacturer were to generate products of similarly poor quality, he'd quickly go out of business. It is astonishing how we can see so well with such bad equipment. In fact, the image is only the starting point for a whole series of neural

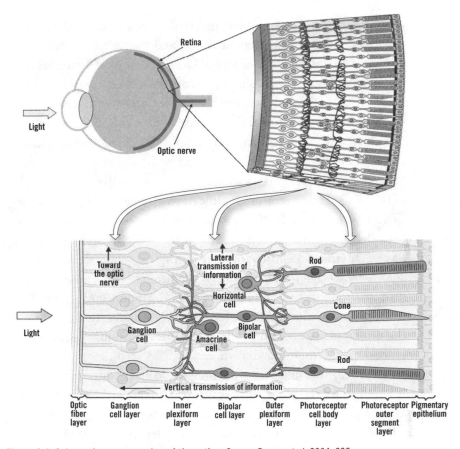

Figure 6.4 Schematic cross-section of the retina. Source: Purves et al. 2004, 235.

processing, to put it properly, aimed at constructing a representation that is stable, rich, aberration-free, and without gaps in the visual world that surrounds us. What the brain processes, what enables us to see what we see, only remotely resembles the image formed by the eye . . . And yet, everything begins with that image. The physical and geometrical aspects of its formation, as well as the physiological characteristics of the retina—the sensitive surface that covers the back of the eye—on which it is painted, determine the extent of what we are able to see" (180–81).

The back (fundus) of the equine eye comprises three superposed envelopes, namely, the sclera (which is continuous in front with the cornea and behind with the dura mater of the optic nerve), the choroid (which contains the tapetum lucidum), and the retina (see plate 1).

The tapetum lucidum, or retinal layer, is a thin reflecting layer no more than about 30 microns thick, composed of extracellular fibers (Ollivier et al. 2004). It occupies the upper portion of the eyeball and ends just above the optic disk. It sends light rays back to the photo-receptor cells of the retina, thus amplifying light and improving twilight and night vision. Owing to its position, it especially reflects ground light, which may aid nocturnal movements (Saslow 2002; Hall 2007). However, this amplification also causes light scattering (Timney and Macuda 2001), which obscures the shapes of objects, thus decreasing visual acuity. The green, yellow, or blue coloring of the tapetum lucidum varies depending on a horse's coat and its age (Noreika 1998). The tapetum lucidum is what causes the eyes of nocturnally active animals to shine. Examples include not only horses but also cats and dogs, for instance, when they are caught in the headlights of an automobile. The inferior fundus, which has no tapetum lucidum, is gray-brown.

The optic disk, or papilla, is slightly ovoid and colored pink-orange. It is also called the blind spot because it has no sensory cells. It is situated on the portion of the retina where the optic nerve leaves the eye. The optic disk is another example of the determining role of the brain in visual perception: If it were to faithfully reproduce the image that forms on the retina, the visual field would be chopped off by a black spot in the upper portion, corresponding to this blind spot.[5]

All told, the retina extends from the pupillary edge of the iris to the optic disk. It also includes a nonsensory area at the back, bounded by the scallop-edged ora serrata. The equine retina is generally very thin: 90 percent of its surface is between 80 and 130 microns thick[6]; it reaches 250 microns only at the optic disk (Ehrenhofer et al. 2002).

5. See the experiment of the abbot Edme Mariotte (1620–1684) enabling a subject to "perceive" a blind spot, which consisted of drawing on a piece of paper two large black dots 5 centimeters apart, closing the left eye, and at a distance of 15 centimeters from the paper, staring at the left dot keeping only the right eye open. Under these conditions, the point to the right cannot be seen because it falls within the blind spot.
6. A human hair has a width of about 50 to 100 microns.

As with most mammals, the sensory retina consists of two layers: a pigmented cell layer, or retinal pigment epithelium, which adheres to the choroid and directs regeneration of the visual pigments in the photoreceptors, and an internal, transparent layer, the neurosensory retina, which is made up of several layers of interconnected cells attached to the retinal pigment epithelium. These latter cells can be classified as follows:

- Outer layers formed of photoreceptor cells and their axonal prolongations. These cells comprise cones and rods (Wouters, de Moor, and Moens 1980; Sandman, Boycott, and Peichl 1996). Both produce electrophysiological signals based on the light energy they receive, a process known as phototransduction. Nonetheless, these two classes of cells are quite different. The cones, which have a high light threshold, only react in situations of full daylight (photopic vision). This photopic apparatus is organized in such a way that it can distinguish fine details and the edges of objects, as well as their colors. The rods, which are several hundred times more sensitive to light than the cones, enable night (scotopic) vision. However, the scotopic visual apparatus is strongly convergent, that is, nerve signals from many rods all converge on the same optic nerve fiber. Consequently, night vision is ill adapted to discerning details and the edges of objects. On the other hand, it is very sensitive to movement. Finally, rods do not enable color vision.
- Intermediate layers of bipolar cells that connect photoreceptor cells to ganglion cells, and associative cells (horizontal and amacrine cells) that control lateral transmission of information between these different cells. These layers help to analyze and encode visual information and (importantly) compress the information by eliminating redundancy. Thus, each ganglion cell receives inputs coming from a certain number of

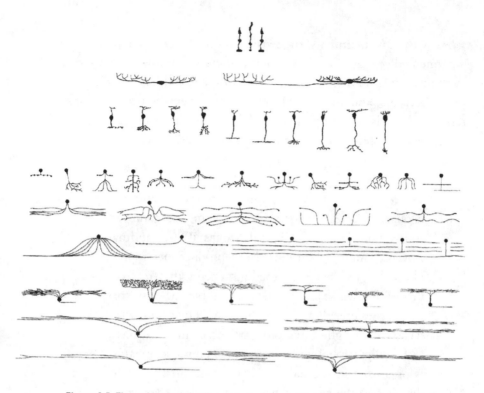

Figure 6.5 The main types of cells in a typical mammalian retina. Top to bottom: photoreceptors, horizontal cells, bipolar cells, amacrine cells, and ganglion cells. Source: Masland 2001, 878.

photoreceptors (by way of the intermediate layer cells), which constitutes its receptor field.

• Ganglion cell layers. These aid both in analyzing and encoding visual information, producing nerve signals in the form of action potentials that are then carried along by the cells' axons. These axons, which make up the innermost layer of the retina, converge toward the optic disk, where they join and emerge from the eyeball as the optic nerve. "The photoreceptors functionally linked to a given ganglion cell are distributed over the surface of the retinal mosaic, in a roughly circular form of varying size depending on their position at the bottom of the eye. This surface constitutes the receptor field, that is, the area in which all visual stimulation provokes a response at the level of the ganglion cell where activity is recorded" (Imbert 2006, 202).

Remember that these classes of cells comprise many types of cells. Indeed, the mammalian retina contains around fifty-five distinct types of cells, each of which regulates a different function (Masland 2001). This suggests the extreme complexity of the analysis and encoding of visual information carried out at this level (see figure 6.5).

I will not go into detail regarding the tasks effected by the retina, but it is worth stressing its functional importance: "The primary function of the retina is to ensure transduction, that is, the transformation of light energy into nerve signals. That notwithstanding, considering its structural complexity as well as the importance of its role in visual processing, it is obvious that, far from being analogous to a simple photographic lens, the retina is a veritable extension of the brain" (Lajoie and Delorme 2003, 84).

Structure of the Retina and Visual Quality

The ganglion cells relay signals coming from the photoreceptors to the central nervous system. Mapping their distribution makes it possible to estimate visual quality within the retina (Evans and McGreevy 2007). In fact, it is the diameter of their receptor field and not that of the photoreceptor cells that determines the spatial resolution of the visual system: The ganglion cells are the real retinal receptors. In humans, the receptor fields of the retinal periphery may be fifty times larger than those of the fovea (Lajoie and Delorme 2003).

Various histological studies of the retina of the horse carried out both thirty years ago (Hebel 1976) and over the last decade (Harman et al. 1999; Guo and Sugita 2000; Evans and McGreevy 2007) reveal the existence of a well-defined, narrow (on the order of 1 to 2 millimeters) visual streak densely populated with ganglion cells that extends horizontally across the ventral periphery of the

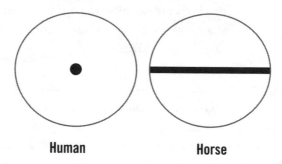

Human   **Horse**

Figure 6.6 Schematic regrouping of cones: where vision is clearest.

retina from end to end, just above (3 to 5 millimeters) the optic disk. This visual streak widens (3 to 4 millimeters) in the temporal arm, coincident with a maximum density of ganglion cells, constituting an area centralis.[7]

In humans, the cones are mainly grouped within the macula, a small circular area 2 millimeters in diameter situated near the center of the retina. The center of the macula itself, the fovea (which measures only 0.3 millimeters across and where vision is clearest), consists solely of cones. In the horse, however, the cones are especially concentrated in the visual streak and even more so in the area centralis, which has very few rods: Within it, a ganglion cell receives inputs from only a few cones, resulting in very high spatial resolution of images (or very fine grain), which enables good discrimination of details and the edges of objects, associated with maximum visual acuity (see figure 6.6). Accordingly, the vision of the world that a horse at rest has does not correspond (as it does in us) to a clear view of the area right in front of it, as high as it is wide, but of a zone relatively limited in height but also panoramic, above and below which visual acuity is poor.

In their study of equine visual acuity, Brian Timney and Kathy Keil (1992) refer to various investigations into the distribution of the retinal ganglion cells in horses and the size of their ocular globe.

---

7. Owing to its shape, the visual streak itself is sometimes called the area centralis striae formis.

Figure 6.7 Photo montage proposed by Alison Harman and colleagues (1999). Photo A corresponds to a photo of a city, taken behind our two observers. Photo B is that of the two observers taken from the front, thus showing the panorama behind them. Photo C represents what a human sees: a very clear central area, surrounded by a blurry area. Photo D gives an idea of what a horse sees, that is, a relatively clear horizontal band, extending from the city to practically behind the observers, with blurriness for everything above and below the band. Source: McGreevy 2004, 40; Harman, 1999, 389.

According to calculated estimates, the visual acuity in the area of the retina where it is best turns out to be between 16.4 and 20.4 cycles per degree.[8] As we will see, these values are somewhat lower than 23.3, which Timney and Keil obtained using a behavioral approach. Alison Harman and colleagues (1999) obtained comparable results, concluding that the visual acuity of the horses they examined was on average 16.5 in the visual streak and that, moreover, it was no more than 2.7 cycles per degree at the periphery, that is, above and below the visual streak. Using photographs (see figure 6.7), these same researchers attempted to compare what a horse sees with what a human sees from the same vantage point looking toward the city of Perth, Australia.

Despite significant interindividual variability (Evans and McGreevy 2007), the order of magnitude density of ganglion cells is consistent across studies (Hebel 1976; Harman et al. 1999; Guo and Sugita 2000; Ehrenhofer et al. 2002; Evans and McGreevy 2007).

8. I will return in more detail to the idea of visual acuity, its definition, and measurement in the next chapter, on equine vision. Note here simply that it can be defined as the number of white and black vertical stripes of equal width per degree of viewing angle.

It generally amounts to several thousand cells per square millimeter for the visual streak and reaches its maximum in the area centralis (around 6,500). At the periphery, in contrast, the density is only a few hundred cells per square millimeter and sometimes less than a hundred.

Marion Ehrenhofer and colleagues (2002) note in addition that most equine ganglion cells are large (with diameters that approach 40 microns), which enables many amacrine cells to converge there and their rapid conduction axons to relay signals regarding subtle changes in light and movement of stimuli. According to these authors, equine sensitivity to moving objects and the fluid character of their vision in much of their visual field have a lot to do with their nervous behavior.

The area centralis is employed in binocular vision, whereas the rest of the visual streak enables high-acuity, panoramic monocular vision (Waring 2003). When a horse raises its head to gaze into the distance in an attitude of alert, the action records the image of what the animal sees on the narrow band of maximal visual acuity on its retina and mobilizes its binocular vision—the best combination for appraising distance. Similarly, when a horse lifts its head sharply and away in response to a hand brought quickly to its forehead, it is seeking to "transfer" the hand from its blind spot to its stereoscopic visual field. In contrast, to more closely examine an offering in the palm of a hand, the horse arches its neck and turns its head to the side to get a better look with one of its two eyes, that is, its nonstereoscopic, monocular vision.

In the wild, the horse generally adopts one of two attitudes: It is calm and quietly grazing, with a clear monocular view of the ground before it while still able to detect unusual movements in its surroundings. Or the animal is alerted and lifts and tilts its head to orient its binocular vision toward whatever it is that has drawn its attention, blurring its monocular vision. The experience is the same for humans when focusing their attention: The periphery of the visual field becomes blurry.

In any event, the existence of the visual streak and the area centralis, together with where the horse's eyes are set in its head, enables the grazing animal to maintain a good view not only of the ground in front of it but also of distant objects off to the side, depending on where it is focusing its attention. Moreover, when the horse is moving calmly, for example, ambling from one place to another on its home range, the relatively low, nearly horizontal position of its neck with a slight poll flexion angles its forehead in a way that enables both eyes to see the shape and irregularities of the terrain before it. Small sideways or vertical movements of the head adjust the animal's vision to take in more of the environment, as needed.

On the other hand, as I mentioned at the beginning of the chapter, the downward orientation of the optic axis of each eye (not forward, as in humans) means that, when mounted and on the bit, with its forehead nearly vertical (and especially if its neck is hyperflexed), the horse cannot see directly in front of it and must lift or tilt its head to do so (Harman et al. 1999; McGreevy 2004). In other words, if the animal's neck is not freed (which does not mean letting go!) at the approach of an obstacle, the horse has no way of using its stereoscopic binocular vision to estimate the distance it still has to go before jumping or how high it must jump. For this reason, a horse may jerk its head a few lengths before jumping to gain additional information, albeit only through monocular vision, especially since once into the jump, the height of the obstacle is obscured by the blind spot below the animal's nose. Obviously, this does not imply that, to see the height of the obstacle for the last few lengths leading up to it, the horse needs to have its "nose in the air," so to speak. For proof, one has only to take the angles described by the limits of the horse's binocular visual field and its forehead in the photographs presented by Alison Harman and colleagues and to transfer them to pictures of horses taken on the approach to an obstacle (see figure 6.8).

Nonetheless, in an article titled "Position of the Head Is Not Associated with Changes in Horse Vision," Bartoš, Bartošová, and Starostová (2008) explicitly call into question the restricted upper portion of the visual field. Such a limitation, they claim, would be an unlikely handicap during competitive interactions between horses, and they cite as evidence the posture of a stallion aggressively

Figure 6.8 Frontal binocular vision of horses approaching an obstacle, during a show jumper competition. Montage of photos taken by M.-A. Leblanc.

approaching another male with a tightly arched neck. I beg to differ: In such a case, the approach, though frontal at times, is still more or less lateral, which easily affords a good view of the partner. Moreover, when this approach is pursued with characteristic signs of aggression, such as intent to bite, or miming biting, the attacker stretches his neck and tilts his forehead. Having said that, the authors note that, in the horse, each eye is equipped with four rectus muscles and two oblique muscles that control the rotation of the eyeball, which suggests the possibility of their maintaining the eyeball in an optimal position with respect to the retina's visual streak, whatever the position of the head. The authors tested their hypothesis by observing the position of the pupil based on different positions of the forehead; in every case the pupil stayed horizontal.

However, Harman (personal communication 2009) observes that, apart from a curious lack of precise data in this article, the authors limited themselves to the position of the pupil, whereas Harman and her colleagues based their conclusions on the use of an ophthalmo-scope, which makes it possible to localize the impact of a light ray on the retina and thus to assess the real extent of the visual field. The simple fact of a given position of the pupil does not necessarily mean that the horse sees from that angle; the retina does not extend to the front of the eyeball. She also observed that the importance of binocu-lar vision for the horse vis-à-vis its vital necessities, such as being able to see the ground it is moving on or nearby food, is basically oriented downward. Finally, as I noted previously, when the horse grazes, using binocular vision it can precisely locate what is right in front of it as well as horizontally detect the arrival of potential predators in the distance. Paul McGreevy (personal communication 2009) adds that, whatever the orbital rotation capabilities of the eyeball, a horse overflexing its neck cannot see where it is going, as its eyelid poses an obstacle. He also makes the point (McGreevy et al. 2010) that the preliminary data gathered to date "indicate that in extreme head positions, the pupil is not parallel with the ground" (185). All the same, McGreevy proposes to approach this entire question in a new way, taking into account the findings of Bartoš and his colleagues, and applying ad hoc opthalmoscopic examinations to a cohort of horses with different head shapes (Evans and McGreevy 2006), since it is known that the distribution of glanglion cells differs with skull length (Evans and McGreevy 2007). To be continued . . .

As is evident in the photos of Perth by Harman and coworkers (1999), humans cannot extrapolate with any confidence what we see with our own eyes to what the horse sees (see figure 6.9). Thus, when loading a horse into a van, a human handler's perception of the open back of the vehicle and that of the horse are not exactly the same. If the horse is not accustomed to the vehicle, it will be more tempted to run out than is commonly appreciated.

Equine reactions to changes in light are themselves worth a little digression, if only for how they influence the daily life of the rider, whether exiting a very dark stall or a forest, negotiating the at times very marked chiaroscuro effects in a steeple chase, where dark and light areas may alternate, or walking the horse in artificially lit areas

Figure 6.9 Representation of what a human and a horse, respectively, see of the rear opening of a van. Source: McGreevy 2004, 43.

at night. The response is particularly striking during rapid passage from a strongly lit area to a very dark one. Under such conditions, a horse may not entirely recover its night vision even after a half-hour (Wouters, De Moor, and Moens 1980).

At the same time, the horse (as I have shown) sees better than humans do in the dark, even if it does not perfectly detect the details or edges of objects. Roughly twenty times more numerous than cones (Wouters and De Moor 1979), rods are distributed throughout the retina and are particularly dense at the periphery. As sensitive to light as they are, they are still excellent receptors for seeing at night or in low light conditions (scotopic vision). However, many rods converge on each ganglion cell; consequently, this massive convergence produces only weak image resolution. At the same time, better perception of objects in the dark does not include good detection of details or edges. This, along with the fact that a horse takes much more time to adapt to the darkness than humans do does not favor situations that involve transitions from bright light to darkness.

The situation may seem paradoxical, but it is analogous to that of other familiar animals, like dogs and cats. This is hardly surprising, given how the horse lives and its need to be on the alert for predators day and night. I have also already noted that in this, as with other animals that are active around the clock, it is aided by a membrane coating on the back of the eye behind the retina (the tapetum lucidum). This membrane reflects light to the rods, thus serving to amplify the light. On the other hand, the horse's slower adaptation to light compared with humans is of no great matter in and of itself, under normal conditions: In the wild, horses are rarely exposed to dramatic light changes.

In other words, although horses see better than humans in constant darkness (at night, for example), they have difficulty dealing with rapid changes of light.

Recall, too, that although rods react to very slight changes of contrast due to their strong sensitivity to light, they are not involved in color perception (except in very weak light), which I will cover in more detail below.

Finally, note that, as emphasized by Serge Rosolen and Florence Rigaudière (2003): "The two groups of photoreceptors are at the origin of peripheral vision, which is characterized by weak spatial resolution and good temporal resolution, which is the basis of sensitivity to movement" (16).

Optical Pathways and Cortical Distribution

As previously pointed out, the axons of the ganglion cells converge in the papilla, or optic disk, where they emerge from the eye to form the optic nerve. The fibers that make up the two optic nerves then become myelinated. The nerves partially cross at the optic chiasm (a process known as decussation), collecting the information from the two retinas right in front of the hypophysis, such that the optic tracts that continue from the optic nerves comprise fibers originating from the homolateral temporal retina on one side and the contralateral nasal retina on the other (see figure 6.10). Around 90 percent (Rosolen and Rigaudière 2003) of the fibers of the optic tracts form relays within the lateral geniculate body of the thalamus that then convey signals to the primary visual area of each cerebral hemisphere. Thus, each left or right side of a visual field projects to the same contralateral visual cortical area.

In humans, around 53 percent of optic nerve fibers pass through the

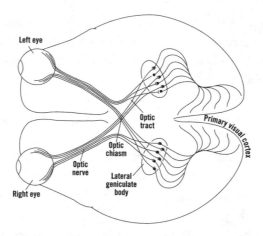

Figure 6.10 Schematic illustration of optic pathways. Source: Pour la Science, November 1979, 81.

opposite optic tract. The percentage is higher in ungulates and in the horse has been reported to reach 81 percent[9] (Barone and Bortolami 2004) and even 85 to 88 percent (Cummings and de Lahunta 1969; Levine et al. 2008).

The few optic tract fibers that do not form relays in the lateral geniculate body of the thalamus serve as reflex (retinocollicular) pathways that project efferences both to the parasympathetic nuclei and to the motor nuclei of the oculomotor nerve at the level of the mesencephalus (in the superior colliculus, which independently receives afferents from the primary visual cortex). In the first case (accommodation pathway), the efferences control pupil retraction in response to increased light, as well as dynamic adjustment through contraction of the ciliary muscle, and aid in photoregula-

---

9. In this context, Robert Barone and Ruggero Bortolami (2004) had the following comment: "The comparison of these percentages in humans, whose very close-set eyes look front and have an extensive binocular field of vision, and animals, whose laterally positioned eyes have a limited binocular field of vision, suggests that the proportion of crossed fibers increases with decreasing binocular field of vision. This appears to be an overgeneralization" (367). However, they write, "in phylogeny, uncrossed fibers represent the more recent acquisition, with improvement of stereoscopic vision giving a more exact sense of depth" (367–368).

tion of neuroendocrine rhythms. In the second case (reflex pathway), the efferences control coordinated movement of the eyes and trigger ocular orientation reflexes as well as muscular reactions of the head (including ear movements) and the neck (head orientation) in response to variations in light intensity or moving objects. Moreover, the superior colliculus sends efferents to the lateral geniculate body as well as to the pulvinar, which may play a significant role in the perception of movement and which is also believed to send efferents to the cortical visual areas involved in detecting movement.

The fibers of the optic tract that do form relays in the lateral geniculate body of the thalamus (retino-geniculate-cortical pathway) ferry basic perceptual information regarding shape, color, perspective, and movement that have already been analyzed and preprocessed upstream and at the level of the ganglion cells and that will now enable complex visual scenes to be recognized and identified. "These ganglion cells also subdivide into two major categories, M (magno) cells and P (parvo) cells . . . Compared with P cells, M cells possess a larger cellular body and axon, larger receptor fields, faster nerve conduction, and greater sensitivity to light contrast. These cells are thus useful in perceiving large objects and detecting weak contrast. Moreover, they respond to rapid movement detected in the large receptor fields of the retinal periphery . . . Each P cell responds to two colors, but its reactions are antagonistic" (Lajoie and Delorme 2003, 88–89). In short, information that codes for shape and color borrows a parvocellular pathway, and information that codes for movement and perspective travels a parallel magnocellular pathway. Both pathways synapse with homologous neurons within the lateral geniculate bodies. At this level, each geniculate neuron is connected to only a few ganglion cells, and visual information—still fairly raw—shows a retinotopic organization that preserves the relative positions within the visual field.

From the lateral geniculate body, the optic tract becomes the optic radiation and (still maintaining its retinotopic organization) follows along the primary visual area of each of the two cerebral hemispheres, located at the occipital end of each hemisphere and called the striate cortex due to its particular conformation. The signals are then transferred to the secondary, nonstriate, visual area, which forms a band adjacent to the primary area. This band maintains connections with various associated areas essentially along two separate pathways: a ventral pathway, which extends toward the temporal region and whose main function is linked to recognition of objects, including shape and color (the "what" pathway), and a dorsal pathway, which extends toward the parietal region, devoted mainly to the localization of objects, that is, spatial position and orientation (the "where" pathway). The secondary visual area is also linked to its homolog in the other hemisphere via the corpus callosum. The lateral geniculate body also receives numerous efferents from the primary visual cortexes that in turn exert major retroactive effects on them.

Without going too deeply into detail, note that visual perception does appear to involve many specialized areas at the cortical level that separately, yet in parallel, process shape, color, movement, orientation, depth, and function (Lajoie and Delorme 2003). This functioning, the product of different interactions with subcortical structures and a variety of retroactive loops, reveals the complexity of the neuronal circuits at work in visual perception, which is only really constructed at the level of the brain.

Chromatic Theory and Color Perception

Color vision is a complex field in and of itself. It calls into play phenomena that involve both physics and biology: "The result of mixing paints is a matter of physics; mixing light beams is mainly biology" (Hubel 1988, chapter 8, 2). In this regard, it is probably worth reviewing a few fundamentals of light and its relationship to color, as well the workings of human vision. As Almut Kelber and colleagues (Kelber, Vorobyev, and Osorio 2003) put it in a review of color vision in animals: "Concepts in color vision are mostly derived from human perception and psychophysics" (82).

In the seventeenth century, Isaac Newton broke up white light into all the colors of the rainbow[10] by shining it through a prism, and reconstituted white light either by passing it through another prism or by reproducing the colors on a disk that, when spun, took on a whitish cast (see plates 2 and 3). At the very beginning of the nineteenth century, Thomas Young proposed that three ocular receptors, sensitive to red, green, and blue, sufficed to permit seeing all colors based on his discovery that white light could be reconstituted through the use of just the three primary colors. Using three beams of light—red, green, and blue—he showed both the reconstitution of white light by combining all the beams, and obtained secondary colors by simultaneously using two beams, so-called additive color synthesis. Similarly, mixing three secondary color paints— cyan, yellow, and magenta—whose pigments act as a filter produces black as well as (again) the three primary colors, a process known as subtractive synthesis (see plate 4). It is subtractive synthesis that gives an object its perceived color. Indeed, the surface of an object reflects only the light rays of the colored pigments in its coating and absorbs all the others. Thus, an object painted blue, lit by a white light, will reflect only blue light and will look blue. In contrast, under red light, its surface will reflect nothing. It will appear black. Young's discovery is the basis of the trichromatic theory of colors developed by physicist and physiologist Hermann von Helmholtz in the middle of the nineteenth century.

However, based on an idea current at the time (and previously advocated by Leonardo da Vinci), namely, that yellow was also a primary color, Ewald Hering observed that red and green and yellow and blue function perceptually in an antagonistic way: There is no greenish red or bluish yellow. Similarly, white and black are a sort of opposite. Hering concluded that there are three antagonistic pairs— red-green, yellow-blue, and white-black—or six basic colors.

---

10. By tradition, Newton divided the rainbow colors into seven main colors: violet, indigo, blue, green, yellow, orange, and red. In reality, indigo is hard to distinguish. In a rainbow it corresponds to a very narrow band of the visible spectrum, which reduces the spectrum to essentially six dominant colors.

As Jean-Didier Bagot summarizes it (2002): "The history of color has been influenced by two major theories: the trichromatic theory proposed by Young (1802) and elaborated by Helmholtz (1867), based on the hypothesis of three receptors (unknown at the time), each having a different spectral sensitivity, and opponent color theory, advanced by Hering (1878), which postulates opposing mechanisms between red and green, blue and yellow, and black and white . . . Trichromatic theory dominated for a long time, for it explained how three well-chosen primary colors could suffice to reproduce the entire range of color sensations and also because it made it possible to interpret certain disorders of color vision as a deficit of one of the primaries. In fact, these two complementary theories are both valid to some extent, the first with respect to cones, and the second with respect to coding by chromatic opponent cells" (160).

At the anatomical and physiological level, three types of retinal cones have been detected in humans based on their spectral absorption of light,[11] and chromatic opponent cells have been identified among the ganglion cells of the retina and in the lateral geniculate body in the thalamus. Color coding begins at the retinal level with the chromatic opponent ganglion cells and continues up to the brain.

According to Bagot (2002): "The chromatic opponent cells are sensitive to certain 'colors' based on the types of cones with which they are associated. In reality, as we have already asserted, the cell is sensitive to a range of wavelengths and not to color, which is constructed by the brain. To simplify, here, too, we refer to color. According to whether a cell is associated with one or several types of cones in its receptor field, its sensitivity to color is as follows:

- cone L → cell sensitive to red (R);
- cone M → cell sensitive to green (G);
- cone S → cell sensitive to blue (B);

---

11. It was not until the early 1960s that the three types of cones could be distinguished based on their light absorption spectrum. Each has a preferential response to long, medium, or short wavelengths (Brown and Wald 1964, and Marks, Dobelle, and MacNichol, cited by Boeglin 2003).

- cone L + cone M → cell sensitive to yellow (Y), that is, red + green;
- cone L + cone S → cell sensitive to magenta (M), that is, red + blue;
- cone M + cone S → cell sensitive to cyan (C), that is, green + blue;
- cone L + cone M → cell sensitive to white (W), that is, red + green + blue.

The connections between cones and cells may activate or inhibit and are antagonistic depending on the opponent color pairs. Cells thus continually combine the activities of different types of cones by subtracting them" (170–71). Such discoveries also show that human color vision involves four main colors: red, blue, green, and yellow. Moreover, these efforts confirm the relevance of previously proposed theories—long after they were first revealed—by establishing their neurophysiological basis in a complementary way.

Newton himself had already figured out that light rays were not intrinsically colored but only provoked a sensation of color in the observer. In fact, color per se is not a physical property of light but a construction of the brain in response (among other things) to wavelength, which is a physical property of light. To a first approximation, perception of color has four main components: the light source, with its constituent distribution of wavelengths; the material that the object or colored surface is made of, with its reflective properties; the eye and its receptors; and the brain (or more generally, the nervous system), including its interpretive system. The brain constructs color based on the nerve signals relayed by the photoreceptor cells, their sensibility to certain wavelength ranges, and the coding that takes plays in other cellular layers of the retina. Note that, generally speaking, according to color theory the eye cannot distinguish individual color components in a mix of different monochromatic wavelengths (in contrast to the ear, which can pick apart the component frequencies within an acoustic wave). Moreover, outside of laboratories, pure color does not exist.

Table 6.1 Approximate wavelengths for the six dominant colors of the visible spectrum.	
Red	620–700 nm
Orange	592–620 nm
Yellow	578–592 nm
Green	500–578 nm
Blue	446–500 nm
Violet	400–446 nm

In physical terms, light is the portion of the electromagnetic wave spectrum that can be seen by an observer. For humans, that wavelength range is typically 400 to 700 nanometers. On either side of this range are to be found, respectively, ultraviolet light and infrared light, which can be seen by other species of animals, as well as X-rays, gamma rays, microwaves, radar waves, and radio waves (see table 6.1 and plate 5)

From both a physical and perceptual perspective, color has three characteristic dimensions:

- Dominant wavelength. This dimension corresponds to the perceptual notion of *hue* or *tonality* (and is what we commonly mean by color: blue, red, green, yellow). Hue refers to the pure color closest to the perceived color.
- Spectral purity. A color's degree of monochromaticity can range from chromatic (100) to achromatic (0). Spectral purity corresponds to the perceptual idea of *purity* or *saturation*. A pure color is 100 percent saturated. "The less saturated a color is, the larger its spectrum" (Boeglin 2003, 109). When a fraction of white light is contained in the color spectrum, the color is more or less bleached or simply washed out. For example, red becomes pinker. At the extreme, with 0 percent saturation, there is no dominant color. Desaturation may tend to white or black.

• Luminance factor. This light intensity parameter is defined by the global energy reflected by a surface, colored or not, compared with a reference white (the luminance factor of a white body is equal to 1, whereas that of a black body is equal to 0).[12] The perceptual equivalent is *luminosity*, or brightness, or clarity (more or less luminous or bright, or more or less dim or dark). Two different pure colors do not have the same luminosity; for example, yellow is less luminous than violet (I will return to this issue regarding the function of spectral sensitivity, in the context of behavioral exploration of color vision). In a black-and-white photo, two different colors may translate into identical grays if the colors have the same luminosity.

The color aspect of an object or a light beam results both from its hue and its saturation, referred to generally as chromaticity. Luminosity, on the other hand, is an achromatic property (see plate 6). Moreover, in everyday language, the association between luminosity and saturation is frequently expressed by a single qualifier: bright + saturated = sharp; bright + washed out = pale; dark + saturated = deep; dark + washed out = grayed down.

It is worth noting, however, as Kelber and colleagues (Kelber, Vorobyev, and Osorio 2003) observe: "There is no evidence that animals perceive hue, saturation, or brightness as separate qualities, or that they categorize colors as yellow, red, etc." (88).

We will return later to color parameters; they have particular methodological implications in the context of the behavioral approach to visual perception.

Photoreceptor cells—cones and rods—are endowed with pigments that preferentially absorb certain wavelengths of light and whose absorption spectra partially overlap (see plate 7). But these absorption spectra

---

12. Whereas luminance is a global measure of intensity, reflectance is a relative value corresponding to the fraction of incident light reemitted by reflection, and depends on the environment.

are also quite different: "relatively narrow for cones, large and extending to the entire visible spectrum for the rods" (Imbert 2006, 185).

Nevertheless, as already indicated, owing to their very different sensitivity to light intensity, cones and rods are essentially active either in daylight (photopic) or at night (in the dark; scotopic), respectively. At twilight, or by the light of the moon (mesopic),[13] the two systems function simultaneously, in which case "visual performance is poor because the luminance is insufficient for appropriate cone functioning and too strong for rods" (Bagot 2002, 127). In other words, cones and rods operate nearly exclusively, which invites an equally exclusive examination of their sensitivity to wavelength.

Eric Warrant (1999) observed that, to the extent that the eye is exposed to light that is a billion times weaker at midnight than at noon, it is hardly surprising that the visual system cannot maintain a high level of performance throughout this enormous spread of light intensity, even though various optic and neural devices help to optimize its performance over as extensive a range as possible.

In chapter 8 we will see that horses, like humans, maintain a perception of color even in very weak light intensity (Roth, Balkenius, and Kelber 2008).

Like most primates, humans have three types of pigmented cones whose maximum sensitivity (Hubel 1995) corresponds to wavelengths of light that are either short (S cone, called blue cone, with a peak sensitivity at around 430 nanometers,[14] situated in the violet); medium (M cone, the green cone, whose peak sensitivity occurs around 530 nanometers, in the blue-green); or long (L cone, the red cone, with a peak sensitivity of around 560 nanometers, located in the yellow-green but extending to the red).

Note, incidentally, that the label S, M, or L cone is far preferable to the more common blue, green, or red cone, which is misleading.

---

13. Kelber and Roth (2006) indicate a light intensity threshold higher than 5 cd/m² for photopic vision and one lower than 0.05 cd/m² for scotopic vision, the intermediate zone corresponding to mesopic vision. Roth, Balkenius, and Kelber (2008) further suggest a light intensity at sundown of around 1.2 cd/m² and by moonlight of around 0.02 cd/m².

14. These values are in fact only approximate owing to variations in measurement methods.

First, the color denomination refers only to the color whose wavelength corresponds roughly to the maximum sensitivity of the cone, whereas each of the cones has pigments whose absorption spectrum (or curve of sensitivity) covers a range of wavelengths relating to many colors, essentially fully encompassing the sensitivity of the three cones, especially the M and L cones. Moreover, as we will see now, no matter the wavelength of light to which it is exposed, a single cone does not suffice to perceive color.

The intensity of the cone response (measured by how often it discharges) is actually based on one variable, the amount of light (number of photons) absorbed and not any specific information relating to wavelength. This quantity of absorbed light itself depends both on the wavelength of the light stimulus, as a function of the absorption spectrum of its pigment (the more the light stimulus approaches the maximum sensitivity of the pigment, the stronger the response) and its intensity. This ambiguity of the cellular code corresponds to what we call the principle of univariance (Rushton 1972). In other words, wavelength and intensity of light stimulus together determine the strength of the cone response; they are indissociable. For wavelength data to be extracted and to result in color vision, comparisons must necessarily be made between the differential stimulation detected by various types of cones (at least two) for the same light stimulus, which de facto is at constant intensity. This supposes, in addition, the intervention of a comparator, which "implies that spectral discrimination and color vision occur at a more central level than that of the photoreceptors" (Imbert 2006, 191). With just one type of photoreceptor, variations in wavelength are only detected as different intensities: "A cone has a large absorption spectrum: thus, it can respond to different wavelengths provided the stimulus is of sufficient intensity. A 'red' cone (in fact, the maximum sensitivity is in the yellow-green, at 558 nanometers[15]) stimulated in the green (490 nanometers) gives the same response

---

15. These cones are nonetheless called red because the wavelengths corresponding to the color red activate them without activating the green cones.

for the two wavelengths if the light intensity of the stimulus is twice as large at 490 as at 558 nanometers. This explains why people with only one type of cone have no color sensation. A retina with two types of cones transmits two values of brightness for each object, and the comparison of these two values allows the brain to distinguish certain colors though with a very limited combinatory analysis" (Calas et al. 1997).

This principle of invariance also posits that, in scotopic vision, where cones are essentially inoperative owing to lack of light, rods cannot contribute to color perception since their pigment (rhodopsin, or visual purple) is sensitive to a single, large wavelength band of light that covers all of the visible spectrum at a maximum of around 500 nanometers. Consequently, "at night, all cats are gray" (although, as we will see in chapter 8, night is really black). David Hubel (1995) writes: "At intermediate levels of light intensity, rods and cones can both be functioning, but except in rare and artificial circumstances the nervous system seems not to subtract rod influences from cone influences. The cones are compared with one another; the rods work alone" (chapter 8, 7).

Like many mammals, horses are dichromats. The horse has only two types of cones (Sandmann, Boycott, and Peichl 1996), whose sensitivity peaks are around 430 nanometers for S cones and 540 nanometers for M/L cones (Carroll et al. 2001; Yokoyama and Radlwimmer 1999; Timney and Macuda, personal communication 2009), that is, wavelengths that for humans correspond to violet and green (see figure 6.11).

Carroll and colleagues (2001) observed that human dichromats with a red-green deficiency "only have two hues, the one most analogous to blue and yellow" (84), whereas trichromats have four basic colors. As they say, one of the most striking differences between human trichromats and dichromats is that the latter are deficient in intermediate hues. When the two extremities of the spectrum are mixed, the result for human dichromats is achromatic (white or gray), perceived either as desaturation or washout of one or the other of the main colors (that is, pale blue or yellow). Moreover, horses,

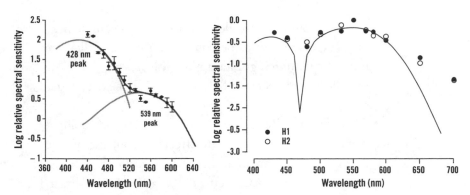

Figure 6.11 Evaluation of the sensitivity of two horse cones. Source: Carroll et al. 2001, 83; Timney and Macuda, personal communication, 2009.

as is generally true of all dichromats, have an achromatic neutral zone, or neutral point, in the middle of the spectrum manifested by a very narrow range of wavelengths, to which the two types of cones respond fairly equally (Jacobs 1981, cited by Geisbauer et al. 2004; Hubel 1995). In horses it is located at around 480 nanometers (Geisbauer et al. 2004; Timney and Macuda, personal communication 2009).

Taking into account the wavelengths measured in equine cones, Carroll and colleagues (2001) compared representations between human trichromatic vision and equine dichromatic vision (see plate 8). However, these representations are obviously somewhat speculative, as the features of cones alone do not tell us much about how the brain of the horse constructs color (see plate 9).

That will wrap up this discussion of the complexity of color perception. As John Boeglin (2003) has noted: "In controlled experimental conditions, one observes a relationship between the dominant wavelength of a stimulus and the color perceived. However, . . . this relationship is far from rigorous in everyday experience" (106). I will also borrow from him a few examples of chromatic phenomena that "a priori may seem illusions or at least exceptions to the rule . . . Some of these phenomena, however, seem more to confirm mechanisms of normal vision" (119), notably the involvement of the brain:

- Simultaneous contrast. For example, "Two complementary colors are said to be contrasting because they are mutually reinforcing: Blue appears bluer when it is close to yellow, and yellow yellower when it is next to blue" (120).
- Assimilation effects. "The juxtaposition of colored surfaces does not only create effects of contrast. In the case of relatively small and repetitive surfaces, contrast gives way to a contrary effect: the assimilation effect, also known as the juxtaposition effect" (121).
- Color constancy. In situations where the visual field is illuminated by a light source with a large spectral band, human color vision is little influenced by variations in light. As Semir Zeki (2005) puts it: "Light is electromagnetic radiation and has no color. The brain constructs color by comparing the wavelength composition of the light reflected by one surface and that reflected from surrounding surfaces. In this comparison, absolute quantities become unimportant and the brain is hence able to construct constant colors" (9). Let me add that it was precisely in showing that perceived color does not depend only on light reflected by a specific object, but also on the color context—that is, light reflected by surrounding objects—and striving to explain it that Edwin Land[16] (1977; 1983) proposed a new theory of color vision: retinex theory.

I will conclude this anatomical and physiological approach to vision, which has necessarily been brief, with another citation from Zeki (1990): "Perhaps more than any other aspect of vision, [the study of color vision] is forcing on us a change in our concept of what the sensory areas of the cerebral cortex do. With its study, we begin to

---

16. Recall that Edwin Land invented the filter glasses and instant cameras manufactured by Polaroid, which he founded and directed.

realize that the cortex does not merely analyze the colors in our visual environment. It actually transforms the information reaching it to create colors, which become a property of the brain, not the world outside. At the same time, however, the brain approximates these constructs—the colors—as much as possible to the physical constants in nature and, in the process, makes itself as nearly independent as possible of the multitude of changes in its environment" (61).

One might well imagine what a colored landscape might look like to a dichromat human whose cones had the sensitivity of equine cones. And yet, we cannot assume that the way in which the equine brain processes the same signals is analogous to the human brain's way of doing it.

7

THE BEHAVIORAL EXPLORATION OF EQUINE VISUAL PERCEPTION: PERCEPTION OF SHAPES AND MOVEMENT

After reviewing the anatomic and physiological characteristics of the visual system of horses most relevant for the purposes of this book, I will now turn to experimental studies whose goal is to "ask" horses themselves what they see.

To do that, I will first describe the principles behind the experiments in question, and then take a more in-depth look at visual acuity, the visual field, nocturnal vision, the visual apparatus (as an integrated system), three-dimensional perception, image recognition, and perception of movement. I will end by considering a question related to the uniqueness of animal vision compared with human vision. Color perception will be taken up in the next chapter.

Experimental Procedures

As Jacques Vauclair (1996b) has noted: "One of the characteristics of the experimental study of animal cognition is the use of experimental techniques that for the most part derive from behavioral approaches, and that for all intent and purposes are borrowed from the area of discriminative learning" (30–31).

Recall that discriminative learning consists of teaching an organism to respond differently to different stimuli. For example, an animal might be tasked with responding positively to striped surfaces and not to respond to nonstriped ones, as I will show below in a study of equine visual acuity.

The relevant classical protocol involves two phases:

- Training phase. During this phase, the animal is taught to "produce an arbitrary response (for example, to press on a lever) when shown a certain stimulus and not to respond when shown another" (ibid., 31). To do this, the arbitrary response to the given stimulus (positive stimulus) is positively reinforced (say, with food) and the same response given to the other stimulus is not reinforced, or negatively reinforced (say, a delay in presenting the next stimulus). With experiments relating to visual acuity, the two stimuli may be presented simultaneously. For horses, in this case, the response frequently consists of indicating the positive stimulus by pressing it with the tip of the nose or by moving toward it; depending on whether the response is to the positive stimulus or the negative one, it is referred to as "good" or "bad." This phase continues until the subject attains a certain level of success (generally 80 percent or more). The goal is threefold: to familiarize the animal with the experimental apparatus; to instruct it in the particulars of the task; and to teach it the behavior it needs to signal its response. During this phase, researchers also ensure that the animal is making its selections based on the discriminative task at hand and not on other cues (for example, left-right type positional cues) and that the experimental design is adapted appropriately. For example, such considerations were important in showing that the height of the

stimulus matters when horses are being taught simple
visual discrimination (Hall, Cassaday, and Derrington
2003). Indeed, the learning was more successful once the
stimulus was presented at ground level rather than on
level with the horse's head, especially when the animal is
required to walk, which forces it to lower its neck (Saslow
1999). This stimulus positioning thus creates better
visual conditions, taking into account the features of the
equine ocular system.

• Test phase, called generalization or transfer testing.
During this phase, the animal systematically undergoes
a series of discrimination tests based on a set experimental protocol. Here, the stimuli are similar but differ
in some dimension, and the parameters of the difference are varied, for example, to evaluate discrimination
thresholds.

Visual Acuity

Visual acuity refers to the eye's threshold of spatial resolution, that
is, its resolving power, namely, the smallest distance allowable for
two separate black points to be detected on a white background. The
basis of measurement is the visual angle necessary to distinguish the
highest resolution or spatial frequency. By international convention,
normal human vision, which serves as a reference, corresponds to a
separator power of 1 minute of arc (or angle).[1] How this measure
is expressed depends on the system of notation, of which there are
several. I will briefly present the main characteristics of the most
widely used among them.

---

1. This definition of visual acuity based on the separating power of the eye is the most useful.
However, it should not be confused with the visible minimum, which is the capacity of the
retina to perceive differences in illumination (for example, the minimum surface of a point
or the minimal thickness of a line that enables them to be seen against a background of
different brightness). The latter is on the order of an arc second and sometimes less.

In France, the traditional measure of visual acuity is based on the decimal fraction corresponding to the inverse of the value of the smallest angle, in minutes of arc, at which a detail can be seen. Thus, in conformity with the international convention cited above, normal vision has a value of 1. If, on the other hand, the smallest angle at which this detail can be seen is 10 minutes of arc, the visual acuity value, expressed as a decimal fraction, is 0.1. The corresponding Monoyer notation is expressed in tenths for acuity between 10/10 (normal vision) and 1/10. Visual acuity above the conventional normal might then be 12/10, 15/10, or even 20/10. For acuity less than 1/10, the numerator remains at 1, and it is the denominator that varies. This notation is simple to use, but it has the disadvantage that the angle difference is much greater between 1/10 and 2/10 than between 9/10 and 10/10.

At the international level, two notations predominate: absolute value, in cycles per degree (based on the measurement of spatial frequency), and relative value, or the Snellen fraction. The latter refers to the distance at which the test is made divided by the distance at which the smallest detail identified subtends an angle of 1 minute of arc.

The notation in cycles per degree relies on evaluation of spatial frequency. This corresponds to the number of parallel black and white bands of equal width presented on a panel at a predetermined distance, which enables calculation of width on the panel equal to a cycle of 1 degree of arc. Because a degree equals 60 minutes, a subject with normal vision will be able to perceive thirty white and black pairs per degree, or thirty cycles.

The Snellen fraction, which is probably the most used indicator worldwide, is based on vision at test distances of 4, 5, or 6 meters or 20 feet. On the basis of a reference distance of 20 feet, the fractions 20/20 20/30 20/50, and 20/100 indicate that a subject tested at a distance of 20 feet can distinguish the same details as a normal subject at a distance of 20, 30, 50, or 100 feet (see table 7.1).

Using an experimental protocol for perception based on spatial frequency, Brian Timney and Kathy Keil (1992) tested the visual

Table 7.1 Correspondence between different units of visual acuity.

Monoyer scale (in/10)	Snellen scale (to 20)	Minutes of arc	Spatial frequency (cycles per degree)
10/10	20/20	1	30
5/10	20/40	2	15
2.5/10	20/80	4	7.5
1.25/10	20/120	8	3.75

acuity of three horses. The experiment is worth reviewing at length because it is a good illustration of the issues mentioned above regarding discriminative learning.

The apparatus "exploited the natural tendency of horses to push at objects with their noses" (ibid., 2290). It consisted of two trapdoor panels (with a fixed horizontal upper edge) that opened slowly on pushing, separated by a vertical divider (whose length determined the minimum distance at which the horse had to make a decision), on which stimuli measuring 20 by 25 centimeters had been affixed (see figure 7.1). The positive stimuli were a set of high-contrast gratings of different spatial frequencies, that is, with broader or narrower stripes. The negative stimulus consisted of a grating with a very high spatial frequency so that the stripes were beyond the resolution acuity of the horse. Pressing against the panel with the positive stimulus gave access to a reward (cubes of concentrated horse food), whereas the panel with the negative stimulus was locked. The lighting for the panels was controlled to ensure satisfactory luminosity of the two stimuli.

During the training phase, the horses were exposed to three different discriminations: white versus black, low-frequency grating versus black, and low-frequency grating versus gray. When position preferences developed, they were corrected by increasing the frequency of the nonpreferred side. After the horses had satisfied the task criterion (twenty-seven correct trials out of thirty consecutive trials), new trials were run with longer dividers (from 40 centimeters at the start to 231 centimeters by the end).

Figure 7.1 Setup for measuring the visual acuity of the horse. Source: Timney and Keil 1992, 2290.

In the test phase, that is, the measurement of visual acuity, a modified experimental method was used to obtain a preliminary acuity threshold. A testing session began with the low-frequency grating. When a correct response was given, the frequency was increased on the next trial until the horse made an error, allowing the researchers to move rapidly to the most difficult gratings.

At this stage, the trials were presented in blocks of five trials of the same difficulty. If the horse achieved at least four or five of the trials, the frequency was increased for the next block. If it achieved only three, the researchers ran a similar set of five trials; if not, the test was restarted at the previous level of difficulty. Usually, testing was stopped after the horse had achieved a sufficient level of success after at least two blocks. Sometimes testing ended because the horse refused to continue.

A final threshold estimate of visual acuity was obtained using a constant stimuli method. To do this, the authors employed five gratings covering a range of spatial frequencies that bracketed the preliminary threshold estimate. These stimuli were presented randomly in blocks of five trials, until all five stimuli had been presented. This

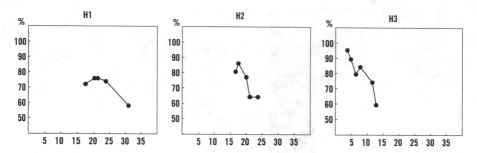

Figure 7.2 Psychometric functions showing, for each of three horses, the percentage of correct responses based on spatial frequency (in cycles per degree). Source: Timney and Keil 1992, 2291.

random method was then repeated until a total of fifty trials had been completed at each frequency.

Analysis of the results showed the visual acuity estimates for the three horses to be 23.3, 21.2, and 10.9 cycles per degree, based on 70 percent correct responses. The results also showed substantial interindividual differences in performance among the three horses. In particular, the first horse responded well to praise and affection, the second was more highly motivated by the food reward, and the third, which was not especially motivated by either, performed its tasks slowly and reluctantly, to the point where it was not possible to obtain complete data. The weak score of this animal may in fact reflect its lack of interest (see figure 7.2).

Timney and Keil considered the best visual acuity obtained—23.3 cycles per degree—to be conservative. It is based on the hypothesis that the horses chose their stimulus only when they reached the end of the divider, that is, a distance of 2.31 meters. If, on the other hand, one assumed that they made their decision before reaching this point, at a distance of 3 meters, the estimated acuities rose, respectively, to 30.8, 27.6, and 14.2 cycles per degree.

The authors also note that the estimates made on the basis of the anatomical features of the ocular system (as I indicated earlier, between 16.4 and 20.4 cycles per degree) are subject to a wide margin of error and thus do not call into question the empirical results of their own study.

By comparison, in work with dogs and cats, Paul Miller (2001) reported very low visual acuity estimates in Snellen fractions: around 20/75 for the dogs and between 20/100 and 20/200 for cats, whereas for horses, 23 cycles per degree corresponds roughly to 20/30. Although, as Miller himself acknowledges, his findings are preliminary and to be treated with caution,[2] the horse does appear to be endowed with significantly better vision than dogs and far better than cats.

It is worth considering breed among the factors influencing inter-individual variability observed in horses. Evans and McGreevy (2007) reported a strong positive correlation between the density of ganglion cells in the visual streak of the horse and nasal length: Horses with a long nose have a greater density of ganglion cells in the visual streak than those with a shorter nose, as do horses with a convex nasal profile compared with those with a concave nasal profile. Thus, thoroughbreds and standardbreds are likely to have better vision than Arabians.

Finally, as I will show in the next chapter, visual acuity is subject to variations in stimulus color (Grzimek 1952).

The Visual Field

Although, as previously shown, interpretation of the anatomical data on the equine eye has led to the longstanding assumption that the animal possesses a very large, nearly circular visual field, it is still the case, as Brian Timney and Todd Macuda (2001) have noted, that estimates made on this basis must be taken with a grain of salt, given that "the optic field does not necessarily correspond to the retinal field" (1568). Moreover, "a complete description of the visually effective field of view requires that it be measured behaviorally" (ibid.).

---

2. To my knowledge, no experimental study has been done on the exact visual acuity of dogs and cats. Hanggi (2006) reports figures of 20/50 for dogs and 20/75 to 20/100 for cats.

Figure 7.3 Left: a horse stationed amid the pylons with its muzzle resting on the release mechanism. Here, the stimili are placed in position C. Right: the horse presses the left paddle with the tip of its nose. Source: Hanggi and Ingersoll 2012, 71. © Evelyn B. Hanggi, Equine Research Foundation.

This is precisely the task recently undertaken by Evelyn Hanggi and Jerry Ingersoll (2012) in a study of three fifteen-year-old geldings: an Arabian, a Missouri fox trotter, and a pinto-draft mix. The researchers tested the horses first on object detection, then on discrimination between two objects symmetrically placed in different lateral or caudolateral (posterolateral) locations, either raised or on the ground.

The experimental apparatus consisted of a semicircular station of 3.66 meters radius, enclosed within a curvilinear, white smooth wall 1.22 meters high. Objects to be detected or discriminated could be temporarily anchored to the wall either at a height of 0.86 meters or level with the ground (see figure 7.3).

Within the station, the test horse was surrounded by four 1.22-meter-high pylons that aligned the animal to face a platform situated 0.91 meters from the ground. Sticking out of the center of the platform was a polyvinyl chloride pipe topped by a marine float. The horse was required to rest its muzzle on the float so as to have only a monocular lateral view of the objects to be detected or discriminated. The float could be lowered by remotely turning the tube, thus constituting a movable release mechanism for the horse.

The platform itself contained a food container centered between

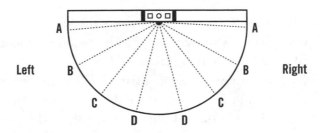

Figure 7.4 Illustration depicting the A, B, C, and D positions relative to the release mechanism. Source: Hanggi and Ingersoll 2012, 71.

two paddles (wood posts with rubber balls attached) that were mounted on spring hinges. The horse had to push one of the paddles with its nose to indicate its choice (see below), after the movable release mechanism had been lowered.

The objects to be detected or discriminated could be hung right or left, at four positions: A, 0.30 meters from the front wall of the station, or B, C, or D, all spaced 1.52 meters apart. The amount of observation time the horse was allowed, before activation of the movable release mechanism, was 3 seconds (see figure 7.4).

The experiment comprised three phases, following an initial learning during which the horses were, among other things, trained to touch a target with the end of their nose, and had become accustomed to standing quietly in a similar station:

- Training phase. During this phase, the horse learned by operant conditioning through successive approximation to press on the paddle situated on the side where a test object appeared. The object was first placed just behind the paddle, then a little farther back, then on the side until it gradually appeared at position A, at a height of 0.86 meters. This training was carried out on both sides, left and right.
- Object-detection phase. During this phase, the horse was presented with two objects placed symmetrically in position A: a plastic and foam racket, attached to the

curved wall (0.86 meters above in the ground in an initial phase), which constituted the positive stimulus; and a cord with nothing at the end, at the same height, which constituted the negative stimulus. The cord served as a control to ensure that the only difference between the two sides was the racket itself. The horse had to press the paddle situated beside the racket. If the animal did so, it heard "Good!" and was rewarded with food automatically delivered to the container placed between the two paddles. If not, the horse heard "No!" and received no food. The stimuli were then withdrawn, the movable release mechanism was raised, the horse put its muzzle back on the float, and a new trial could begin with a new stimulus presentation. The success criterion for a given position (for example, in A, 0.86 meters from the ground) was 80 percent correct responses out of twenty consecutive trials. When the horse satisfied this criterion, it was tested in the following position at the same height (B, then C, then D). If it obtained a score only between 70 and 75 percent during the first series of twenty trials, it then proceeded to new series of twenty trials to determine whether its performance improved. If, on the other hand, the score was less than 70 percent, the session was terminated, and the procedure was resumed at the following position during a next session until all the positions had been tested. The procedure was identical in the second phase, with the stimuli hanging at ground level. As for the other test phases, various methodological precautions were taken, such as randomly determining the left versus right positions of the stimuli.

• Discrimination phase between two symmetrically situated objects, first in a high position, then in a low one, for each of the positions A, B, C, and D. This phase followed the same procedure as the object-detection phase. The racket was still the positive stimulus, but a rubber ball attached to the end of the cord functioned as a negative stimulus. Thus, the good response no longer simply assumed detection of the presence of an object (the racket), but the ability to discriminate the good object (the racket) from another object that was also present (the ball).

At the end of the experiment, it was found that all three horses satisfied the success criterion required for object detection in any of positions A, B, C, or D and for a stimulus either above or level with the ground.

In contrast, for discrimination between the two objects, the racket and the ball, although all three of the horses also succeeded when the stimulus was in position A, B, or C, whether above or level with the ground, they all failed when the stimulus was in position D. Statistical analysis carried out on the results also showed that, although there was no significant difference when the racket was presented with the cord alone or with the ball for positions A, B, and C, there were significantly more errors when the presentation involved the racket and the ball in position D, whether above or level with the ground. As the authors note, notwithstanding measurement uncertainties relating to the experimental apparatus, and to interindividual differences between the horses tested (size, position of the eyes, and so forth), "this suggests that objects appearing laterally from approximately 90° up to approximately 138° can be discriminated by a forward-facing horse but those that appear nearly behind the horse (162° or more) are difficult to identify . . . In other words, these horses were able to *notice* an object appearing from the side in all tested positions and heights, including nearly behind them, but could only *discriminate* between objects within a certain range of lateral vision" (Hanggi and Ingersoll 2012, 74–75).

The researchers also note that there were practically no significant differences between the performances of the three horses tested and that, overall, the height at which the stimuli were positioned did not matter.

Finally, considering the different characteristics of equine vision, their sensitivity at the periphery of the visual field to the detection of movement, and their adaptive character, especially with regard to detection of predators, the authors also remark that "a horse may react adversely to the sudden movement of a human appearing toward the rear due to its physiological makeup rather than a lack of perceptual ability or cognitive comprehension" (ibid., 76).

Night Vision

Anyone who has ever ridden a horse at night in the country cannot but notice the ease with which his or her mount moves, even though it be pitch black. In fact, horses are well known to have substantially better night vision than do humans, but for a long time, no one really bothered to measure it.

Recently, Evelyn Hanggi and Jerry Ingersoll (2009b) set about remedying this gap in knowledge by subjecting four horses to an experimental task of visual discrimination of geometric shapes in a gradually darkening environment. With a touch of its nose, the horse had to indicate which of two simple shapes—a circle or a triangle—it had previously been trained to recognize as a positive stimulus by rewarding it with grain. As had been observed previously, this type of learning task presents no particular difficulty in and of itself (Hanggi 1999a).

During a preliminary training phase, under normal illumination, two of the horses learned to discriminate a black triangle on a white background as a positive stimulus, and the two others a black circle on a white background (see figure 7.5). This learning was achieved fairly easily, as the horses required only between two and four sessions of twenty to thirty trials to accomplish it. Moreover, when they were presented immediately afterward with a white circle or triangle on a black background, the horses responded correctly, showing (as in other experiments) their capacity for generalizing.

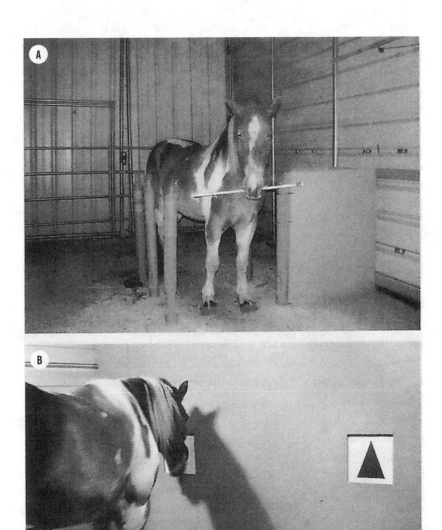

Figure 7.5 (A) Station where a horse awaits the presentation of a stimulus. (B) A horse indicating its choice by touching the stimulus with the tip of its nose (here, in full light).

Source: Hanggi and Ingersoll 2009b, 47. © Evelyn B. Hanggi, Equine Research Foundation.

In the following test phase, the experimental apparatus consisted of a 2.5-by-3.6-meter wall located in a windowless metal building with a light-sealed door. The wall contained two square openings for presenting 227-square-centimeter cards showing a circle or a triangle, as described above. The two openings were separated by a distance of 1.2 meters and were situated at a height of 80 centimeters. Facing the wall, at a distance of 2.4 meters, was a sort of station that the experimenters could close or open by raising or lowering a horizontal bar through remote control from a room behind the apparatus. This station was intended as a holding area for the horse at the start of each discrimination task.

The experimenters used a Sky Quality Meter, a high-precision tool used in astronomy to measure the brightness of the night sky,[3] to precisely measure the room illumination. In addition, they employed night vision cameras to observe the horse from the room behind the apparatus.

At the beginning of each test session, the horse was allowed to adapt to the starting level of darkness for about fifteen minutes, until it could move easily in the room without bumping into anything. In fact, as I showed in the previous chapter, the time required for a horse to adapt to darkness is substantially greater than that for humans, to the extent that, in going from a strongly lit area to pitch dark, a horse may not totally recover its night vision even after half an hour. Once it was clear that the horse could move about the room, it was placed in the station, and the test session began.

The stimuli were presented simultaneously, and after 5 seconds, the bar was lowered to free the horse. The animal moved toward the stimuli and made its choice. It remained there until it heard "Good!" or "No!" It then received 15 grams of a mix of grains if the choice was correct. After having eaten, or hearing "No!" it returned to the station for a new trial (recall that the horses had been previously trained, in the context of other experiments, to return on their own to the station after each trial).

---

3. The unit of measure for this instrument is magnitude per square arc second (mag/arsec^2), which is a logarithmic unit convertible to cd/m^2 (candela per square meter).

The success criterion was 80 percent correct responses for two consecutive runs of twenty trials (p = 0.01). Once this score was obtained, the test continued with a reduction of the level of light in the room, which at the start had been illuminated at a level akin to daylight. Sessions were run four days per week on average; each lasted forty-five to ninety minutes, depending on the horse, and comprised thirty to sixty trials.

In the case of black shapes on a white background, four horses satisfied the success criterion until the level of light in the room was lowered to 3.35×10^{-5} cd/m², corresponding to a moonless night lit only by stars (without additional light from a nearby town). Only beyond 2.71×10^{-5}, that is, illumination corresponding to light within a dense forest under a moonless sky, did the responses become random.

With white shapes on a black background, only three horses underwent further testing (the fourth horse could not be tested owing to time constraints), starting as before with a relatively low level of illumination comparable to the light from a full moon. The results were similar. Despite their failure to discriminate in conditions similar to those in a dense forest under a moonless sky, the horses could still navigate easily, without bumping into anything.

For comparison, experimenters placed in the same conditions as the horses had difficulty with discriminating tasks, around 4.30×10^{-4} cd/m² (overcast full moon). Nor could they do it when the light was lowered to 6.81×10^{-5} cd/m² (starlight). Finally, they could not distinguish anything around them beyond 4.30×10^{-5} cd/m² (attenuated starlight).

The horses thus showed excellent scotopic vision, much better than that of humans, which not only enables them to distinguish the shapes of objects around them but also to move around in conditions of profound darkness. As the authors note, it is only when the dark is almost total, as in a dense forest with a moonless sky, that the animals lose their ability to distinguish objects, although they can still navigate in familiar surroundings.

Finally, Hanggi and Ingersoll assert that, although the tapetum lucidum may reduce the animal's capacities of visual discrimination to some extent, it does not appear to hinder horses from using scotopic vision to distinguish simple shapes of average size. On the other hand, their ability to make out finer details or smaller objects under the same conditions is still an open question.

The Visual Apparatus: An Integrated System . . .

In describing the characteristics of the visual field, I referred to a longstanding belief that a horse shown the same object at two different times, first with one eye, then with the other, cannot subsequently recognize the object and may even react to it with fright. This belief is inconsistent with the existence of the corpus callosum, which assures constant communication between the two cerebral hemispheres. To disprove this thesis experimentally, Hanggi (1999b) subjected two horses to a visual discrimination task by covering one of each pair of eyes to test their capacity for interocular transfer of learning.

The stimuli, measuring 30 by 30 centimeters, were placed vertically one over the other in two of four panel openings in a wooden board. The horse stood in front of the board. When the animal's left eye was covered with an eyepatch, the stimuli appeared in the top and bottom openings on the right side of the panel. When the right eye was covered, the stimuli appeared on the left side. These stimuli consisted of four pairs of different white and black shapes, with roughly equal amounts of black and white (see figure 7.6).

The horse had to press the choice of stimulus with its nose. When it chose correctly, a container at the bottom of the panel automatically filled with food. The success criterion for learning for the entire experiment was 90 percent correct responses out of two consecutive runs of twenty trials.

Following an initial phase of training, during which the two horses were familiarized with the experimental apparatus and the test procedure, the experiment was conducted in two phases.

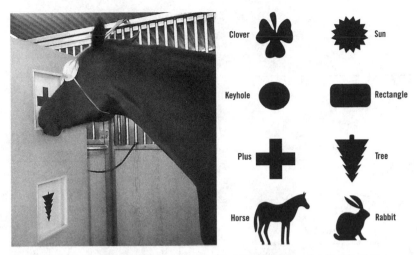

Figure 7.6 Left: learning setup for monocular visual discrimination of right side up or upside down. Right: pairs of shapes presented. Source: Hanggi 1999a, 519. © Evelyn B. Hanggi, Equine Research Foundation.

In the first phase, one of the horses, Ellie, was tested on four stimulus pairs. The discriminative learning was carried out with the left eye only for pairs 1 and 4, and with the right eye for pairs 2 and 3. The horse was then tested with the other eye for each of the four pairs, that is, with the right eye for pairs 1 and 4, and with the left eye for pairs 2 and 3. For pairs 1, 2, and 4, the test showed successful transfer from the two initial consecutive runs of twenty trials. Pair 3 ("Plus" versus "Tree") required only a second run of twenty trials to succeed.

The second phase of the experiment involved both horses and brought into play reversal discrimination, that is, the positive stimulus became negative, and vice versa.

Thus, the horse that participated in the first phase of the experiment, Ellie, now had to discriminate the first pair ("Clover" versus "Sun") with reversal of the positive and negative stimuli vis-à-vis her previous learning. This second training was done with the right eye, and the discrimination task with the left eye.

The other horse, Coco Bean, which had not participated in the first part of the experiment, was first shown the fourth stimulus

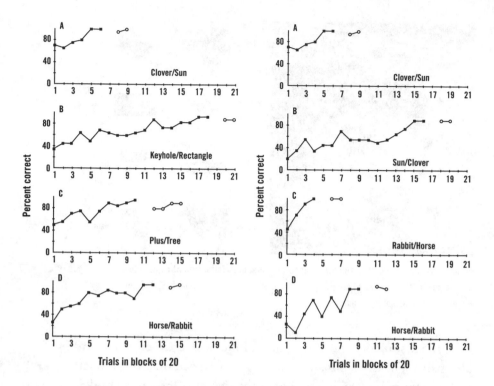

Figure 7.7 Left: Ellie's learning curves for the first four discriminations. Right: learning curves for discriminations with reversal learning (Ellie: A and B; Coco Bean: C and D). The black squares correspond to the trained eye, the white circles to the transfer test of the other eye. Source: Hanggi 1999a, 520–21.

pair in an initial training (positive stimulus "horse" versus negative stimulus "rabbit") with the right eye, followed by discrimination with only the left eye. Then the horse was shown this same pair with reverse reinforcement (positive stimulus "rabbit" versus negative stimulus "horse") with the left eye, followed by a discrimination with the right eye alone. The results of this second phase of the experiment were conclusive from the two first consecutive runs of twenty trials (see figure 7.7).

Thus, taken together, these experimental results confirm the corpus callosum-based hypothesis, namely, that no matter which eye a horse uses to view an object, it can recognize it instantly with the other eye.

And what of the anecdotal accounts by riders attesting that this or that horse, returning from an outing, reacted with fright to the sight of an object that previously inspired no fear? The explanation must be sought elsewhere, for example, in the angle from which an object is seen. I will return to this point in discussing object recognition.

... In the Context of Cerebral Hemispheric Specialization

In chapter 4, on the brain, I touched on the existence in many animal species of a certain functional specialization of each of the cerebral hemispheres that translates into perceptual or motor preference for one side or the other. Recent work has focused on confirming this lateral bias in the horse and characterizing its nature and extent.

In the area of visual perception, five studies have shown strong response in the horse to emotional stimuli presented in the left visual field, which also happens to overlap quite a bit with the monocular visual field of the left eye. Indeed, on the one hand, the lateral setting of the horse's eyes translates into a narrow binocular visual field and wide monocular fields largely independent of each other. On the other hand, its optic fibers are for the most part crossed (Barone and Bortolami 2004). Thus, nearly all the visual data of the left eye project to the right hemisphere, which is predominant in many species, in detecting novelty and emotional reactions.

In three of the recent studies, the horses were presented with objects. The last two, on the other hand, involved interaction with humans.

The first of these studies, carried out by Claire Larose and colleagues (Larose et al. 2006) involved 65 young horses (35 males and 30 females), aged two and three years, including 19 trotters (handled equally on both sides because they are harnessed) and 46 French saddlebreds (trained to be mounted from the left). These horses were shown a new object (a barred cage, 100 by 80 by 80 centimeters, with a shiny red tape round it) placed in a familiar environment. The idea was to arouse various neophobic reactions in the animals and to measure them using an index of emotionality (Wolff, Hausberger, and Le Scolan 1997).

In addition to evaluating individual behavior patterns, the authors observed whether, at intervals of 10 seconds after being shown the object, the horses looked at the object with their right or left eye and whether they sniffed or smelled it with their right or left nostril (the authors did not, however, report any detailed results on this aspect), as well as what the position of the axis of their head was with respect to the object.

Although the subjects as a group showed no side preference with regard to the number of times an object was looked at or which nostril smelled or sniffed it, and although neither sex nor age appear to have an influence in this respect, two clear observations emerged concerning visual exploration in relation to the index of emotionality.

On the one hand, the authors observed a stronger tendency to look with the left eye when the subject was more emotional.

On the other hand, they observed a much higher frequency of right-eyed looks from less emotional French saddlebreds than from more emotional ones, whereas all the trotters tended to use their left eye to glance at the object. These behaviors translated into a clear significant correlation between use of the left eye and emotionality for the French saddlebreds, but not for trotters. Finally, no significant correlation in the emotional index was found between individuals of either breed. In any case, the French saddlebreds glanced at the object more often with the right than the left eye, whereas the tendency was the opposite for the trotters (see figure 7.8).

Although the first observation confirms a relative predominance of the right hemisphere for emotionality, which is consistent with findings for many other species, the second observation raises an interesting question about the possible influence of learning on the expression of laterality, which is also examined in the studies discussed below, focusing on equine interactions with humans (Farmer, Krueger, and Byrne 2009; Sankey et al. 2011b).

The authors remark that French saddlebreds, normally handled and mounted from the left side, consequently may become less reac-

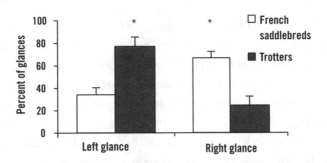

Figure 7.8 French saddlebreds look at the object more frequently with the right eye than with the left, whereas the tendency is reversed for trotters. Confidence interval: $*p < 0.05$. Source: Larose et al. 2006, 362.

tive to stimuli appearing on that side, which suggests a reversal of cerebral hemispheric dominance based on experience. Trotters, on the other hand, which are harnessed on both sides, may exhibit a lateral bias more characteristic of the species, which is a tendency to explore novelty with the left eye.

Consequently, the authors propose that the custom of mounting from the left corresponds to a lateral bias as much for humans as for horses. For humans, who are generally right-handed, it is easier to use the left leg as a postural pivot while the right leg is balanced over the saddle (independent of the oft-cited explanation that swords are carried on the left). For horses, inasmuch as they are more reactive on this side to unsettling stimuli, the approach of a rider from the left enables the animals to evaluate his or her emotional posture (known or unknown, friendly or menacing).

In keeping with this hypothesis, note also that the practice of saddling up from the left appears to be universal across the ages. Toward the end of the fourth century, the nomadic peoples of Central Asia and Siberia and the Chinese were known to use stirrups. But even earlier, at the very beginning of that same century, the Henan and Hunan of China (Cartier 1993; Lazaris 2005) used "proto-stirrups consisting of a triangle, on the left side of the mount only, suspended from a strap attached either to the front part of the saddle tree arch or to the flap" (Cartier 1993, 33), which would have been mounting stirrups.

Nicole Austin and Lesley Rogers (2007) compared the reactions of thirty horses (eighteen males and twelve females, two to twenty-seven years old) to the presentation of a novel stimulus. A person 5 meters away opened a blue-and-white umbrella 105 centimeters across and pointed toward the neck of a horse that was eating food that had been scattered on the ground before it, in a grazing position. The person then raised the umbrella while walking rapidly toward the horse's head region, either directly from the front, from the left side, or from the right side. The researchers asked whether the flight reactions provoked differed depending on the direction of the stimulus and whether the reactions tended toward the right or left when the same stimulus was presented in the binocular field. Out of twenty-four horses tested on both sides, seventeen fled a greater distance, thus apparently reflecting greater emotional tension, when the stimulus was presented on the left, which implies a right hemisphere response, which is consistent with the experiments of Larose and colleagues. In contrast, only four of the horses shown a stimulus on the right fled as far, and three other horses showed no preference. However, the horses reacted differently depending on whether the stimulus was first presented on the left (eleven horses) then on the right (thirteen horses), or vice versa. In the first case, they responded more strongly to the stimulus presented on the left, whereas no difference in reactivity was observed between sides when the stimulus was first presented on the right (see figure 7.9).

The authors asked whether the initial presentation of the stimulus on the right might not somehow be associated with inhibition of the flight response by the left hemisphere. In such a scenario, the horse would learn that the stimulus presented no danger, and this information would then be transmitted to the right hemisphere.

Moreover, no lateral bias was observed with respect to the way in which the horses initiated their flight, right or left, when the stimulus was presented in the binocular field; in contrast, horses that turned to the right fled a greater distance than those that turned to the left.

A third study, conducted by Alice de Boyer des Roches and colleagues (de Boyer des Roches et al. 2008) sought to determine

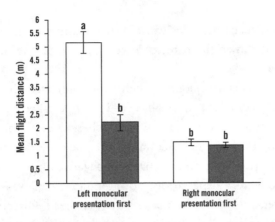

Figure 7.9 Mean distance of flight depending on whether the stimulus was first presented to the left (n = 11) or right (n = 13). White bars: presentation to the left; gray bars: presentation to the right. The difference between the bars indicated was significant (p < 0.05) for each of the bars marked b. Source: Austin and Rogers 2007, 470.

whether visual exploration of an object by a horse is done more with the right or left eye depending on emotional valence—positive, negative, or neutral—of the object. The study also involved investigation of a possible asymmetry in the olfactory exploration of the object.

Accordingly, three objects were selected:

- a gray plastic bucket normally used to feed the horses and thus attractive, that is, positive value;
- an orange plastic cone, like the ones used in roadwork, which the horses had never encountered and that was considered to have a neutral value; and
- a white short-sleeved shirt, laid on the ground; normally worn by stud farm veterinarians, it was potentially associated with stress-inducing situations, and thus had a negative value.

The group of subjects consisted of thirty-eight adult mares (aged five to twenty-one) in the last month of gestation, which avoided possible biases related to sex or to cyclic changes in behavior; all the mares had the same care regime (human contact, veterinary visits,

183

and feeding). Because the idea was to test the animals' first reaction, each mare was shown the three objects, at a rate of one per day, in random order.

The horses' general exploratory behavior confirmed the assumed emotional valences of the three objects: The time taken to approach the positive object (the bucket) was much shorter than the time required to approach the new, neutral object (the cone), which in turn took much less time to approach than the negative object (the shirt).

Vis-à-vis olfactory exploration, no laterality effect was observed for the bucket, insofar as the mares put their heads directly into it as soon as they reached it. For the two other objects, comparisons made regarding preferential use of one or the other nostril depending on the object revealed no significant differences. Most of the mares tended to use the right nostril first rather than the left to sniff both the cone and the shirt. In addition, the animals exhibited a slight tendency to use the right nostril more often to explore the novel object (the plastic cone) than the shirt, an observation I will come back to briefly in discussion olfactory perception.

With respect to visual exploration, the authors made the following observations:

As a whole, the results showed no laterality bias with respect to which eye was used first to glance at an object. Nonetheless, in taking their first look, significantly more ($p < 0.01$) mares used their right eye for the cone than for the bucket. Thereafter, the mares showed no preference in using their monocular and binocular fields of vision for the cone, whereas they used the binocular field of vision more to examine the bucket and a monocular field of vision to investigate the shirt. Thus, the use of monocular and binocular fields of vision clearly varies depending on the object: mainly binocular for the bucket (positive); shared for the novel object, the cone (neutral); and largely monocular for the shirt (negative) (see figure 7.10).

When the mares used their monocular field, they tended to preferentially use their right eye (processed by the left hemisphere) for the new, neutral object (the cone) but more frequently their left eye (processed by the right hemisphere) rather than the right to look at

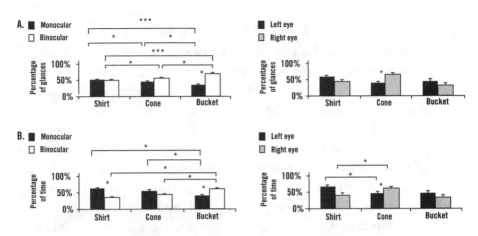

Figure 7.10 Number of glances (percent) and time elapsed (percent) in the monocular, binocular, left monocular, and right monocular visual field. Confidence interval: $*p < 0.05$; $**p < 0.01$; $*** p < 0.001$. Source: De Boyer des Roches et al. 2008, 489.

the shirt, which suggests a side bias in visual exploration of disturbing objects, consistent with the results of the two previous studies (Larose et al. 2006; Austin and Rogers 2007). Citing Lesley Rogers and Richard Andrew (2002), the authors reiterate that in many species of vertebrates, the right hemisphere appears to be specialized in processing stimuli that arouse negative emotional responses.

Conversely, when the mares visually explored the cone, they made greater use of their right eye than the left eye, which suggests that the left hemisphere is involved in classifying novel objects with neutral value. Accordingly, the authors conclude that their "data confirm the role of the left hemisphere in assessing novelty in horses like many vertebrate species" (de Boyer des Roches et al. 2008, 490), adding that their "results seem therefore to confirm that lateralization of visual and olfactory responses could be independent, as previously shown for olfaction and motor bias in horses" (489). I will return to this point in the next chapter. For their own part, Rogers and colleagues (2004) propose that "[m]any species share a general pattern of using the right hemisphere to attend to novelty and execute rapid responses, whereas the left hemisphere is used to categorize stimuli and control responses requiring consideration of

alternatives" (S420). In any event, the question of novelty, its implications in cognitive and emotional processes, and its connection to
laterality appears quite complex and deserves to be explored further.

De Boyer des Roches and colleagues also point out that the fact
that mares prefer their binocular field for the positive object (the
bucket) might be taken to mean that they are seeking a better three-
dimensional assessment. Yet insofar as they show no such binocular
preference for the other objects, and in particular for the stimuli with
negative value, the more plausible hypothesis is that stimuli with
positive value are mainly processed by both hemispheres.

In concluding, they remark that their findings that "negative
emotions are relatively lateralized whereas positive emotions are less
lateralized raise interesting issues about the weight of their involvement in survival mechanisms" (de Boyer des Roches et al. 2008, 490).

Revisiting the hypothesis formulated by Claire Larose and
colleagues (2006), according to which the custom of saddling up
from the left is associated with lateralization in both horses and
humans, Kate Farmer and colleagues (Farmer, Krueger, and Byrne
2009) explored visual lateralization in the horse in interactions with
humans, in particular horses trained to be led and mounted from the
left as well as horses trained to be led and mounted from both sides.
Considering, among others, the studies just discussed, these researchers note that it is not clear whether "the lateralization observed with
inanimate objects also applies in connection with humans, nor
whether it is restricted to emotional situations" (230).

Consequently, they embarked on a series of four experiments with
the goal of studying different aspects of the question. They used a population of 55 horses. Among them, 40 (divided into two subgroups) had
been trained in the traditional manner, being saddled and mounted
from the left, and 15 (divided into two subgroups with 10 horses
common to the two subgroups) had been trained to trust humans that
led and saddled them from both sides, as well as to be desensitized to
human activity and to unknown objects situated on either side.

In the first experiment, which involved fourteen traditionally
trained horses, the animals had to traverse a specified route to find

a reward, either alone in the test area (a so-called neutral situation) or in the presence of an unknown human, who remained passive. In the presence of this human, the horses showed a significant tendency to preferentially use their left eye, except for one animal that used its right eye significantly more. In the neutral situation, this tendency was less marked, and only four horses used their left eye significantly more. The authors note that, consequently, the lateralization preference observation cannot be exclusively linked to emotional factors, but might also be connected to evaluation of the environment.

The second experiment concerned twelve horses trained on both sides that had to complete the same task while in the presence of a passive human who was either familiar or unknown. In the presence of the familiar human, nine of the horses showed the same, but nonsignificant, tendency to use their left eye. One horse expressed this same tendency significantly; another horse showed a significant tendency to use its right eye. In the presence of the stranger, although ten horses showed a certain tendency to use the left eye, this same tendency was significant for only three of the horses; the other two animals, in contrast, showed a significant tendency to use the right eye. Overall, these horses appear to have shown a left eye preference in the presence of a stranger equal to that of traditionally trained horses in the neutral situation, and less in the presence of a familiar human. The authors also note that, for the horses showing a right eye preference, surprisingly, no lessening of the tendency was recorded in the presence of a stranger from one experiment to the other, which suggests that the nature of the training does not appear to influence the choice—left or right—of the preferred eye.

The third experiment involved 26 traditionally trained horses and 13 horses trained on both sides, in two interactive situations with a familiar human: Based on the join-up and hook-up techniques popularized by Monty Roberts and Buck Brannaman, respectively, these techniques both consist in part of having a trainer chase the horse before allowing it to approach by turning his back. Although, for the traditionally trained horses, insufficient data preclude individual analysis of the results, 23 of the animals more frequently positioned themselves so that the experimenter was in their left visual field. For horses trained on both sides, 12 showed such a preference, and for 11 of them, the preference was individually significant. Considering that the experimenter was very familiar with the horses (care, feeding, work) and thus that there was no reason to assume that he was a source of negative emotion for them, the authors conclude that their observations are consistent with those of de Boyer des Roches and colleagues (2008). In other words, preferential use of the left eye may be associated with positive and negative emotions and not uniquely with anxiety and negative emotions.

As a fourth experiment, the authors carried out a comparative analysis of the results of the second experiment, in the situation where the passive experimenter was familiar to the horse, and the third (interaction with familiar experimenter) for the ten horses trained on both sides. Paired comparison showed a stronger tendency to use the left eye in interactive situations, which the authors attribute as most likely due to the interactive situation itself, considering that there was no reason to assume that the horses were feeling more emotion or fear vis-à-vis the experimenters in one or another situation since the humans were familiar in both cases.

In concluding, Farmer and colleagues note first that, among a population of fifty-five domestic riding horses used in the experiments, a strong preference for the left eye emerged for scanning a person present or, in the absence of a person, the environment. Forty-eight horses showed a consistent preference for the left eye, five for the right eye, one showed no particular preference, and one varied its preference depending on the situation. As noted by the authors: "Horses appear to prefer to keep 'high priority' stimuli in one visual field (the left, for most horses)" (Farmer, Krueger, and Byrne, 237). Second, when horses interact with a human experimenter, the experimenter becomes the highest priority, irrespective of the way in which the horse was previously trained. Finally, "When a human is present but passive, the human is a priority for conventionally trained horses who regard them as a risk, but less for the bilaterally trained horses, apparently because they are more comfortable in the presence of humans" (ibid.). Moreover, the authors note that their own observations with respect to the stronger response of horses to stimuli presented on the left (Austin and Rogers 2007), and consistent with the hypothesis proposed by Larose and colleagues (2006) regarding the custom of mounting from the left, suggest that the general preference of horses for humans to be on the left is not simply the result of habit and training, but rather related to their emotions and perceptions.

Carol Sankey and colleagues (2011b) also tackled the issue of equine interaction with humans. However, they chose to focus on a comparison between young, untrained (naïve) horses and trained horses to evaluate the effect of training on lateralization of responses to human approach.

To this end, the researchers used two groups of young horses. One group comprised 16 one-year-old animals (10 females and 6 geldings) with no experience of humans except for mandatory veterinary checks and delivery of food in winter. The other group comprised 23 two-year-old horses (15 females and 8 geldings) who, from the age of one year, had been trained for two months to remain immobile on command, by a verbal order that was given them from the left, and then to accept a range of routine handling procedures, such as being fitted with a halter and a surcingle and giving their feet. These groups had been raised in similar conditions during their first year: remaining near their mother until weaning at six months, then being maintained in stalls in groups of five or six during winter, until the arrival of spring.

In short, the experimental procedure consisted of having one and the same experimenter approach each of the horses in the two groups in seven different ways. Distanced 8 meters from the horse at the start, the experimenter approached slowly, arms behind her back, gazing at her target zone, which was either the shoulder, the flank, the croup (both right and left sides), or the front. Once near the horse, the experimenter extended her left arm toward the animal and softly touched the target zone.

The experimental results were based on the success or failure of the approach, according to whether it resulted in touching the horse (or not), as well as on the animal's positive (remaining quiet, looking at the experimentor, or sniffing her) or negative (avoidance, escape, threat of biting, or kicks) reaction on contact.

For the naïve horses, almost all of them (fifteen out of sixteen) accepted being touched at least once, with on average two to three zones touched out of the seven target zones. Approaches from the right more frequently resulted in contact (35.4 percent) than did those from the left (29.2 percent). Some zones provoked flight reactions from the horses, namely, the shoulder (right or left) and the left flank. In contrast, no place was particularly easy to reach. The authors also remark (see table 7.2) that, although not statistically significant, out of the four horses that permitted approach at shoulder height to the point of touching (as is generally in the case in saddling a horse), all the animals reacted negatively to left-side approaches, whereas three of them reacted positively to the right side.

For trained horses, almost all (twenty-two out of twenty-three), just as for the naïve horses, accepted being touched at least once, in such cases four to five zones touched on average out of the seven target zones. However, unlike the case of naïve horses, marked preferences were expressed in favor of certain zones; thus, the horses allowed approach without trying to flee or expressing threats when the shoulders (either right or left), the forehead, and the right croup were involved. Moreover (see table 7.2), most of the horses showed positive rather than negative reactions during contact, whatever the area touched, which is particularly true of the eighteen horses that permitted touching on either shoulder (see figure 7.11).

Comparison of the results between the two groups of horses shows clearly significant differences ($p = 0.001$) for greater mean tolerance to contact for trained horses and for a larger proportion of these horses to accept contact on one or the other shoulder, as well as the right croup. The authors also observe that it is equally striking to observe that, although there is no behaviorally significant difference for contact on the right shoulder and croup, the results do show a strongly significant difference in the behavioral responses of the two groups with respect to the left shoulder ($p = 0.001$) and the croup ($p = 0.002$).

Table 7.2 Number and percentage of one- and two-year-old horses accepting to be touched on the different zones and their positive or negative response to contact and statistics (binomial tests).

Approached zone	One-year-old (naïve)					Two-year-old (trained)				
	Success rate	P	Positive reactions	Negative reactions	P	Success rate	P	Positive reactions	Negative reactions	P
Left shoulder	4/16 (25%)	0.04	0	4	NA	18/23 (78%)	0.005	18	0	< 0.001
Left flank	4/16 (25%)	0.04	2	2	NA	12/23 (52%)	0.5	10	2	0.02
Left croup	6/16 (38%)	0.2	1	5	0.1	10/23 (43%)	0.3	8	2	0.05
Front	10/16 (63%)	0.2	6	4	0.4	19/23 (83%)	0.001	17	2	< 0.001
Right shoulder	4/16 (25%)	0.04	3	1	NA	18/23 (78%)	0.005	18	0	< 0.001
Right flank	8/16 (50%)	0.6	3	5	0.4	14/23 (61%)	0.2	11	3	0.03
Right croup	5/16 (31%)	0.1	3	2	0.5	18/23 (78%)	0.005	12	6	0.1

Source: Sankey et al. 2011b., 466

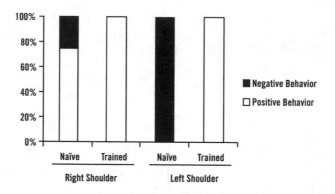

Figure 7.11 **Percentage of horses reacting either positively or negatively to a human touching their left/right shoulder.** Source: Sankey et al. 2011b, 466.

For the authors, the observation that there is a clear asymmetry in horses aged one year, which show substantially more reticence and more negative reactions when approach and contact are made within the left monocular visual field and, conversely, fewer negative reactions and more acceptance on the right, "support[s] previous studies conducted in adult horses, that showed that the left monocular visual field (information processed by the right cerebral hemisphere) was associated with a higher reactivity . . . and fits in the theory of valence (right hemisphere mainly processing negative emotions and left hemisphere mostly processing neutral or positive emotions" (466).

However, the observations that two-year-old horses show no negative behavioral responses when they are approached from the left and that in fact they express no asymmetrical behavioral response and accept approach and contact just as well on one side as on the other, brings up the question of the possible influence of age. Yet, with respect to different existing studies, and taking into account the weak age difference between the two groups, the authors conclude that the differences observed are most likely due to the training the two-year-old horses received.

Continuing their analysis in this vein, the authors venture that the results of preceding studies (Larose et al. 2010) show the potential importance of training methods, and in particular the use of positive and negative reinforcement, on the perception of work sessions and more generally on the relationship between humans and horses (Sankey et al. 2010b; Sankey et al. 2010c). This might explain the decrease in negative responses for the left side in two-year-old horses whose training was accompanied by positive reinforcement, the left eye being generally associated with situations of fear and the right with more positive events. Moreover, horses are spontaneously less reactive on the right side than the left side in the presence of a worrisome object (Austin and Rogers 2007), which is consistent with the weaker emotional reactions on the left side observed here in trained horses.

The authors add that, although Farmer and colleagues (Farmer, Krueger, and Byrne 2010) recently reported contrary findings, namely, training horses in interactive situations resulted in a strong preference for looking at the experimenter with the left eye, this is most likely due to the training method used. For Sankey and colleagues, in fact, chasing the horse before allowing it to approach may have been a source of stress and could have provoked negative emotions toward the experimenter. To their mind, the positive relationship existing between experimenters and horses is not necessarily sufficient to compensate such an effect, as negative events appear to leave stronger traces in memory than positive ones (Fureix et al. 2009).

Ultimately, for Sankey and colleagues, it does appear that visual preference for fixating an object may provide a clue to the way in which the object is perceived. Considering the study by de Boyer des Roches and colleagues (2008), according to which horses use their left eye more to look at an object with negative emotional value than a neutral or positive object, Sankey and colleagues hypothesize that "the type of work and how it is perceived by horses, rather than work itself, impacts differently on horses' perceptual laterality" (467). Finally, they note the practical implications of their findings: "Young or highly reactive horses would certainly learn faster and be more relaxed if first approached on the right side. Beginning their training by handling them on their most reactive side (i.e., left) may in some cases exacerbate their emotional reactivity. Besides, knowing that interocular transfer occurs in horses . . . , at least from the right eye to the left one (i.e., from left to right hemisphere); . . . they could progressively learn to accept approach and handling on the left side. Once a positive relationship is established, this study suggests that the laterality bias may fade to allow horses to accept human contact equally on both sides" (ibid.).

Perceiving the Third Dimension

Depth perception, which tells us how far we are from objects, or how far one object is from another, as well as their actual size, relies on many sources of perceptual information, both monocular and binocular, as shown by Andre Delorme and Jacques Lajoie (2003) (see table 7.3).

Table 7.3 Main sources of data on depth perception.				
Data source	Visual/ nonvisual	Absolute/ relative	Monocular/ binocular	Pictorial/ cinematic
Accommodation	Nonvisual	Absolute	Monocular	—
Convergence	Nonvisual	Absolute	Binocular	—
Familiar size	Visual	Absolute	Monocular	Pictorial
Occlusion	Visual	Relative	Monocular	Pictorial
Relative height	Visual	Relative	Monocular	Pictorial
Relative size	Visual	Relative	Monocular	Pictorial
Texture gradient	Visual	Relative	Monocular	Pictorial
Linear perspective	Visual	Relative	Monocular	Pictorial
Brightness gradient	Visual	Relative	Monocular	Pictorial
Motion parallax	Visual	Relative	Monocular	Cinematic
Variable size	Visual	Relative	Monocular	Cinematic
Binocular disparity	Visual	Relative	Binocular	Cinematic
Source: Delorme and Lajoie 2003, 278.				

In particular, the spectacle of horses jumping over obstacles suggests that they possess a marked ability to judge distance and height: Think of how they are able to place their approach in stride and to adjust their jump to a few centimeters over the obstacle.

To clarify the nature of the information the horse uses, Brian Timney and Kathy Keil asked whether horses are sensitive to monocular depth cues of a pictorial nature (1996) and also whether their binocular field of vision is genuinely stereoscopic (1999).

Depth Perception Based on Pictorial Monocular Cues

Depth perception in the horse was demonstrated by testing two horses with the Ponzo illusion, which as I mentioned earlier causes two horizontal bars set within converging lines to look as though they have different lengths although they are in fact exactly the same.

The experimental apparatus was analogous to the one the same authors used previously to measure visual acuity. A pair of trapdoors whose opening potentially gives access to a reward presents two views: One (the negative stimulus) shows two straight horizontal bars of equal length placed one over the other; in the second image (the positive stimulus), the upper bar is longer than the lower bar. Following the initial discrimination tasks, the two stimuli were separated by a 2-meter-long divider establishing the minimum distance at which the horses had to carry out their task.

The experiment unfolded in three phases:

- a training phase, during which the horses learned to discriminate between the two images, based on the relative length of the two bars;
- a measurement phase to determine the animals' discrimination threshold; and
- a test phase, during which the images presented were photographic representations: one showed white bars of equal length situated within two railway tracks converging toward a vanishing point, taken at ground level, with the features of the rural landscape adding perspective;

Figure 7.12 Photographs of two scenes used in the critical phase of discrimination. Source: Timney and Keil 1996, 1122.

the second showed two white bars of equal length super-posed on a pastoral scene of a pond and a creek, but with no convergence or perspective cues (see figure 7.12).

During the test phase, the scene containing the white bars within the railway tracks—which created the illusion that the upper bar was longer than the lower bar—was fairly systematically chosen by the two horses, as opposed to the image that created no illusion (see figure 7.13).

When, as a control, the illusion was removed by shortening the upper white bar in the scene with the railway tracks and lengthening it in the pastoral scene, the horses' choices reversed. Finally, when the horses were exposed to the two railway scenes, they chose the one where the upper white bar appeared longest, showing their sensitivity to the Ponzo illusion.

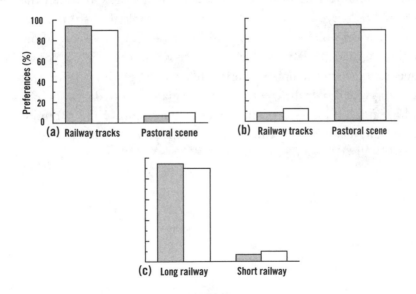

Figure 7.13 Discrimination task results (pairs of bars correspond to two horses). (a) Preference for the tracks or the pastoral scene with white bars of equal length. (b) Preference for a longer upper bar for the pastoral scene and a shorter bar for the tracks. (c) Preference when the two tracks are the stimulus, but with one longer upper bar for one of the stimuli. Source: Timney and Keil 1996, 1126.

Binocular Vision and Stereopsis

Although horses' eyes are set to the side, as we have seen, they have appreciable binocular vision toward the front, on the order of 60 degrees. This binocular visual field gives rise to a retinal disparity. In effect, owing to the interpupillary distance, the images that form on each of the two retinas are not identical; they are disparate. The index of retinal disparity is the basis of binocular vision. As Vilayanur Ramachandran and Diane Rogers-Ramachandran (2009) explain it: "The differences (called retinal disparities) are proportional to the relative distances of the objects from you. Try this quick experiment to see what we mean: Hold two fingers up, one in front of the other. Now, while fixating on the closer finger, alternately open and close each eye. You'll notice that the farther the far finger is from you (don't move the near finger), the greater the lateral shift in its position as you open and close each eye. On the retinas, this difference in line-of-sight shift manifests itself as disparity between the left and right eye images . . . A simplified example shows this effect clearly. When you look at the pyramid, the right eye sees more of the right side than the left eye does, and vice versa; it is a simple consequence of geometric optics. Notice that the images in the two eyes are correspondingly different; the inner square is shifted right or left. This retinal disparity is proportional to the height of the pyramid. The brain measures the difference and experiences it as stereoscopic depth" (20–21) (see figure 7.14).

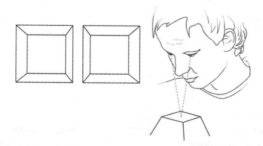

Figure 7.14 **The two eyes do not see exactly the same truncated pyramid. The disparity increases with the height of the pyramid.** Source: Ramachandran and Rogers-Ramachandran 2009, 88.

Moreover, write André Delorme and Jacques Lajoie (2003), according to Béla Julesz (1971), there are two forms of stereopsis, local and global: "The first relies on local features that are clearly discernible and thus easily 'fused,' as in most traditional stereograms. The second, which is unique to random-dot stereograms, appears as a solution to an initial ambiguity . . . The stereoscopic form is perceived by virtue of a 'global process' that selects among the matches of possible points those that have the greatest density of correspondences. All the other potential matches are thus inhibited" (Delorme and Lajoie, 293–94).

To find out whether horses' binocular visual field provides them with genuine stereoscopic vision, Timney and Keil (1999) tested two horses[4] using a pair of visual discrimination experiments, one focusing on local stereopsis and the other on global stereopsis.

Local Stereopsis

Horses are able to recognize monocular depth cues that enable them to discriminate using one eye. Nonetheless, researchers have hypothesized that, if the animals could use both eyes, they would still have the same monocular cues but would have the additional benefit of binocular depth cues related to retinal disparity. As a result, their depth perception threshold should be better.

Using an experimental apparatus analogous to that used in the preceding experiments, Timney and Keil initiated a training phase comprising three discriminative learning tasks involving each of the following stimulus pairs in succession. As before, following the initial discrimination, the two stimuli were separated by a 2-meter-long panel, which established the minimum distance at which the horse had to decide.

The positive stimulus (see figure 7.15, left) consisted of a planar surface with black dots and a square drawn in the center, flush with

---

4. These two horses merit special mention for their contribution to equine science. Indeed, they are the same horses that participated not only in the preceding experiment and the investigation of visual acuity but also the tests carried out by Timney. Moreover, as we will see, they were involved in experiments in color vision.

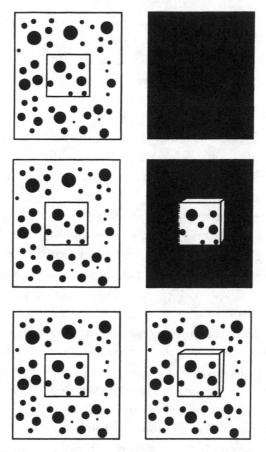

Figure 7.15 Stimulus pairs used for three successive discrimination tasks. Source: Timney and Keil 1999, 1863.

the background. For the first pair, the negative stimulus was a black planar surface; for the second pair, a black background with a square like the one in the positive stimulus with a depth separation of 20 centimeters from the background; and for the third pair, a background identical to that of the positive stimulus with the same square as for the second pair (see figure 7.15, right).

Once this phase of training was successfully completed (the success criterion was twenty-seven correct choices out of thirty successive trials), a measure of the depth detection threshold was

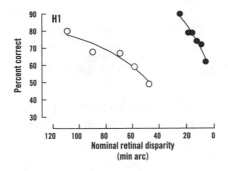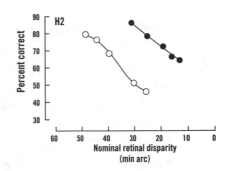

Figure 7.16 Results of depth detection based on retinal disparity for each of two horses. Black circles: binocular vision. White circles: monocular vision. Source: Timney and Keil 1999, 1864.

done by varying the distance of the negative stimulus square relative to the background.

All these steps were first performed with binocular vision for each of the two horses. Later, the steps were repeated with monocular vision. The researchers observed a significant deterioration of performance: on the one hand, a marked increase in the number of trials required to reach the success criterion for the final phase of training, although the horses had accomplished this same task using binocular vision several hundred times; on the other hand, a marked decrease of the threshold for depth discrimination. Timney and Keil also noted significant interindividual variation in this threshold, which might reflect a differential ability to take advantage of monocular depth cues (see figure 7.16).

Thus, in a situation presenting local monocular depth perception cues, the use of both eyes enables horses to better estimate depth, which supports the hypothesis, namely, that based on the same monocular cues, the animals also appear to benefit from binocular depth cues provided by retinal disparity.

Global Stereopsis

To better understand stereoscopic vision in horses, a second, more rigorous experiment was carried out with the aim of definitively establishing whether the horse possesses stereopsis.

201

The experimental apparatus was similar to the one just described. Following initial discrimination tasks, the two stimuli were separated by a 2-meter-long divider that set the minimum distance from which the horses had to decide.

The stimuli consisted of photographic reproductions of anaglyphic random-dot stereograms arranged as in the stereopsis testing patterns of Julesz (1971). An anaglyph is formed by two superposed images of two complementary colors representing the same scene or the same abstract design, but with a slight offset. By placing a filter in one of the two colors over one of the subject's eyes and a filter in the other color over the other eye, each eye can only perceive the visible elements through the respective filter, which creates a depth effect due to retinal disparity.

At a distance of 2 meters, based on an interpupillary distance of 20 centimeters, the stereograms had a disparity of around 19 minutes of arc, corresponding to an effective depth of 12 centimeters.

Two patterns were used: a negative stimulus with 100 percent correlated elements that (for a human observer wearing glasses equipped with a green filter for one eye and a red filter for the other) made a square stand out in the center of the photograph and a positive stimulus with only 40 percent correlated elements that, under the same conditions, did not produce an apparent depth effect. These two patterns constituted the standard stereogram pair.

The horses were first retrained with the third pair of stimuli from the preceding experiment (including "true" depth for the negative stimulus). Once they succeeded at the discrimination task under the specified conditions, they were fitted with glasses with red and green filters and continued the same training to determine whether they could sustain their level of performance. They were subsequently exposed thirty-five times to the standard stereogram pair, alternating with random exposures to the preceding test, and received a reward regardless of their selection.

Next, a series of control sessions was carried out to ensure that the horses were not discriminating based on nonrelevant cues. These experiments began by reexposing the horses to standard pairs of

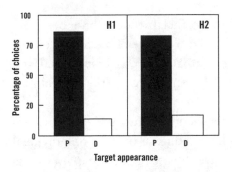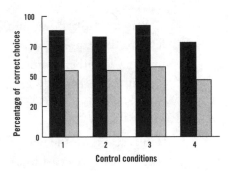

Figure 7.17 Left: Results of the choices of each of the two horses with planar (P) and depth (D) stereograms, with a reward in either case. Right: results for control tests. Black bars: standard pairs of stereograms; gray bars: control pairs where one stimulus is arbitrarily designated correct. Source: Timney and Keil 1999, 1865.

stereograms, but rewarding them only for choosing the "flat" stimulus (with no depth effect) until they had satisfied the success criterion. They were then tested in blocks of twenty trials by alternating standard stereogram pairs with successive pairs constructed in the following way: both stereograms flat; both eyes covered with a green filter; both eyes covered with a red filter; and both stereograms protruding. For each control block session where the horses had no basis for discriminating, one of the stimuli was randomly designated "good" and rewarded.

In the discrimination task with the standard stereogram pairs, the horses chose correctly in 90 percent of cases (see figure 7.17). In contrast, in control situations their selections were equally distributed between the two stimuli. Thus, horses do indeed appear to possess not only local but global stereopsis.

Image Recognition

Images are often used to study social or individual recognition within species. This is especially the case with certain domestic ungulates, such as sheep (Kendrick et al. 1995, 2001) and cows (Coulon et al. 2007).

Yet, as Dalila Bovet and Jacques Vauclair (2000) have observed, few studies explicitly focus on the equivalence of an object and its image: "In other words, it is not at all obvious that animal and human subjects really consider 2-D stimuli to be the 3-D objects that they represent" (143).

With regard to the horse, the scientific literature on this question is scant. To the best of my knowledge, in the area of individual recognition, only two studies involve the use of photographs (Tanida et al. 2005; Stone 2010). At the date of writing, the first of these has yet to be published. In the matter of equivalence between objects and their images, the only research published to date appears to be that of Evelyn Hanggi (2001).

In carrying out her study, Hanggi tested three horses using the same experimental protocol she used in her work on interocular transfer of learning (see "The Visual Apparatus: An Integrated System . . ." above).

The experimental principle consisted of having the horses discriminate between two objects (3-D), or between two photographs of the same objects (2-D), then testing them on transfer of learning either from the objects to the photographs or from the photographs to the objects, depending on the situation.

The object pairs used in the discriminative learning prior to a transfer test based on their photographs were the following: a plate, a candle, a measuring scale, a dog-shaped frame, a bottle, a hedgehog, a football, a Frisbee, a Cookie Monster Beanie toy, and a Beanie Snuffaluffagus toy.

For each trial, the pairs of objects or photographs to be discriminated were shown to the horse for no more than 5 seconds, after which the animal approached and pressed its nose against the place where the chosen stimulus had appeared until the horse heard "Good!" and was given 15 grams of grain (if it made the right choice) or "No!" if it did not.

The success criterion for learning was nineteen correct responses out of twenty trials (p = 0.01). Once learning had occurred for a given stimulus pair, overlearning sessions were undertaken to reinforce the response.[5] These were followed by a transfer test comprising twenty trials. Only after this entire process was complete for a given stimulus pair did the horse proceed to discriminative learning with another pair.

Two of the horses (Coco Bean and Bodie) completed learning tasks, each of which was followed by a transfer test. But the third horse, Cache, only completed four tasks due to time constraints. Moreover, at the end of the experiment, Coco Bean underwent reversal learning for photographs of the bottle and the hedgehog, followed by a transfer test with the objects themselves, to more thoroughly test the robustness of transfer.

Because the horses showed a preference for one of the elements of each pair, the nonpreferred stimulus was chosen as the positive stimulus, except for the first discrimination (plate versus candle), for which the positive stimulus was the preferred stimulus. In each of the four learning tasks for which the positive stimulus was not the preferred stimulus, the horses scored at or below chance on ten out of twelve sets for their first twenty trials, and their performance became significantly higher than chance in the course of learning; in contrast, the animals' success was well above that predicted by chance for all the transfer sessions (again, each comprising a block of twenty trials), apart from two. A significant difference between the first twenty learning trials and the twenty transfer trials was also observed ($p = 0.01$). Two of the horses completed all the transfers successfully, an outcome also achieved by Coco Bean for the transfer reversal set. The third horse, Bodie, delivered less consistent results (including transfers, where it achieved three out of five, as well as the difference between the initial trials of the two situations, only the first being significant). While acknowledging that Bodie might not have recognized the relationship between 3-D and 2-D stimuli, Hanggi nonetheless suggested that the results might also reflect the horse's behavior during the sessions: Bodie was the dominant horse

---

5. Note, however, that it is not evident that overlearning has such an effect (Doré and Mercier 1992).

in the group and was more frequently distracted by the noise made by the horses in the neighboring pasture.

Thus, it appears that some horses are able to recognize objects both in and based on photographs. Hanggi notes in this context that, although the horses made errors in the twenty training and transfer tests (while still obtaining significant average scores), one cannot conclude that they simply learned the transfer tasks quicker since each new learning task, following a nearly instantaneously successful transfer, reflected a typical learning curve. Nor does it seem likely that the findings resulted from a process of categorization, insofar as the stimuli varied greatly in terms of size, shape, color, and so forth, and do not appear to independent observers to have any discernible link. Hanggi further concludes that although horses show that they can establish a relationship between initial stimuli and transfer stimuli, it still remains to be determined what features (the whole or parts of the whole) are important in this type of recognition.

Finally, let me add that, in the context of a study on long-term memory in horses, Hanggi and Ingersoll (2009a) reproduced this experiment six years later with two of the three horses, Coco Bean and Bodie, replacing the photographs with computer liquid crystal display (LCD) screens bearing the same stimuli as the original experiment. Straight off, without any prior learning and with no interim exposure to the stimuli presented since the initial experiment, the horses performed excellently; as before, Bodie's results were inferior to Coco Bean's, and his performance dropped markedly in the third of five pairs. This suggests a possible influence of primacy and recency in retaining information (see table 7.4). In any event, images projected onto an LCD screen posed no particular recognition problem compared with photographs.

This finding was confirmed by two other experiments reported in the same publication and involving long-term memory of categories and concepts, which similarly used LCD screens in place of photographs.

Sherril Stone (2010) has studied the capacity of horses to learn

Table 7.4 Retention of discriminative learning after six years, with photographs replaced by projections on an LCD screen.

Horse	Stimulus (S+/S−)	CR	Total	% CR	Trial 1	Binomial (Z)
Coco Bean	Plate/Pillar	10	10	100	CR	3.162[a]
	Scale/Frame	9	10	90	CR	2.530[a]
	Hedgehog/Bottle	5	10	50	CR	0.000
	Football/Frisbee	9	10	90	CR	2.530[a]
	CM/Gus	9	10	90	CR	2.530[a]
Bodie	Plate/Pillar	10	10	100	CR	3.162[a]
	Hedgehog/Bottle	8	10	80	CR	1.897[b]
	Bottle/Hedgehog	2	10	20	CR	−1.897
	Football/Frisbee	8	10	80	CR	1.897[b]
	CM/Gus	8	10	80	CR	1.897[b]

CR: Correct response. CM: Cookie Monster. Gus: Snuffaluffagus. [a]$p < 0.01$. [b]$p < 0.05$, binomial test. Source: Hanggi and Ingersoll 2009a, 455.

to discriminate pairs of human faces based on photographs of the images, and their ability to generalize this learning by spending more time, in an experimental setting, in an open field near whichever of the two persons whose face was associated with a reward during learning. Despite the somewhat contradictory results, the conclusions of the study are generally positive.

Object Recognition

Maintaining separate relations with conspecifics, identifying prey and predators, recognizing sources of food, locating landmarks in their living space, and so forth are all mechanisms of survival for many species, which explains the fact that the capacity to recognize and identify 3-D objects is widespread in the animal kingdom.

This capacity relies on preexisting mental representations. In particular, it assumes that, to be recognized, the perceived object may be confronted with a permanent representation of its shape in memory, and be assimilated into it, even when the angle from which it is seen gives rise to a different "visual image" from the one stored in memory. Thus, animals endowed with this capacity must be able to recognize that an object presented from different angles is nonetheless the same object.

These observations raise two questions: What is the nature of representations stored in memory? And what is the process by which animals compare the perceived object with the remembered one?

Much of the research work on object recognition derives from two general classes of theories (Spetch and Friedman 2006).

The first, based on global appearance, relies on recognition by pattern matching, that is, prototypical visual forms of objects stored in long-term memory with different angular perspectives. In comparing the perceived object to a certain number of plausible candidates stored in memory, the object is identified as the one closest to memory, through a process of mental rotation matching. The time required for matching and the level of success then depend on the angle formed between the perceived object and the selected pattern. In particular, it has been shown in humans (Shepard and Metzler 1971) and under certain conditions[6] in baboons (Vauclair, Fagot, and Hopkins 1993) that, when subjects are given a matching task between an image of a complex geometric object and the image of the same object rotated, the time required to match the two images is a linear function of the angular difference between the two objects, as it would be for a rotation executed in the physical world.

The other class of theories is based on a structural description that "explicitly represents objects as configurations of attributes (typically parts) in specified relations to one another" (Hummel and Biederman 1992, 483). Irving Biederman (1987) named these attributes *geons* (from geometric ions). They are basic 3-D shapes that correspond to a part of an object (concave or convex contours,

---

6. The results of the work with baboons are relevant to the question of cerebral hemispheric specialization, mentioned elsewhere, since they were obtained only when the stimulus was in the right visual hemifield, thus involving the left hemisphere.

angles) and that represent its structural invariants. In comparing the geons of the perceived object and their arrangement to those of the object stored in memory, the object is identified or not based on correspondence (or lack of it). Rotating an object may conceal some geons or reveal others, which may impede recognition, thus affecting the time and degree of recognition. In contrast, there is no direct relationship between these elements and the angle of rotation of the perceived object. Pigeons, for example, can recognize an object presented from different angles, but their response time is unconnected to the object's angle of rotation (Hollard and Delius 1982).

Observing that no studies had been done previously on recognition of objects after rotation in horses, Hanggi (2010b) carried out a series of experiments. These experiments extended the hypothesis, first proposed during her study of interocular transfer of learning (Hanggi 1999b), according to which a horse's fear response, on returning from a walk, to an object that had not frightened it before might be linked to a change in perspective from which the object is seen, which could impede recognition. To find out, Hanggi tested four horses using tasks requiring them to discriminate among four stimuli, using three groups of stimulus pairs in succession, consisting of plastic children's toys, each element of the same pair having the same color: a wheelbarrow and a mower, a tractor and a truck, and a dinosaur and a lizard.

During an initial phase, a preference test was conducted with the first pair, and the nonpreferred element for this pair was selected as the positive stimulus and rewarded. The second stage involved learning, during which the horses learned to choose the positive stimulus, each of the two elements of the pair being position upright, with their front aimed to the left. Finally, once the learning was complete with this arrangement of stimuli, the horses were given discriminations in which the stimuli were rotated in various ways and randomly interspersed with tasks for which the stimuli were arranged as in the learning phase. These trials were continued until each type of rotation had gone through twenty trials. The types of trials used (eleven) involved orienting the object to the left, to the right, directly facing or back to the horse, and with the top or bottom of the object oriented upward, downward, or directly facing the horse.

Once these three steps had been run through for the first pair of stimuli, they were restarted for the second pair (tractor and truck), after which the third pair (dinosaur and lizard) was implemented in the same way. For each of these object pairs, the learning produced typical learning curves that were generally comparable for the four horses.

With respect to the rotations of the objects tested, the horses' performance on all rotations was significantly above chance. However, analysis of individual results showed differences: Although there were rotations for which the horses all scored significantly above chance, the same was not true for others, which some horses recognized and others did not. For most of the rotations, there was no significant difference associated with type of stimulus.

Finally, although there was no significant difference between the rotations in which the top of the stimulus was visible and those in which the bottom was visible, there was, however, a very significant difference between these two categories of rotations that was confirmed at the individual level. The horses clearly performed better when they had some view of the top of the object as opposed to the bottom.

Hanggi suggests that this finding might be connected to the arrangement of the stimuli in the learning phase, which gave the horses the chance to acquire a maximum of information about the characteristics of the objects. Although the horses also had the chance to draw near to the objects to make their choice, acquiring additional information based on slightly different angles, they had no overall view. And in extreme positions, such as when they were directly facing an object, the information available was limited.

For Hanggi, the results of these experiments suggest object recognition based on attributes: geons. The fact that, generally speaking, the horses could always recognize objects under extreme conditions, even when it was the bottom that was visible, implies that they continued to identify the primary characteristics.

In any event, the findings also show that changes in perspective by themselves do not explain the fright reactions of the horses mentioned earlier. Perhaps, writes Hanggi, it would be worth considering that, although horses have excellent long-term memory, learning and memory rely on attention. Thus, horses may startle at the sight of objects that had previously escaped their attention, either owing to their own behavior, or to the demands of their rider.

Perceiving Movement

Although no research has been devoted specifically to equine perception of movement, as exists for the dog, for example (Miller and Murphy 1995), or the cat (Pasternak and Merigan 1980), the investigations conducted by Oskar Pfungst (2000–1907) on Clever Hans, the "calculating" horse (see chapter 2), in the early 1900s do shed some light on equine visual sensitivity to movement.

Recall that in 1904, in Berlin, a horse named Hans was alleged to be able to spell words and to calculate sums by tapping its hoof. The episode caused a stir, but it turned out (Pfungst 2000) that Hans was relying on visual cues to produce his answers, such as small movements of the head or trunk made by the questioner. In her eponymous book titled "Hans, the counting horse," Vinciane Despret (2004) states that Hans's evaluators "all made the same little movement just after interrogating the horse: they inclined their head and trunk slightly forward. This movement is so slight that it could not be perceived except in a comparison test . . . The exact measure of the amplitude was between a quarter of a millimeter and a millimeter and a half! When the movement had occurred, the horse began his work of counting. Similarly, when the horse reached what was to be the last tap of the foot, Pfungst saw the questioner make a slight upward jerk of his head, even less perceptible than before, raising his head to its initial position after the last tap" (51–52).

Such observations do suggest that horses have very fine perception of movement. As I showed previously in discussing the composition of the retina vis-à-vis visual quality, this perception preferentially involves peripheral vision. It is thus especially well adapted to scanning the environment even when horses are grazing.

Nonetheless, I hasten to add that, although Despret's critical analysis of Pfungst raises interesting epistemological questions,[7] it cannot be indiscriminately applied to equine ethology. The only "authority" she cites is Jean-Claude Barrey, better known for his personal interpretation of horse behavior disseminated in equestrian magazines or over the Internet than for his familiarity with the scientific literature that is the basis for equine ethology. Of course, unconfirmed speculations are hardly surprising when it comes to interpreting equine behavior.

I will confine myself to a short illustration regarding the visual field: "Naturally, this made Hans an exceptional horse because he had delocalized an ability—isopraxis, which enabled him to read in his own muscular response that of his rider—on another level (the visual level), where horses do not normally excel . . . Naturally, horses use vision, but in Barrey's apt words, 'he never believes his eyes; he always verifies with his nose'" (Despret 2004, 129–30). Without uselessly dwelling on isopraxis, these opinions appear to betray a certain lack of information regarding the equine visual system.[8] With no substantiation whatsoever, they propose that smell can substitute for a supposedly deficient visual apparatus, whereas nothing (apart from pure speculation) even suggests that smell is anything but a fairly obvious source of information that is different, though complementary to vision. Moreover, as this proposition derives from experiments on the comparative effect of blocked vision and smell on the

---

7. As indicated in chapter 2, an additional source on this subject, not cited by Despret, is the proceedings of a colloquium organized by the Academy of Sciences of New York in 1980 on the "Clever Hans Phenomenon," which contains contributions from noted scientists.

8. Particularly telling, in light of the research cited above, is the assertion made by Barrey in a 2007 conference talk, which contains the fragment cited by Despret: "In the horse, whose eyes are positioned relatively to the side, images are too different to be superposable, and consequently, the horse has no binocular vision . . . The disadvantage of this system is that it provides no depth perception" (http://pulsaye-station-recherche.ifrance.com/ex_pages/text-therapias.htm, accessed 5 January 2009).

sexual behavior of stallions (Anderson et al. 1966), vision may well predominate over smell. Indeed, the perception of visual stimuli appears to be very important for the expression of precopulatory behavior and for sustaining sexual excitement. The absence of smell, on the other hand, seems to have only minor effects, such as suppression of the flehmen effect. Nor is flehmen particularly linked to a sexual context, contrary to popular belief, as I will show later in the chapter on chemical perception. In any event, under environmental conditions, perception is basically multimodal, which I have already shown and will underscore in later chapters.

The Equine Visual Environment: Seen as a Whole or the Sum of Its Parts?

In addition to the different aspects of visual perception in the horse already mentioned, there is one that, today, it must be said, raises more questions than it answers. At issue, according to Temple Grandin (Grandin and Johnson 2005), is the way that animals see their environment, not globally, as humans do, by creating a conceptual representation, but by processing all the details: "That's the big difference between animals and people, and also between autistic people and nonautistic people. Animals and autistic people don't see their *ideas* of things; they see the actual things themselves. We see the details that make up the world, while normal people blur all those details together into their general concept of the world" (30). In developing this concept of animal vision, Grandin, who is also a respected scientist, simultaneously draws on her own experience of autism and on her professional success as an inventor of livestock handling devices.

Grandin is what is called a high-functioning autistic. Afflicted from a very young age with Asperger's syndrome, she finds it extremely hard to grasp the mental states of others, such as emotions and intentions. Yet she has impressive intellectual ability. And as is common with autistics, she thinks in pictures rather than words. "I think in pictures. Words are like a second language to me. I translate both spoken and written words into full-color movies, complete with sound, which run like a VCR tape in my head. When somebody speaks to me, his words are instantly translated into pictures" (Grandin 1995, 3). When she was nearly fifteen, at the urging of one of her teachers, she transformed her "particular considerations of farm animals and machinery to a general interest in biology and all science. And here Temple, still quite abnormal in her understanding of ordinary or social language—she still missed allusions, presuppositions, irony, metaphors, jokes—found the language of science and technology a huge relief. It was much clearer, much more explicit, with far less depending on unstated assumptions. Technical language was as easy for her as social language was difficult, and it now provided her with an entry into science" (Sacks 1993, 117). In fact, Grandin subsequently obtained a Ph.D. The author of many scientific articles (and autobiographical accounts), she is currently professor of animal science at the Colorado State University and also directs a successful enterprise that she founded.

For Grandin, "when an animal or an autistic person is seeing the real world instead of his idea of the world that means he's seeing detail. This is the single most important thing to know about the way animals perceive the world: animals see details people don't see. They are totally detail-oriented. That's the key" (Grandin and Johnson 2005, 31), which states explicitly what she implies elsewhere: "To understand how the environment is affecting an animal's behavior, you *have* to look at what the animal is seeing" (19).

Actually, Grandin's thesis goes beyond perception: "Autistic people can think the way animals think. Of course, we also think like the way people think—we aren't *that* different from normal humans. Autism is a kind of way station on the road from animals to humans, which puts autistic people like me in a perfect position to translate 'animal talk' into English. I can tell people why their animals are doing the things they do" (6–7).

Referring explicitly to the work I have just cited, specialists in neuroscience and animal behavior (Vallortigara et al. 2008), while applauding Grandin's contributions to the understanding of autism and animal well-being, also criticized her ideas comparing the skills of autistic savants to those of animals. Since the farm animals and domestic pets she studied experience living conditions that can be stressful, their mental processes might be different from those of wild animals. Consequently, noted the authors, their own questions would address the broad context of animal behavior in a natural environment.

They noted in particular that, whereas humans with extraordinary aptitude (especially for music, mathematics, and art) often possess deficits in other cognitive domains, for animals, it is quite different. Although some species do in fact show remarkable capacities—for example, Clark's nutcracker[9] which can remember thousands of places where it has hidden its food reserves—these capacities more likely correspond to adaptive specializations specific to the species that are unconnected to deficits in other domains. With respect to "detailed thinking," and referring as much to experimental work in comparative psychology as to hemispheric specialization in humans and animals, they suggest that, as with humans, animals focus either on details of a visual scene or on the whole depending on the perceptual context.[10] Moreover, endowed with hemispheric specialization similar to that of normal humans, they are not in any event overwhelmed by details, unlike people with autism, in whom there may be dysfunction of the left hemisphere.

Asked to respond to the critique of Vallortigara and his colleagues, Grandin ventured that the main point of disagreement had to do with the concept of details, that is, exactly how details are perceived by humans, who think in language, compared with animals, who think in sensory data, and whose store of memory is by nature necessarily more detailed. The main similarity between animal thought and her own was the lack of verbal language. She also questioned the notion that certain species of animals with extremely developed aptitudes, such as the exceptional memory of Clark's nutcrackers, were deficient in other domains, and that, by way of validation, further experiments needed to be done with birds such as Clark's nutcrackers. That notwithstanding, she found that the points raised by her critics clearly confirmed the reality of animal cognition, was delighted that her book had stimulated so much discussion, and hoped to encourage more research on animal cognition.

In fact, one can only conclude that, as things stand, however appealing it seems at first blush, Grandin's thesis remains a fairly speculative one.

9. Clark's nutcracker (*Nucifraga columbiana*) is a small corvid that feeds on pine nuts.
10. Although they concede that some species, for example, pigeons, exhibit an interest in details in certain contexts, probably due to specific adaptations.

8

THE BEHAVIORAL EXPLORATION OF EQUINE VISUAL PERCEPTION: THE QUEST FOR COLOR PERCEPTION

From the earliest experiments on color vision in horses, carried out by Bernhard Grzimek (1952), much ink has been spilled on the subject, owing especially to the problems caused by having to control many critical parameters (hue, purity, brightness, and conditions for stimulus lighting), the limited number of horses tested, and the very few studies devoted to the subject. Not only must research contend with complex methodological programs, but these studies are by definition long term, and as I have already emphasized, the horse is a heavy, expensive, hard-to-handle animal that is less amenable to experimentation than many other species. In recent years, the never-ending quest for color perception in horses has nonetheless continued, and significant advances have been made. Examples include a body of conclusions that appear to be coming together, with more consistent data than previously between the experimental behavioral and anatomo-physiological approaches.

Before reviewing these experiments, I hasten to add that the question is not whether the horse sees this or that particular color in the same way we do—that would imply subjective experience of color perception about which we can admittedly only speculate. Rather, the question is simply (not that it is really that simple, as I will show) to know whether the horse is at all able to discriminate colors based on their chromatic characteristics (as opposed to degrees of grayness), to determine which colors it can distinguish from gray, and to understand how it organizes the spectrum of colors that it perceives.

Brightness: A Vexing Dimension

One crucial aspect, of course, is to ascertain (from the discriminations between gray and color that horses must make) which factors are associated with chromaticity and which with brightness (achromaticity).

As I previously mentioned, color has three characteristic dimensions: hue (associated with dominant wavelength), purity or saturation (associated with spectral purity), and brightness (associated with the luminance factor). The first two dimensions determine the chromatic properties of color; the third is by nature achromatic (see plate 6).

What makes the problem especially complicated is that the eye is not equally responsive to all wavelengths (similarly, as we will see, the ear is not equally responsive to all frequencies). The eye's response to monochromatic light of a given intensity—its spectral sensitivity—varies strongly as a function of wavelength. This general sensitivity of the eye is itself connected to the relative spectral sensitivity of each of the cones. To explain how that works, let us take the example of human vision.

As I described earlier, the three cones of the human eye—blue, green, red, or more precisely, S, M, and L—are each sensitive to a specific range of wavelengths, and this sensitivity varies widely even within this range. Moreover, the maximum response amplitude is not the same for all of the cones (see plate 10). The spectral luminous efficiency function, or spectral sensitivity, described by the spectral sensitivity curve of the eye, defines the eye's relative sensitivity along the continuum of wavelengths of the spectrum. The function makes it possible to establish a relationship between the subjective sensitivity of the eye and the purely physical luminance value.

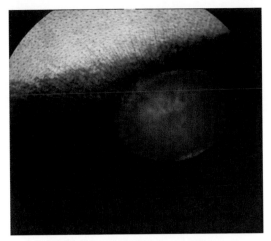

Plate 1 Fundus of the eyeball. The tapetum lucidum in green and the optic disk (slightly ovoid) in light red.
Source: Franck Olivier, University of Florida.

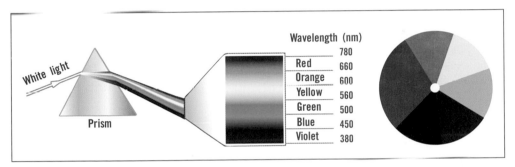

Wavelength (nm)
780
Red — 660
Orange — 600
Yellow — 560
Green — 500
Blue — 450
Violet — 380

Plates 2 and 3 Decomposition of white light by a prism (left) and the disk used by Newton to recompose white light (right).

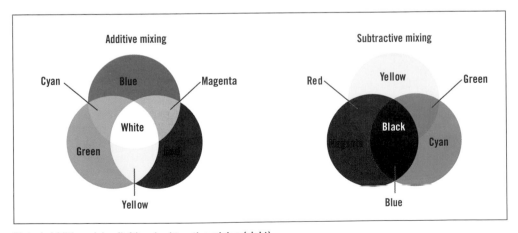

Plate 4 Additive mixing (left) and subtractive mixing (right).

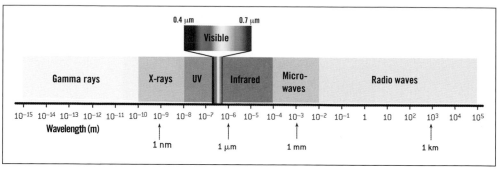

Plate 5 Electromagnetic spectrum.

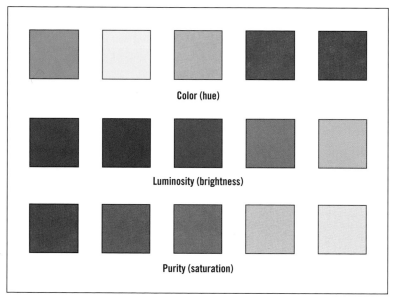

Plate 6 Hue (color), luminosity (brightness), and purity (saturation).

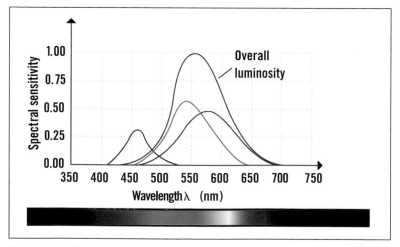

Plate 7 Normalized spectral response of human cones and rods.

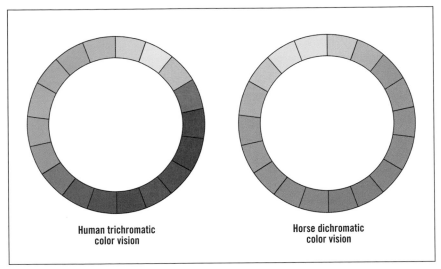

Human trichromatic
color vision

Horse dichromatic
color vision

Plate 8 These color wheels illustrate the difference between normal trichromatic human vision (left) and the horse's dichromatic color vision (right). Dichromatic color vision is the result of a reduction in the number of types of cones from three to two. This in turn enormously reduces the number of colors that can be seen. Source: J. Carroll et al., 2001.

Plate 9 Four images—two unaltered (A, B) and two altered (C, D)—show the effects of dichromatic color vision for the horse. The altered images—made using a computer algorithm—show what the colors in the original image look like to a dichromatic horse whose visual pigments have the spectra determined in Carroll et al.'s study. The images were also adjusted to simulate the horse's decreased spatial acuity and thus provide an indication of the visual experience of the horse (though without showing the horse's visual field). Source: J. Carroll et al., 2001.

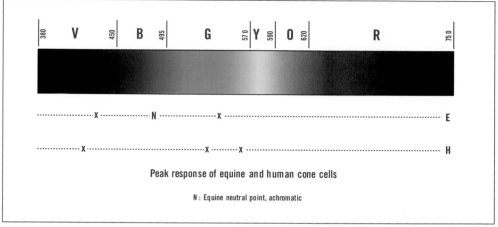

Peak response of equine and human cone cells

N : Equine neutral point, achromatic

Plate 10 The peak sensitivity of equine and human S cones occurs at roughly the same wavelength; in contrast, the wavelength of peak sensitivity of equine M/L cones is midway between that of human M and L cones. The color discrimination of horses is greatest on either side of their neutral (achromatic) zone (N).

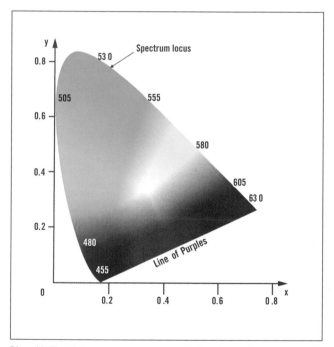

Plate 11 The visible spectrum according to the chromaticity diagram of the International Commission on Illumination (1931) in two dimensions, under isoluminance.

Normal vision

Protan vision

Deutan vision

Tritan vision

Plate 12 Comparison between human normal vision (top) and protan, deutan, and tritan vision (descending). Source: Burnham, Hanes, and Bartleson, 1963.

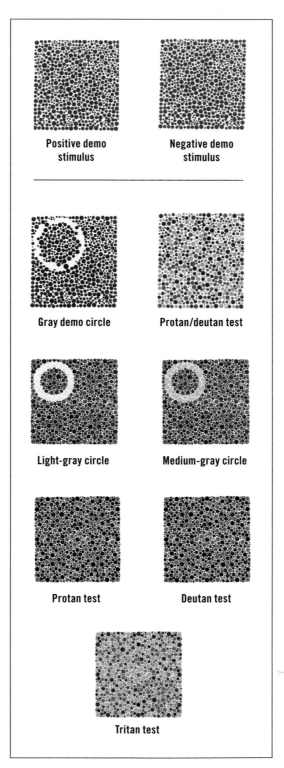

Plate 13 Panels inspired by Waggoner's CVTMET (1994), used by Hanggi et al. (2007). Source: © Evelyn B. Hanggi, Equine Research Foundation.

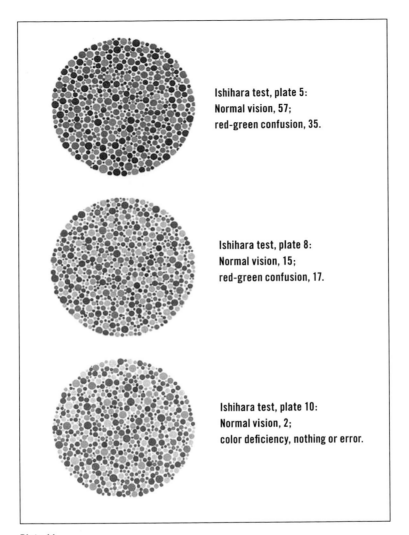

Ishihara test, plate 5:
Normal vision, 57;
red-green confusion, 35.

Ishihara test, plate 8:
Normal vision, 15;
red-green confusion, 17.

Ishihara test, plate 10:
Normal vision, 2;
color deficiency, nothing or error.

Plate 14

l (nm)	400		450		500		570		590		610		750
Color		violet		blue		green		yellow		orange		red	
n (10¹⁴ Hz)	7.5		6.7		6.0		5.3		5.1		4.9		4.1

Plate 15 At equal light intensity, green and yellow appear brighter than blue and red. In the range around 460 to 530 nanometers, horses can discriminate colors fairly well. Outside that range, their discriminative abilities appear more or less limited for the same wavelength spread. Recall that the neutral point, where gray and blue cannot be discriminated, occurs around 480 nanometers.

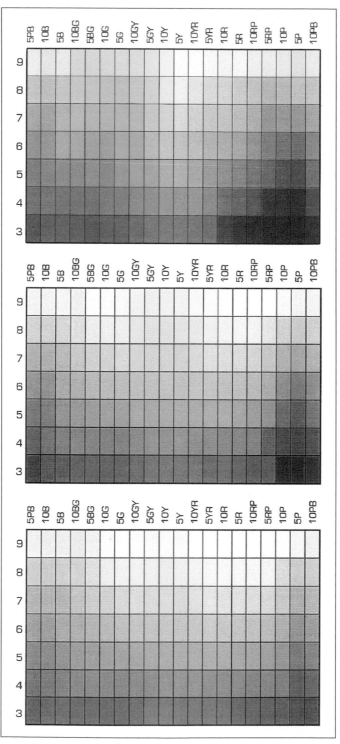

Plate 16 Rough comparison of color perception between a normal human trichromat (top) with a deuteranopic dichromat (middle) and a protanopic dichromat (bottom). Hues are labeled along the abcissa, and brightness along the ordinate. Source: Bonnardel 2005.

To determine this function for the human eye, in 1924 the International Commission on Illumination (CIE) launched a statistical study based on a survey with subjects about the brightness sensations they experienced vis-à-vis each color (interindividual variability exists). The CIE Spectral Luminous Efficiency Function thus established had the effect, in particular, of revealing a peak sensitivity at a wavelength of 555 nanometers for photopic vision. Examining the curve also reveals that, in photopic vision, the sensitivity of the human eye to monochromatic light at a wavelength of 490 nanometers represents only 20 percent of its sensitivity at 555 nanometers. Thus, the first case requires five times the intensity of the second for both to be perceived as having the same brightness. In other words, at equal intensity, blue is perceived as being much less bright than green or yellow. Similarly, light at 660 nanometers (in the red range) must have an intensity around ten times greater than at 560 nanometers (in the green-yellow range) to be perceived as equally bright (see plate 15).

When testing animals for color discrimination, in the absence of such a function it is impossible to know a priori whether they are responding to brightness cues or to color. One cannot ask an animal not endowed with language to give a verbal answer to the question whether it sees color or gray. Thus, showing experimentally that an animal can discriminate yellow and blue, for example, without having taken certain precautions, leaves open the possibility that this animal has chosen simply because it perceived yellow as brighter than blue.[1] Two strategies exist for taking this problem of brightness into account in studying animals, both of which have been applied to horses, as reported by Brian Timney and Todd Macuda (2001).

The first method is based on the principle used by Karl von Frisch (1955) in his pioneering experiments on color vision in bees: "As we cannot foresee in which particular shade of brightness our

---

1. Recall that in scotopic vision, the spectral sensitivity curve of the human eye is clearly shifted toward the weakest wavelengths, with a sensitivity peak toward 507 nanometers, which explains how, in the dark, we perceive blue better than, for example, yellow. The opposite is the case in photopic vision.

blue paper would appear to a totally color-blind bee's eye, we have to find out whether the bee can distinguish it from every possible shade of brightness. In order to do this, we have to use a whole series of gray papers, leading, in fine gradation, from pure white to absolute black. If we place a clean blue sheet without food in the midst of such a series of gray papers, arranged in an arbitrary order, in front of bees previously fed on blue, they will still fly towards the blue sheet as if quite sure of their goal, and settle on it . . . This shows that they can, indeed, distinguish the blue as a color even from the entire gamut of gray shades" (64–65). This method thus consists of comparing chromatic stimuli with a group of achromatic stimuli of varying brightness in the hope of neutralizing the brightness cues, as we will see more specifically in the first experiments presented below.

The second method is based on the principle of the animal's previously established ability to discriminate brightness, using two roughly equal achromatic stimuli and varying the luminance of one of them. This makes it possible to determine the range of uncertainty within which the animal does not perceive the brightness difference, and thus to subsequently test its ability to discriminate between color and achromatic targets within this range where the brightness difference is not perceptible. Because the brightness cues have been made irrelevant, only color is left as a cue for discrimination. Consequently, to the extent that the animal can discriminate between achromatic and chromatic stimuli within this margin of indecision, it is basing its decision on color alone.

We will see which of these methods is brought to bear in reviewing the studies carried out to date on equine color perception. Proceeding in chronological order will allow us to follow the evolution of research in the area, most of which has developed over the past decade, and in particular in the last few years. Here and there I will go a little deeper than previously into the presentation of experiments and their results, as their interpretation is very complex and requires a nuanced approach.

To begin, I will provide an example of the first strategy—combining chromatic and achromatic stimuli of varying brightness—for the simple reason that this method is the one used by the authors of the first three studies.

A Pioneering Study (Grzimek 1952)

Although it was published in 1952, this study was actually done toward the end of the Second World War, in the northern part of Berlin, between December 1944 and March 1945, when the city was being bombarded and Soviet troops were approaching. The experiments took place in a large covered arena, on a soft surface blanketed with wood chips such that, according to Bernhard Grzimek (then a veterinary officer), the animals were suitably protected from the elements and from the disturbances going on outside.

To carry out this first experiment on equine color vision, the author tested three horses. However, only the results of two of them (Zuckung and Farinette) were counted owing to the skittishness shown by the third horse on certain days and inconsistent results. The underlying principle, based on von Frisch's (1955) protocol for bees, consisted of using operant conditioning and a food reward to train horses to discriminate color from shades of gray. To address the problem of brightness, the animals were required to choose between a given color and a range of grays from light to dark, such that if the animal saw the color as gray, it would confuse it with one of the other gray shades. According to the hypothesis, there had to be at least one gray whose signal was sufficiently close to the color so that, if the horse discriminated the color from all the grays, it could not be relying on achromatic brightness cues.

The color composition was based on pigments from a standard color system (the Prase color system) arranged around a disk containing forty-eight color swatches. The main colors included yellow, citron, green, blue, violet, purple, red, and orange. In addition, pure

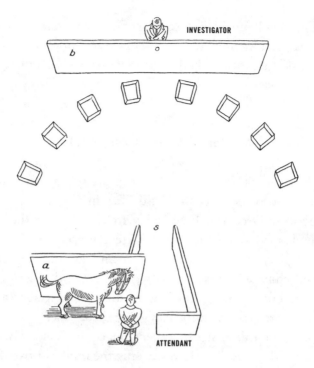

Figure 8.1 Grzimek's experimental apparatus for color recognition. Source: Grzimek 1952, 24.

colors could be lightened or darkened by adding white or black. The gray palette contained twenty-seven shades.

During the training phase, the horse was first led by the bridle to a wooden box, placed on the ground, which was open at the top and contained oats for it to eat. Then the animal repeated the procedure on its own. Next, two boxes were placed in front of the horse, one bearing a colored card 23 by 27.5 centimeters, and the other a gray card of the same size (see figure 8.1). If the horse went toward the box with the color card, it was allowed to eat the contents. If it headed for the other box, it was caught and commanded "Back!"

Once the learning was completed, in the test phase the horse had to identify the box with the color card amid an ever-increasing number of boxes arranged in a semicircle and bearing gray cards of different shades. These boxes also contained oats, but the horse was prevented

Table 8.1	Summary of results recorded for each of two horses for different colors.					
	Zuckung			Farinette		
Color	Nb B	Nb trials	+%	Nb B	Nb trials	+%
Blue	6	330	80.9	10	290	79.3
Green	10	150	83.4	10	208	90.8
Yellow	10	162	98.6	10	145	94.4
Red	6	396	70.4	6	374	54.5
Nb B: Number of boxes. +%: Percentage of correct choices. Source: Grzimek 1952, 35.						

from eating them. In other words, it was in a situation analogous to that of a subject who must respond to a multiple-choice question where each choice has only one right answer, and that answer is rewarded. Given the number of gray shades (twenty-seven), several steps were required to ensure that each color was compared against each gray.

For yellow and green, the horses scored high until they were faced with a choice between ten boxes presented simultaneously. In contrast, for red and blue, it was not possible to reach this level; positive results decreased sharply beginning with six boxes (see table 8.1).

Moreover, although green and yellow recognition training was relatively simple, for blue it was significantly more difficult, and even harder for red, a finding observed for both horses even though they had been trained with a different order of colors.

Grzimek concluded that horses have some degree of color vision and that, like us, they can discriminate the four main colors, that is, yellow, green, blue, and red. However, for horses, yellow is easiest to distinguish from gray, then green, blue, and finally red, which appears to cause them some difficulty. In addition, light red was easier to discriminate than more saturated reds. Conversely, saturated blue was easier to discriminate than lighter blue. Finally, when the horses were occasionally presented with several colors at the same time, each time they preferred yellow to the three other colors.

In a second set of experiments, the horses were given a choice between two gray boxes, one of which had a perpendicular, colored line in front, whose hue and width varied. These two boxes were

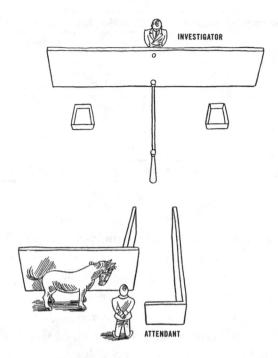

Figure 8.2 Grzimek's experimental apparatus for visual acuity based on color. Source: Grzimek 1952, 37.

on opposite sides of a partition 3.30 meters long, thus obliging the horses to decide from this distance (see figure 8.2).

The researcher found that the horses' visual acuity varied depending on the distance, since at a distance of 3.30 meters the horses chose correctly in 66 percent of trials involving a yellow perpendicular line 0.5 millimeters wide. In contrast, a blue line had to be 2 centimeters wide for the horses to attain the same score.[3]

In their work on color recognition, Macuda and Timney (1999) argued that varying the luminance of gray boxes more or less randomly to keep the horses from using achromatic cues does not necessarily work: "While such an approach should reduce the potential salience of brightness cues, there remains the possibility,

---

3. The latter result is similar to the visual acuity values obtained for the visual streak and the peripheral retina, respectively (Hall et al. 2006).

especially in an animal whose spectral luminosity function is not known, that the range of distractor luminances does not include a true brightness match. As a consequence, brightness could still play a role" (302).

The results obtained by Grzimek made very little stir at the time insofar as, anatomically and physiologically speaking, there was practically no information on the vestigial or functional character of equine cones. Moreover, it was not clear what adaptive value color vision might have for horses. Proof that equine cones had a genuine function would have to await the end of the 1970s (Wouters, de Moor, and Moens 1980).

An Inconclusive Replication (Pick et al. 1994)

David Pick, Greg Lovell, Suzanne Brown, and Dan Dail (1994) subsequently tried to replicate Grzimek's experiment on the basis of a two-choice discrimination, but with only one horse. The experimental apparatus consisted of an open-sided barn shaped like a T and fitted with plywood panels sized 1.22 square meters, one gray, the other colored. Not knowing the spectral sensitivity function of the horse, the authors varied the reflectance values of their stimuli, both colored and gray. The gray stimuli were divided into six equal increments between white and black, thus giving rise to five grays. Three levels of blue (462 nanometers), green (496 nanometers), and red (700 nanometers) were mixed so that their reflectance value, verified by spectral analysis, matched that of the three mid-reflectance grays. The horse could thus be tested on the different combinations of brightness levels. Various methodological precautions (so-called double-blinding) were also taken to prevent intervening visual, auditory, olfactory, tactile, and positional cues.

The researchers were unable to confirm Grzimek's conclusions. Although an initial red-gray discrimination was obtained fairly easily with only 102 trials, including the required controls, the first attempt at green-gray discrimination had to be abandoned after 541 trials. It took 243 presentations, including controls, to achieve blue-gray discrimination. A second attempt at green-gray discrimination was also abandoned after 334 trials. A second attempt at red-gray discrimination confirmed the first set of results, after which a third attempt at green-gray discrimination succeeded after 361 trials. The authors note, however, that this result might simply be due to chance given the number of trials. They attribute their failure to reproduce Grzimek's results to several causes, including deterioration of the horse's visual system owing to age (nineteen years), or the possibility that Grzimek had controlled better for reflectance. Although they had also intended to test yellow, they abandoned the idea when they found that they could not obtain a sufficiently saturated yellow to match the reflectance of the lightest midreflectance gray stimulus. Moreover, they suspected that Grzimek had used yellows that reflected more light than his grays.

Ultimately, the most probable explanation for their results seemed to them to be that horses possess two classes of cones, sensitive to red and to blue, in addition to rods, as is the case with other ungulates, and that, since dichromacy had already been demonstrated in other species of mammals, horses too are dichromats.

An Apparent Confirmation of Grzimek's Results (Smith and Goldman 1999)

A study by Susan Smith and Larry Goldman (1999) had the advantage of a larger number of subjects: They used five horses that they chose from an initial population of eleven, six of which could not adapt to the experimental environment. Smith and Goldman tested their horses using discriminative learning based on a choice between two translucent, swinging panels. One of the panels was illuminated by a projector through a bright-colored, dominant-wavelength filter (red, 617 nanometers; yellow, 581 nanometers; green, 538 nanometers; blue, 470 nanometers). The other panel was illuminated through a gray filter. The filters met precise standards with respect to their brightness characteristics (see figure 8.3).

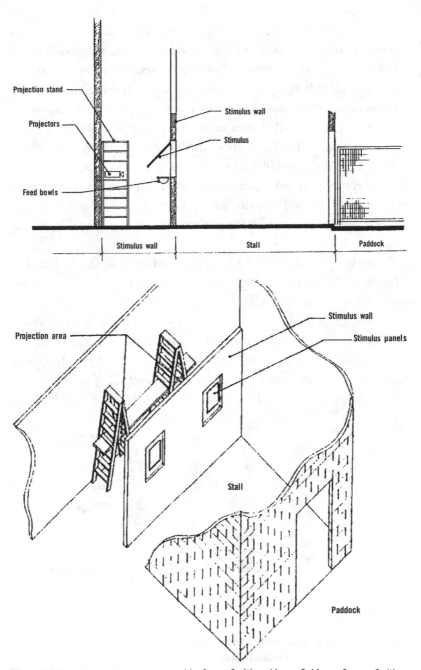

Figure 8.3 Experimental apparatus used by Susan Smith and Larry Goldman. Source: Smith and Goldman 1999, 16–17.

On completion of the training phase, during which the horses learned to press their nose on the color panel, giving them access to a bowl of food, the animals had to discriminate between each of the colors paired with a gray and equated for intensity with a light meter. The success criterion was 85 percent correct responses for sessions of forty trials (significantly better than chance at p < 0.0001).

Because the spectral sensitivity of the equine eye was unknown, and to limit the possibility of the horses using brightness cues, the researchers then paired colored lights with three different intensities of gray (the gray used initially, a slightly brighter gray, and a slightly darker gray). The gray shades were then presented randomly with the colored light.

Three horses (AK, AL, and NK) were tested on all four colors, but a fourth (AZ) was tested only on two colors, and a fifth (PK) only on one, owing to time constraints (see figure 8.4).

Two of the horses successfully discriminated color versus gray for each of the four colors. A third (NK) succeeded with red and blue, but failed with yellow and green. A fourth (AZ) discriminated green and yellow, the only colors on which he was tested. The fifth horse (PK) was tested only with blue, which he was able to distinguish from

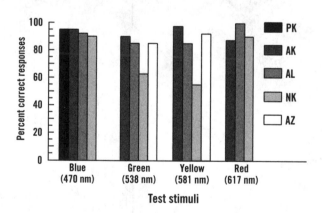

Figure 8.4 Percentage of correct responses for the final session of each color. Discriminations were made with three shades of gray, except for NK, for whom green and yellow were matched with just one shade of gray, which the horse failed to discriminate. Source: Smith and Goldman 1999, 22.

gray. According to the authors, these results (which posed no obvious difficulty in discriminating colors either during learning or final performance) confirmed those of Grzimek, that is, that horses can distinguish red, yellow, green, and blue, different shades of gray, and consequently that they possess color vision that is sensitive to short, medium, and long (S, M, L) wavelengths. They attribute the failure of one of their horses to discriminate yellow and green to a physiological inability to discriminate medium wavelengths, which could also explain the failure of the horse tested by Pick and colleagues (1994). Smith and Goldman conclude that, although their results are consistent with trichromatic vision in normal horses, their findings could also be interpreted in terms of a dichromatic system with a neutral point (as defined in chapter 6) that was not detected by the stimuli used.

New Uncertainties Centering on Brightness (Macuda and Timney 1999)

Macuda and Timney (1999) tackled the question of neutralizing brightness cues by adopting the second strategy introduced at the beginning of this chapter. As I described previously, it consists of determining the horse's ability to discriminate brightness (which comes back to establishing the brightness discrimination function for achromatic stimuli) and then making chromatic discriminations within the achromatic range of uncertainty, where the brightness difference is imperceptible and, consequently, neutralized.

The experimental procedure was inspired by the method used previously by David Peeples and Davida Teller (1975) to show the existence of a certain form of color vision in two-month-old babies.[4] A rectangular white bar was embedded in a white screen, and the intensity of the bar was systematically varied to establish the infants' luminance discrimination function. The researchers observed that babies regularly stared at the white bar except when its intensity approached that of the screen,[5] which enabled them to determine the intensity range corresponding to an inability to discriminate between the white bar and the white background, that is, near-chance performance. Next, they replaced the white bar with a red bar, which they also varied systematically within a wide range of intensities. They observed that babies stared regularly at the red bar no matter its brightness, thereby indicating that they discriminated it consistently and thus perceived it as colored.

Proceeding in the same spirit, Macuda and Timney (1999) adapted the experimental apparatus that they had already used in other research on equine vision that I touched on earlier. The swinging trapdoors were specifically designed: They had openings (21 by 31 centimeters) at their center equipped with two thick sheets of Plexiglas with black projection material sandwiched between them, lit from the back (see figure 8.5). The lighting was furnished by two computer-controlled projectors, to the rear of the apparatus. The stimuli projected onto the panels were each divided into three equal parts, separated by a 1.25-centimeter, vertical, black bar. Using standard filters, the side portions of the positive stimulus panel were neutral gray, and the brightness and color of the central portion varied. In contrast, for the negative stimuli, the three parts of the panel were all a neutral gray (see figure 8.5).

In this way, the researchers established luminance discrimination functions, which were similar for their two horses, showing that they could not discriminate achromatic stimuli whose luminance difference between the side and central parts of the panel was less than 0.2 log units.[6] Moreover, if the horses could not recognize color stimuli as colored, the results were likely to be the same for the chromatic discrimination functions, which in fact was clearly the case for the

4 Note, incidentally, that, in the field of cognitive ethology, it is worthwhile examining the protocols used to investigate the cognitive abilities of infants, which share their verbal incapacity with nonhuman animals. Additional examples include the study by Hanggi et al. (2007) and, in the following chapter, that of Proops et al. (2009). Conversely, developmental psychologists may find interesting the experimental protocols used in ethology. An example is that provided in a totally different field by the use of a test spot on the face of the chimpanzee studied by Gordon Gallup (1970), which was originated in infants by René Zazzo (1979) in his work on mirror images.

5. This corresponds to the method known as preferential looking.

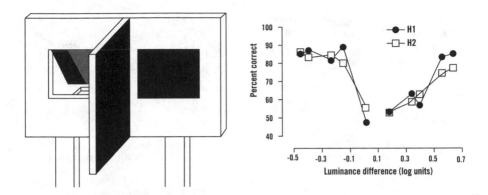

Figure 8.5 Left: Schematic view of the apparatus. Right: luminance discrimination functions. Source: Macuda and Timney 1999, 303–4.

yellow stimulus and somewhat less so for the green. This latter finding suggests that the horses were able to extract at least some chromatic information. In contrast, for the blue and red stimuli, the horses continued to discriminate with no notable difference below this limit, indicating that they did perceive the stimuli as colored (see figure 8.6).

These observations, which appear to confirm that horses are dichromatic at the very least, are consistent with those of Pick and colleagues (1994). They also suggest that the findings of Grzimek (1952) and Smith and Goldman (1999) might be attributable to an unsuccessful attempt to completely eliminate the effects of brightness. Macuda and Timney also note that, because no data were available on the spectral luminosity function of the horse at the time of their experiment, they had to measure the luminance of colored stimuli based on the human luminosity function. However, having subsequently measured this function (Macuda

---

6. The authors note that, as shown in figure 8.6, this limit applied to situations of gradually decreasing brightness differences in the central part of the panel, that is, when the central part was brighter than the sides. In contrast, in situations of gradually increasing brightness differences, where the central part of the panel was darker than the sides, the (now higher) threshold reached 0.5 log units. The authors also note that a similar asymmetry was observed in infants.

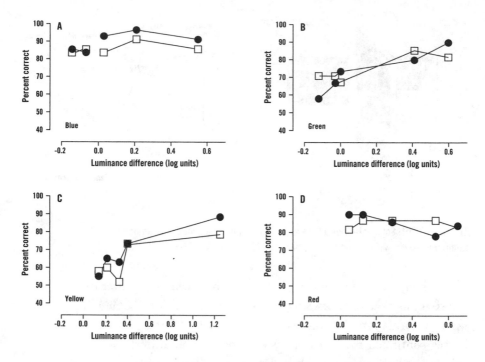

Figure 8.6 Chromatic discrimination functions of four colors for each of two horses. Source: Macuda and Timney 1999, 305.

and Timney 1999), they could say with confidence that the differences in spectral luminosity observed for red and blue did not affect their results.

In a review of the relevant literature published several years later (Timney and Macuda 2001), these same authors maintained that, although studies to date had established that horses had little difficulty discriminating red and blue from gray, the same was not true of yellow and green, which some horses could discriminate and others not. A decline in performance regarding these colors as the luminance match is approached suggests that color discrimination in horses in this region of the spectrum is weak.

Thus, observations made both during these different experiments and those stemming from anatomical and physiological studies (Carroll et al. 2001) conclude that horses are indeed dichromatic, although agreement is mixed regarding which colors they are able to identify. Now, dichromats cannot discriminate between an achromatic luminous source and a chromatic luminous source when both classes of cones are equally stimulated. Such a condition is met by a specific wavelength (or a narrow wavelength region) called the neutral point, which, as we have seen, is found only in dichromats (Jacobs 1981).

The Evidence for a Neutral Point (Geisbauer et al. 2004)

The questions of brightness discrimination in horses, the existence of a neutral point in their color spectrum, and the determination of its wavelength were taken up by Gudrun Geisbauer, Ulrike Griebel, Axel Schmid, and Brian Timney (2004). In their investigations, these authors used two horses that they tested for two-choice discrimination using an experimental apparatus analogous to that employed by Timney in his previous experiments on vision (see figure 8.7). The brightness test involved thirty gray panels measuring 30 by 30 centimeters, with values from G1 (brightest) to G30 (darkest). The panels were inserted into either of two trapdoors that swung inward to give access to a food reward. The positive stimulus was the brighter of the two gray panels presented.

The training phase was implemented with the brightest gray (G1) and a dark gray (G20). The G1 trapdoor gave access to the food (pieces of apples or carrots), and the G20 trapdoor was locked. The learning criterion was reached with 90 percent correct choices in thirty-one trials. During the test phase, the horses had to discriminate between two shades of gray for each of the thirty steps of gray. Once the learning criterion was reached, the brightness difference between the two grays was successively reduced until the horse could no longer discriminate between them, indicating that it had reached the brightness discrimination threshold (or brightness difference threshold) for the shade of gray tested.

A Weber fraction, or the relative difference threshold, was calculated for each threshold as K = $\Delta R^*/R$, where ΔR^* is the difference

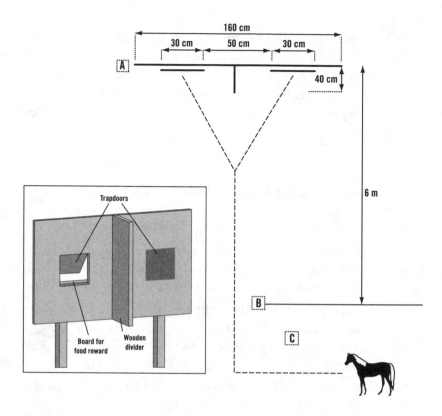

Figure 8.7 Experimental apparatus used by Geisbauer et al. Source: Geisbauer et al. 2004, 661, 664.

in relative reflection between the distinguishable grays and R the relative reflection of the starting stimulus. The authors note that the mean Weber fraction for one of the horses was 0.45 and for the other 0.42, and that this result was not markedly different from that reported by Macuda and Timney (1999) in their study of brightness discrimination, namely, around 0.3 log units, which corresponds to a Weber fraction of 0.5. In addition, they note that this Weber fraction is somewhat greater than that for humans, which reportedly ranges from 0.11 to 0.14. They note, too, that species that are active during both night and day, such as horses, dogs, fur seals, and manatees, are visual generalists. As such, they are well adapted to both scotopic and photopic conditions and appear to have much higher brightness discrimination thresholds than diurnal species such as humans.[7]

Following the brightness discrimination test, the authors performed an equal brightness test between an achromatic stimulus and a chromatic stimulus that presented each of the thirty grays (still the positive stimulus) against three blue-green shades. The researchers observed a gray region between the light and dark grays where the horses could not choose between the achromatic and chromatic stimuli—that is, a region where they perceived color and gray as equally bright.

The researchers then carried out a neutral point test, which required them to retrain the horses. Here, the discrimination criterion was color, and the color stimulus was the positive stimulus. During the test, the three blue-green shades used previously were paired with grays of equal brightness, as determined in the preceding test. The success criterion was 90 percent correct choices, or eighteen correct choices out of twenty trials. If the horse was able to discriminate both color and all the different grays presented, one could conclude that color was perceived as an intrinsic quality.

Although the two horses succeeded in discriminating color versus gray in four to six sessions for two of the blue-green stimuli, they both failed with the third color and showed either strong right or left preferences or refused to continue the test. Thus, they could not discriminate this particular blue-green from certain grays in terms either of color or brightness, suggesting that this stimulus excites both cones equally and consequently cannot be distinguished from an achromatic stimulus (white or gray). The authors cite the unpublished doctoral work of Macuda (2000), who concluded that the existence of a neutral point in the horse corresponds to a wavelength of around 480 nanometers. Indeed, the spectral distribution of the blue-green color that could not be discriminated showed a maximum of 480 nanometers, which appeared to confirm the existence

---

7. For purposes of comparison, Weber fractions obtained for dogs are roughly intermediate, around 0.22 to 0.27 (Pretterer et al. 2004).

of a neutral point toward this wavelength and thus that horses are dichromats.

Color Preferences (Hall et al. 2005)

In approaching the question of color vision from another angle, Carol Hall, Helen Cassaday, Chris Vincent, and Andrew Derrington (2005) began by asking whether horses showed color preferences, based on Grzimek's (1952) conclusion that yellow was easier to discriminate than gray, and asking whether such preferences were related to the way in which some colors might appear "more colorful" to horses.

To find out, they tested six horses, previously trained to discriminate colors from grays, using two-choice tasks between color pairs, in groups of eight: blue, blue-green, green, yellow, orange, red, black, and white. The stimulus consisted of identical boxes in the eight colors, arranged on the ground in front of the horse and separated by 1.20 meters one from the other. A given color was matched once to each of the other colors, which resulted in twenty-eight pairs. The spectral reflectance of each stimulus was also measured under experimental conditions by spectroradiometry, and the cone excitation ratio for each color was calculated (Govardovskii et al. 2000, and Brainard 1997, cited by the authors). These ratios were then compared with that for the grays, and the differences used to evaluate the chromaticity of the stimuli perceived by the animals. A correlation was then sought between this measure and the number of times a given color was chosen (see figure 8.8).

The researchers observed that yellow was chosen significantly more often ($p < 0.05$) than any of the other colors, and that red was chosen significantly less often ($p < 0.05$) than any of the other colors, with the exception of black and blue-green. A significant positive correlation ($p < 0.05$) was also found between the excitation differences of the cones (between color and gray) and the number of color choices.

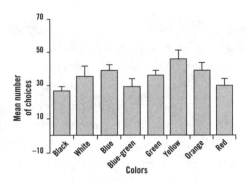

Figure 8.8 Mean number of choices of each color made by six horses (+ 1 standard deviation). Source: Hall et al. 2005, 42.

The authors conclude that some colors are more likely than others to attract horses and that this influences their behavior. They also note that some stimuli that humans perceive as very colorful (like red) do not have the same effect on horses. The authors recommend taking into account differences in color perception between humans and horses to avoid misleading interpretations of certain aspects of equine behavior.

Do Horses Perceive the Entire Color Spectrum? (Hall et al. 2006)

Revisiting the varying results of previous work on color vision (Grzimek 1952; Pick et al. 1994; Macuda and Timney 1999; Smith and Goldman 1999), Hall and colleagues (2006) made a number of observations. They noted that four colors (red, blue, green, and yellow) had been tested and that the experimental apparatus used by Grzimek differed strikingly from that of other researchers in the height at which it was placed: in Grzimek's case at ground level and for the other studies at nose height (Pick et al. 1994; Macuda and Timney 1999) or at 1.22 meters (Smith and Goldman 1999). According to Hall and colleagues (and as I have already mentioned), for horses, visual discrimination learning, including brightness discrimination, appears to be more effective when stimuli are placed on the ground than when they are placed level with the head (Hall, Cassaday, and

Derrington 2003). This observation might explain in particular the good discrimination of green obtained by Grzimek, in contrast to the results reported by others. Hall and colleagues thought it probable that the placement of the stimulus played a nonnegligible role in discrepancies among the results of previous studies. They also remarked that visual acuity varies as a function of stimulus color (Grzimek 1952): close to the value obtained using an image located on the visual streak for yellow and that obtained at the peripheral retina for blue. Moreover, to ensure the visibility of all colors presented, they must be of a size that, at the presentation distance, results in a visual angle greater than 0.5 degrees. Finally, given that the peak sensitivity of the equine cone photopigments is estimated to be 429 nanometers for one cone and 545 nanometers for the other (Macuda 2000), one can derive the spectral sensitivity curves and predict how easily a horse should be able to discriminate particular shades of gray. Thus armed, one could compare the data obtained by behavioral experiments with the predicted effect of color on the visual system.

Accordingly, Hall and colleagues decided to test a group of six horses with an array of colors against gray. Inspired by an experimental protocol used by Karl von Frisch (1955) in his study of color vision in bees, they devised learning tasks to discriminate between colors and gray based not only on one of the four colors red, green, blue, and yellow, but on a group of fifteen colors, namely, violet, indigo, purple, blue, powder blue, cyan, blue-green, greenish, green, lime green, khaki, yellow, orange, red, and pink-purple (C1 to C15 in figures 8.9 to 8.11).

Specifically, the horse had to choose one of two 37.5-square-centimeter cards constituting the top flap of two identical boxes, each card subdivided into eight panels (see figure 8.9). Among the 256 shades of gray made available by the software used to generate them, eight were selected ranging from light to dark gray. The card designated as the positive stimulus contained a color panel placed randomly on one of the eight positions. The other seven panels represented seven different shades of gray randomly occupying the other available positions. The eight negative stimulus panels, on the other hand, were all different grays and randomly placed.

 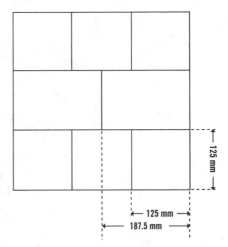

125 mm

← 125 mm →
← 187.5 mm →

Figure 8.9 Left: Karl von Frisch's experimental apparatus. The bees, trained to blue, go directly to the blue panel looking for food and not to any of the gray panels. Right: the apparatus used by Hall et al. A color is placed randomly in one of the panels. The other panels are different grays. Source: Von Frisch 1955; Hall et al. 2006, 440.

The boxes with the stimuli were placed on the floor to aid the visual discrimination of the horses. The top flap of each box was sloped at an angle of 60 degrees from the vertical, and the boxes were separated by a gap of 1.50 meters. The starting line for the horses was 6.5 meters from the boxes, which translated into a visual angle of the card of 3 degrees and of each individual panel of 1 degree.

The spectral reflectance of each of the colors and grays was measured, under experimental conditions, with the stimuli in the same position as in the experiments, with the aid of a spectroradiometer. Wavelength values between 380 and 720 nanometers were recorded in 1-nanometer steps.

During the training phase, the horses were divided into two groups of three. For one of the groups, each horse was trained with colors of gradually increasing wavelength. The second group followed a reverse procedure. When a horse erred, the same stimulus pair was presented to it again until it succeeded, but the initial choice was scored as incorrect. The performance of the horses was assessed four different ways: number of trials taken to reach the learning crite-

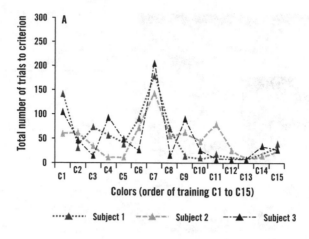

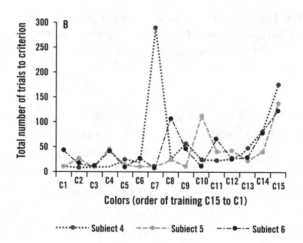

Figure 8.10 Total number of trials to reach the learning criterion for each horse in groups A and B. Source: Hall et al. 2006, 445.

rion, accuracy (percentage of correct choices beginning with the first attempt), the number of repeated errors, and the variation in the speed of approach.

All the horses completed the color discrimination training, reaching the learning criterion of ten consecutive correct choices (p < 0.001) more or less easily depending on the color. Blue-green (color C7) required the greatest number of trials and orange (C13) the least. Distinct interindividual variations were also observed, as well as a

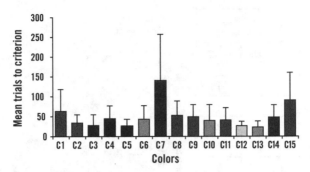

Figure 8.11 Mean number of trials taken to reach the learning criterion (95 percent confidence interval). Source: Hall et al. 2006, 445.

learning effect during discrimination training with the initial colors. It also appears that individual performance with blue-green differed within the two groups of horses, depending on whether training began with short or long wavelengths. In the first case (group A), performance was weak for all three horses, whereas in the second case (group B), only one of the horses had trouble (see figures 8.10 and 8.11).

At the conclusion of this phase of training, two transfer tests were carried out. First, the fifteen colors were randomly divided into five groups of three colors, and the horses were tested on each of these groups by the same procedure as for the training phase, save that the three colors in any group were displayed randomly within a session of twenty trials. Each of the colors was shown at least five times. The performance measure was the same as for the training phase, that is, ten consecutive correct trials. Second, still employing the same procedure, the authors used six new colors to make up two new groups of three colors: The first contained green, yellow, and brown, and the second blue, violet, and pink. One of the two previous groups of three horses was tested with these two new groups beginning with the first, and the other with the second.

All the horses succeeded in both transfer tests: Once they had learned to discriminate chromatic from achromatic, independent of any specific color, they were able to generalize the rule and apply it to all colors, not only the ones on which they had already

Color set	Component colors	Order of presentation	
		Group 1	Group 2
S1	C4, C9, C14	1	1
S2	C3, C10, C13	2	2
S3	C6, C11, C12	3	3
S4	C2, C8, C15	4	4
S5	C1, C5, C7	5	5
NS1	NC1, NC2, NC3	6	7
NS2	NC4, NC5, NC6	7	6

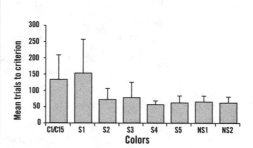

Figure 8.12 Left: component colors and order of presentation of color sets used in the transfer tests. Right: mean number of trials taken to reach the learning criterion of the final color trained (C1 or C15) for pretrained color sets and for the novel color sets (confidence interval 95 percent). Source: Hall et al. 2006, 444, 446.

been trained, but also new colors. Thus, none of the combinations resulted in any obvious lowering of performance, including for the two new color groups. Moreover, blue-green, which had been the most difficult color during the training phase, caused no decrease in performance for the group it was part of: Once a color had been trained, it became as easy to discriminate from gray as any other color (see figure 8.12).

Thus, although some horses succeeded more easily than others, all the horses were able to use chromatic cues to carry out their discriminations, all along the spectrum, including the five greens. As Grzimek (1952) showed in his experiments, horses can discriminate green and gray when the stimuli are placed at ground level, as well as yellow and gray. Hall and colleagues note that horses are selective herbivores and that color may provide them with cues that are complementary to smell, taste, and touch cues in their choice of food. Nonetheless, consistent with the preceding studies, reaching learning criterion for one particular color, blue-green (C7), required more trials with less reliable performance. The problems reported in this and earlier studies suggest that C7 was probably close to the neutral point of the horse and appeared less colorful than the other shades of green. The authors also note that the reflectance spectra of the colors used in their study cover a broad range of wavelengths, as is common for most natural objects, and so it was possible that some chromatic information was still available to the horse. The discrepancies between findings could also be due partially to interindividual differences, as well as to the effects of previous training, as appears to be the case in the results described above.

In comparing the different spectral reflectance tests carried out on each of the stimuli and the spectral sensitivity curves resulting from estimates of the spectral sensitivity of equine cone photopigments, the authors observe that the colors whose S to M/L cone excitation ratios differed most from those of achromatic stimuli (that is, those predicted to be the most colorful for the horse) were also the ones the horses could most easily distinguish from gray (based on repeated errors and latency of approach).

On balance, horses are clearly capable of using chromatic information from the entire spectrum (except for the neutral point), including colors whose wavelengths we perceive as green or yellow. Moreover, for the authors the close link between behavioral and physiological data corresponding to cone excitation indicates that cone excitation calculations could help to predict the ease with which horses will discriminate colors and to provide additional proof of equine dichromacy. Finally, Hall and colleagues suggest that such predictions could provide a base for comparing perception of color in horses with that of human trichromats, thereby increasing understanding of equine behavior.

Two points made by the authors in their discussion merit a closer look, as we will see in the study that follows. First, they ask whether the panels that we perceive as gray, that is, achromatic, really appear that way to horses, and whether the animals may not simply see a panel of a different color. Citing a review by Almut Kelber and coworkers (Kelber, Vorobyev, and Osorio 2003) of color vision in animals, they note the extensive use of gray stimuli in studies of color vision in other species that do not report the phenomenon, concluding that such a hypothesis would seem to be an implausible explanation of their results. Second, referring to the performance for blue-green (C7) in their own experiments, Hall and colleagues note the variability of the results: better when the learning began with long wavelengths and then moved to medium ones, less good when learning began with short wavelengths. They suggest that M/L wavelengths, including C7, look alike to horses, but appear different from S wavelengths. This would be consistent with the way in which a dichromat perceives the visible spectrum: two hues divided by a neutral area (Sharpe, Stockman, Jagle, and Nathans 1999, cited by the authors).

The Neutral Point: Break or Continuity? (Roth, Balkenius, and Kelber 2007)

In wondering how dichromatic animals, including horses, perceive their color space, Lina Roth, Anna Balkenius, and Almut Kelber (2007) turned their attention to the role of the neutral point.

Dichromatic vision is founded on comparing the excitation of two classes of cones. As such, it amounts to two-dimensional color space—the dimensions being hue and brightness—with saturation not taken into account. At constant brightness (isoluminance), this space reduces to a single dimension, hue. Within this one-dimensional space, colors that stimulate only S cones are presented at one end and colors that stimulate only M/L cones at the other. Between these two ends is a neutral point, where the two classes of cones are stimulated in the same ratio, corresponding to a wavelength that cannot be distinguished from those of achromatic stimuli (for example, gray).

The authors point out that, in trichromatic space, gray shades are interpreted as qualitatively different from chromatic colors, such as red and green, which also holds for animals, as Carl Jones, Daniel Osorio, and Roland Baddeley (2001) showed in a study of domestic chicks. Because ultraviolet light was excluded from the experimental conditions, chicks (which are tetrachromats) were effectively left

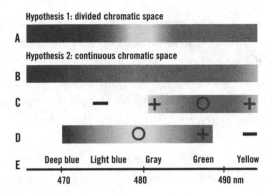

Figure 8.13 Two hypotheses regarding dichromatic color space, and the principle behind two initial experiments. The signs + and − correspond to positive and negative stimuli, and the circles represent the test colors. Source: Roth et al. 2007, 2796.

with trichromatic color vision. In such a situation, colors with identical brightness can be visualized in a triangle with a gray point at its center (see plate 11).

At the conclusion of these experiments, the authors observed that the chicks were able to generalize their learning to intermediate colors based on discriminations for which they had received positive reinforcement (for example, generalizing to purple after training to yellow and red as positive stimuli). In contrast, chicks positively trained to blue and yellow, which are on opposite sides of the gray point, could not generalize over it. In other words, the chicks did not perceive gray as an intermediate color between yellow and blue, but processed the latter colors as being in different categories (like human trichromats).

That notwithstanding, Roth and coworkers (Roth, Balkenius, and Kelber 2007) note that, in the one-dimensional color space of dichromats (that is, under constant brightness), the role played by the neutral point in perceiving color is unclear. Two alternative hypotheses have been advanced to explain it (see figure 8.13).

According to the first hypothesis, the neutral point divides color space into two color categories. In such a case, dichromats would perceive gray at the neutral point, and they would perceive it as different from chromatic colors, as chicks do.

In contrast, the second hypothesis assumes that the one-dimensional color space of dichromats comprises a continuous scale of hues where the color at the neutral point is not qualitatively different from other perceived colors. According to this hypothesis, dichromats may be able to generalize between colors whose wavelengths are on opposite sides of the neutral point. To test these hypotheses, the authors carried out three experiments with seven horses, two participating in the first experiment, three in the second, and two in the third. The experimental apparatus was similar to that used previously by Timney in his studies on vision, where stimuli were projected onto two swinging trapdoors (Timney and Keil 1992; Timney and Keil 1996; Macuda and Timney 1999).

The first two experiments relied on the same principle as in the work done by Jones and colleagues (Jones, Osorio, and Baddeley 2001) with chicks.

In the first experiment, the two horses were first trained to two positive stimuli whose wavelengths were on the same side of the neutral point (seen as yellow and gray by human trichromats) and to one negative stimulus on the other side of the neutral point (seen as blue by human trichromats). Both reached the learning criterion of 75 percent correct choices over three consecutive days. However, performance with the gray positive stimulus, which was closer to the negative blue stimulus, was lower than that with the yellow positive stimulus, which differed most from gray.

For the test phase, the authors added a green stimulus test to the discrimination pairs, with a wavelength that was intermediate between the gray and yellow positive stimuli.

When they were tested with combinations of the discriminations green versus gray, green versus yellow, gray versus yellow, and (as a control) gray versus a novel blue, the horses systematically preferred the longer wavelength color, thus showing that they could generalize from the positive stimuli training (see figure 8.14).

In the second experiment, the three horses were trained to two positive stimuli with wavelengths located on opposite sides of the neutral point (seen as green and blue by human trichromats) and to one negative stimulus (seen as yellow by human trichromats).

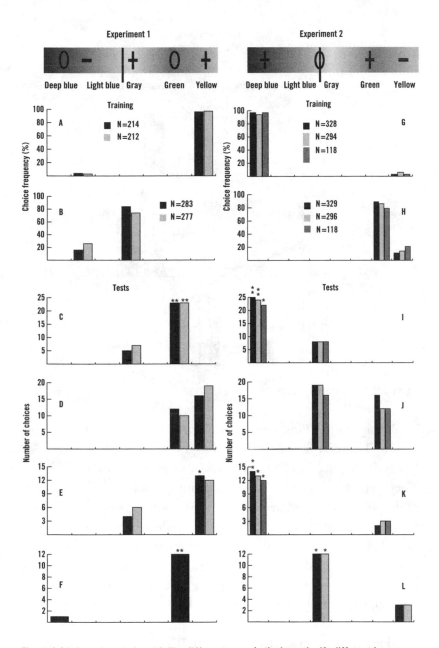

Figure 8.14 Experiments 1 and 2. The different grays in the bars signify different horses. Only one horse participated in the final test of experiment 1, and only two horses completed the final test of experiment 2. Significance levels: *p < 0.05; **p < 0.01. Source: Roth et al. 2007, 2798.

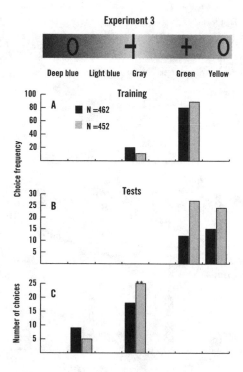

Figure 8.15 Experiment 3. Source: Roth et al. 2007, 2799.

During the training phase, the horses achieved all the learning criteria. However, as in the preceding experiment, they performed less well in tests involving colors whose wavelengths were closer. They were then tested with combinations of choices similar to those of the preceding experiment, with the addition of a gray stimulus test whose wavelength corresponded to that of the neutral point.

As before, the horses chose in a relative fashion, preferring shorter wavelengths, including in the final test where they had to discriminate gray and yellow. Thus, it appears that the neutral point is not processed differently than other colors for dichromatic horses.

The authors then did a third experiment to see whether the horses could learn colors in a relative manner. They trained the two other horses to a positive green stimulus and a negative gray stimulus, which involved choosing the color corresponding to the longest wavelength (see figure 8.15).

In a first test, the horse had to choose between the green positive stimulus and a novel stimulus, yellow, that is, with a longer wavelength. The horses chose both colors equally often, which suggests that they did not discriminate between them and that they may have interpreted both as positive stimuli.

A second test required the animals to choose between the negative, gray stimulus and a novel color, blue, with a shorter wavelength. The horses had great difficulty deciding, but the gray was nonetheless chosen more often than the blue. Thus, although the horses had previously learned that gray was negative, it also had the longest wavelength and was the one the horses chose.

Apart from demonstrating relative color learning, this experiment adds weight to the suggestion that horses process gray the same way as any other color in their color space.

In addition to the observation that performance level in all the experiments appears to be linked to proximity or remoteness of the wavelengths of the stimuli involved, the authors also note that the conclusions agree with those of Jan Hemmi (1999) concerning an Australian marsupial, the tammar wallaby. They also point out that Thomas Wachtler, Ulrike Dohrmann, and Rainer Hertel (2004) showed that dichromatic humans use the term *green* for colors at the neutral point, which is yet another indication that the neutral point does not split the color space and that they could perceive a chromatic color there. The same is probably true of all dichromats.

The authors note as well that different colors that humans perceive as green may in fact be quite close or far away from the neutral point of dichromats, which might partly explain the discrepancies of previous studies regarding color vision in horses.

Finally, in their view, to the extent that dichromats do not perceive gray as qualitatively different from red, blue, green, or yellow, a better test of color vision would be to use colors that differ from those of the neutral point, rather than the classic gray versus color tests.

Equine Dichromacy: A Qualification (Hanggi, Ingersoll, and Waggoner 2007)

More recently, noting the divergent results of studies based on discrimination of colors and gray (in particular, due to difficulties in controlling for brightness), Evelyn Hanggi, Jerry Ingersoll, and Terrace Waggoner (2007) tried a new approach. They adapted the pseudoisochromatic plate test, a color vision test first published in 1917 by Shinobu Ishihara based on the ability to discriminate one color from another and still widely used around the world to detect deficiencies in color vision (dyschromatopsia).

Recall briefly that individuals affected by dyschromatopsia (dyschromats) cannot distinguish the three primary colors red, green, and blue, either because one of the three associated photopigments is abnormal or dysfunctional (causing an abnormal sensitivity to this or that color) or absent (see plate 12). The latter category of individuals subdivide into protanopes, deuteranopes, and tritanopes,[8] according to whether they are red, green, or blue blind, that is, they lack the photopigment associated with the L, M, or S cone. In any event, all three are dichromats.[9]

The thirty-eight plates that make up the Ishihara test consist of a mosaic of dots of a small number of different colors, arranged in an apparently random fashion. Each hue appears in several different sizes, saturations, and brightnesses, and these differences are identical for each hue represented. The mosaic contains a shape (a number, for example): The uniformity of the hue is what makes it possible to recognize the shape in the plate. But the dots that make it up have different saturations or brightnesses. A dyschromat who does not see the hue will not be able to identify the shape because it has no homogeneous saturation or brightness.

---

8. These terms derive from the Greek prefixes *prot(os)*, *deuter(os)*, and *trit(os)*, meaning first, second, and third, with reference to the trio red, green, and blue, with the suffix *anopie* or *anopsie* (from the Greek *an-opsis*), which means "absence of vision."

9. Around 1 percent of men are protanopes, and just as many are deuteranopes. But these proportions are a hundred times less in women. Tritanopia is extremely rare and occurs in only 0.008 percent of the human population.

An adaptation of this test (called the Color Vision Testing Made Easy Test, CVTMET), developed by Waggoner (1994), initially aimed at evaluating deficient vision for individuals of all ages, including preschool-age children for whom the Ishihara test is less reliable. In fact, the CVTMET assumes only that the subject is able to identify simple geometric shapes or objects (such as a circle, square, star, car, dog, house, or boat) rather than to test more complex cognitive tasks, such as identifying or tracing numbers.

For the purposes of the experiment I am about to describe, the authors used a modified version of the CVTMET to test four horses. It consisted of laminated cards (plates) 18.5 by 20 centimeters in size based more or less on the principles of Ishihara (see plates 13 and 14), as follows:

- for the positive stimulus, a circle formed by a mosaic of dots of a certain color, with varying levels of size, saturation, and brightness, on a background of dots of another color all with similar levels of size, saturation, and brightness, and
- for the negative stimulus, the same background without the circle.

The colorimetric characteristics of the plates were based on CIE standards, and verified using a colorimetric spectrometer.

The plates were slid into 30-square-centimeter openings[10] positioned at equal heights in a 1.9-by-2.4-meter wooden board. The horse stood at a distance of 2.4 meters from the board when the stimuli were revealed. These were visible for 5 seconds, then covered, at the same time as the horse was released to approach the top of the panel on its own and touch a stimulus with its nose. It then received a stimulus consisting of verbal conditioned reinforcer—"Good!" or "No!"—and in the first case a food reward, after which it returned to

---

10. Although height was not specified in this article, it can be estimated nonetheless to be around 8 to 90 centimeters, based on the photographs in figure 8.16.

its starting position. Note that the horse had previously been trained through operant conditioning to position itself at the station to limit the risk of distracting cues caused by operator intervention (see the experimental apparatus in figure 8.16).

During the training period, the horses were shown cards where the positive stimulus was a large circle of orange dots on a blue-gray dotted background, which any human (including the color deficient) could discriminate. Once the first discrimination was learned, a control, which was successful, was carried out with a plate where the circle was gray.

The first part of the experiment consisted in showing the horse plates that tested protanopia and deuteranopia, where the positive stimulus was a dotted brown circle and a dotted green background: For protans and deutans, brown and green blend into one another. The four horses failed this test. In a second phase of the experiment, the plates had smaller circles. A first step, during which the horses were presented with plates where the positive stimulus had either a dotted light-gray circle or a dotted medium-gray circle and a dotted gray background, showed that the horses had no trouble generalizing their prior learning. They were then subjected to a series of three tests:

- a protan discrimination test, where the positive stimulus was a dotted red-violet circle on a dotted gray background (red-violet looks gray to protans);
- a deutan discrimination test, where the positive stimulus was a dotted blue-violet circle on a dotted gray background (blue-violet looks gray to deutans); and
- a tritan discrimination test, where the positive stimulus is a circle of dotted yellow-green on a dotted gray background (yellow-green looks gray to tritans).

Although all four horses successfully completed the tritan test, none scored significantly above chance (80 percent correct responses) either for the protan or the deutan test (see figure 8.17).

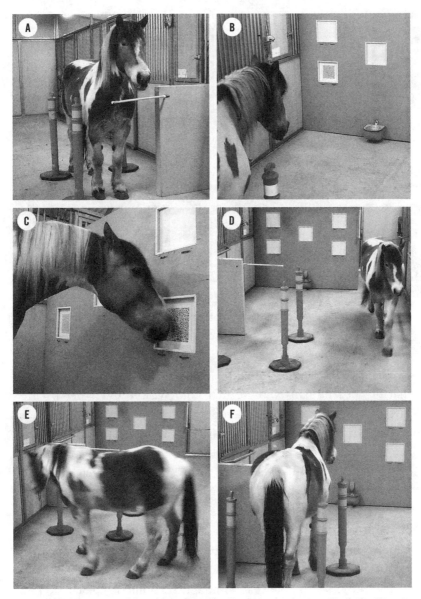

Figure 8.16 Testing apparatus, station, and experimental procedure used by Evelyn Hanggi, which included learning a chain of behaviors taught by operant conditioning. At the end of the learning, the horse worked independent of the handler. In the experiment described in the text, only the two lower panels were used to project stimuli. Source. Hanggi 2007, 67.
© Evelyn B. Hanggi, Equine Research Foundation.

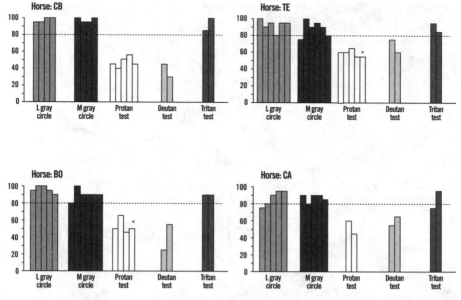

Figure 8.17 Success rate for different test plates and for each of the four horses, based on runs of twenty trials. Source: Hanggi 2007, 70.

The authors concluded that the color vision of horses differs from that of both humans with regular color vision and other trichromats. Horses are neither tritanopes nor dichromats with pigment S (blue) deficiency. Rather, they are either anomalous trichromats[11] (protanomalous or deuteranomalous) or dichromats (protanopes or deuteranopes). In other words, horses have red-green deficiency but not blue deficiency. Yet a human anomalous trichromat would normally be able to pass either the protan or deutan test, whereas a human dichromat would succeed at neither because his color deficiency is more severe. In this respect, the results of the horses in this study do indeed suggest dichromacy.

The authors also note that, although trichomatic color vision has obvious advantages with respect to dichromatic vision, the evolutionary success of dichromat mammals[12] confirms that dichromacy

---

11. An anomalous trichromat is relatively insensitive to one of the three primary colors owing to an abnormal or dysfunctional photopigment. Depending on the particular photopigment, the deficiency is referred to as protanomaly, deuteranomaly, or tritanomaly.

in general suffices for dealing with many visual tasks within diverse habitats. As for the horse, it has no need to identify colorful foods or to sport flashy colors to attract sexual partners. Instead, what the animal needs is vision that is good enough under scotopic conditions especially to detect predators and to find food. And in fact, according to Simon Verhulst and Frans Maes (1998), dichromats are better at this than trichromats in such conditions.[13] In an earlier publication, Hanggi (2006) had reported that, in situations where horses are not limited to reacting to a single category of visual stimulus, they are capable of using visual cues other than color to solve a range of discrimination and avoidance tasks. For example, when several green objects are placed on a green plastic tarp, a horse brought to within 5.5 meters of the tarp had no difficulty locating the objects and approaching to examine them. The same horse even managed to pick out an object measuring only 2.54 by 5.08 by 1.59 centimeters that was itself covered with the same plastic as the tarp on which it was placed.

Moreover, as the authors of the present study point out, and which is not surprising considering the many preceding studies mentioned (in particular Timney and Keil 1999), horses move about easily in nature through the most varied terrains, or jump green hedges and brown bars in competitions where the ground is green or brown, thus needing to rely on brightness cues, variations in hue and depth, and so forth, to compensate for deficiencies in their color vision.

A New Experiment in Chromatic Discrimination (Blackmore et al. 2008)

Returning to conflicting results regarding the wavelengths that horses are able to discriminate, Tania Blackmore, Therese Foster, Catherine

---

12. Among eutherian mammals, only some primates—Old World monkeys, or catarrhines—are trichromats, with roughly two exceptions; New World monkeys, or platyrrhines, are dichromats, except for around half the females of most species, which are trichromats (Dominy, Svenning, and Li 2003).

13. This hypothesis, based on an analysis of human perceptual light thresholds after adaptation to dark, was subsequently questioned after taking into account human behavioral results (Simunovic, Regan, and Mollon 2001).

Sumpter, and William Temple (2008) note that some behavioral studies have used reflective stimuli, that is, painted surfaces, and other projected stimuli, that is, light passed through color filters.

The authors emphasize that, in the latter case, the properties of the stimuli are more easily controlled: It is, for example, possible to employ standard filters, used in lighting or photography, whose characteristics (such as wavelength) are known and which enable better brightness matching with achromatic filters. This makes it easier to describe the properties of the stimulus for the purposes of experimental replication.

On the question of the height at which stimuli are placed, the authors state that, although Carol Hall and her colleagues (Hall, Cassaday, and Derrington 2003) showed that height may play a role, it does not explain the differences observed in the results.

Consequently, the studies with the best control conditions appear to be those of Macuda (2000) and Macuda and Timney (1998 1999), who used projected stimuli, and that of Evelyn Hanggi and colleagues (2007), using pseudoisochromatic plates.

To carry out their own experiments, Blackmore and her colleagues tested four horses with stimuli back-projected onto two semitransparent "windows" (screens), one of which was illuminated with a color slide and the other with a gray one (see figure 8.18). Each of the stimuli was emitted by one of two projectors equipped with a carousel containing filters. The entire apparatus was controlled by a computer. The screens, measuring 30 by 40 centimeters and separated by a distance of 34 centimeters, were situated at the level of the horse's head and embedded in a panel painted black. In front of each screen was a table with two levers, one under each window, and between them a container that could be automatically filled with food.

At the end of each learning phase, the horse had to press its nose on the lever under the window where the colored stimulus was projected. In this case, the container was intermittently filled with food, according to a random interval schedule of reinforcement that offered better resistance to learning extinction than a systematic reinforcement schedule.

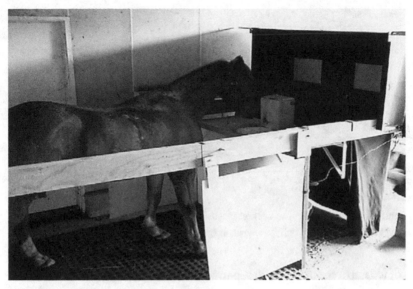

Figure 8.18 Experimental apparatus used by Tania Blackmore and colleagues. Source: Blackmore et al. 2008, 390.

Various methodological precautions were taken to avoid the risk of "parasite" stimuli. For example, the filter carousels were changed between each session to avoid the influence of auditory stimuli, and the left-right position of the color stimuli varied in a nearly random order (with no more than three consecutive presentations in the same position), while maintaining an overall balance.

The filters used for the color stimuli included three shades of blue (470, 474, and 482 nanometers), green (532, 533, and 545 nanometers), yellow (579, 582, and 583 nanometers), and red (609, 611, and 615 nanometers).

For achromatic stimuli, gray filters (neutral density) were chosen so their brightness matched as closely as possible with the color filters. Gray filters that were brighter and dimmer than each color filter were also selected. However, gray filters of varying brightness were not available for the blue medium (470 nanometers) or for the light yellow (579 nanometers), and consequently the matches were made with the closest gray filters.

For each color tested, the nine matches between color and gray

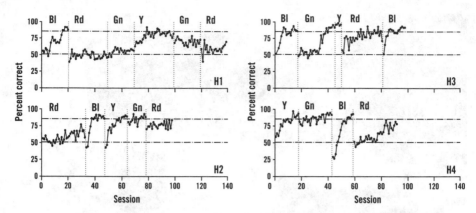

Figure 8.19 **Percentages of correct responses of all horses (H1, H2, H3, H4) on all color conditions: Bl (blue), Rd (red), Y (yellow), Gr (green).** Source: Blackmore et al. 2008, 392.

(that is, three color shades paired with three dimmer or brighter grays) were presented twenty times per session in a random order that was different for each horse.

As indicated above, correct responses were randomly reinforced with food; nonetheless, each of the correct responses was consistently followed by an electronic feedback beep.

The order in which the four colors were tested was different for each of the four horses; a color test was terminated when any of the following conditions were satisfied: Either the horse tested achieved 85 percent correct responses in five consecutive sessions, or it completed twenty sessions without a significant improvement in score. In addition, some horses were tested more than once for the same color to confirm the replicability of the results. The findings (see figure 8.19) show that all four horses, H1, H2, H3, and H4, achieved 85 percent accuracy in five consecutive sessions for wavelengths corresponding to blue (470, 474, and 482 nanometers), after 20, 14, 16, and 16 sessions, respectively.

In contrast, no horse satisfied this criterion for wavelengths corresponding to red (609, 611, and 615 nanometers), even after many sessions, although some achieved 85 percent correct responses for particular sessions. H2 improved after a second red test and thereafter consistently achieved 75 to 80 percent. H3 and H4 also achieved

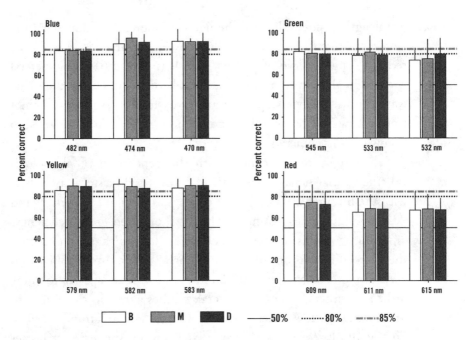

Figure 8.20 Average percent correct for each color for the nine pairings based on average accuracy over the last five sessions: M (matched), B (brighter), D (dimmer). Source: Blackmore et al. 2008, 393.

a high level of success, but not 85 percent within five consecutive sessions. Only H1 failed to achieve the criterion for wavelengths corresponding to green (532, 533, and 545 nanometers) and yellow (579, 582, and 583 nanometers). Its level of performance for green was consistently around 55 percent accuracy, despite a second test, and 80 percent for yellow, although it exceeded 85 percent in four separate sessions.

Furthermore, the average number of correct responses for each color based on nine slide pairings was calculated based on the last five sessions for each of the horses (see figure 8.20). These results show no clear or consistent differences related to the comparative brightness of the gray and colored slides. This suggests that, here, horses do not rely on brightness cues and that they discriminate based on the wavelength of the color stimuli.

The average percent correct was high across the yellow and blue wavelengths, though less for the lightest blue (482 nanometers, the closest to green of the blues). The lowest accuracy was with the red wavelengths. Two of the red stimuli (611 and 615 nanometers) resulted in the lowest accuracy across all the stimuli. For the green wavelengths, the worst performance was for the stimulus farthest from yellow (532 nanometers) and the best for the stimulus closest to yellow (545 nanometers).

All of these results concur with the conclusion of physiological studies that horses have a dichromatic visual system, as well as with the findings of Hanggi and colleagues (Hanggi, Ingersoll, and Waggoner 2007) suggesting that horses show a red-green deficiency.

Citing previous studies, Blackmore and colleagues note that there is a consensus on the ability of horses to discriminate blue and gray. For green, on the other hand, some studies appear to show difficulty; others do not. In general, studies agree on the ability to discriminate yellow, except for that of Macuda and Timney (1999). Finally, results for red are fairly mixed, but do suggest that horses are not totally incapable of discriminating it, though they may have difficulty doing so.

Considering the possible influence of the order of presentation of the different colors, the authors note that, in their experiments accuracy levels decreased only slightly in changing from yellow to green and vice versa. In contrast, performance decreased markedly with other color changes (except for red to green for H1, whose accuracy was low). They emphasize that each of the stimuli was chosen to represent the complete spectrum of wavelengths for each color, but that the filter used for the brightest green (545 nanometers) appeared yellowish to humans.

With respect to this point, recall that, although laser rays consist of a single wavelength, the same is not true of white light passing through a colored filter. Indeed, these rays comprise a varying range of wavelengths—their bandwidth—for which the peak of the spectral distribution curve corresponds to their dominant color, that is, the one specified by the filter.

In superimposing the spectral distribution curves of the green and yellow filters, the authors observed that there is an overlap in the area around 500 to 550 nanometers, which confirms that the filters transmit some shared wavelengths. The spectral values of the green and yellow filters were also plotted on a CIE diagram, which revealed that the low values of two of the green filters were close to the yellow region of the visible spectrum. That could explain the continuing high accuracy in the change from green to yellow. In other words, the horses may have been able to discriminate green because two of the filters contained yellow hues.

This observation is important, and I will return to it, for it underscores the need to take into account the spectral curves of the stimuli, and in particular their bandwidth, whether the stimuli are reflective or projected.

Blackmore and colleagues also note that some of the disparities in study results may be due to differences in the criteria for judging whether a given color has been correctly discriminated or not. For example, Pick and colleagues (1994) and Hall and colleagues (2006) judge their horses to have succeeded in discriminating gray on completion of ten consecutive responses. In contrast, Smith and Goldman (1999) required 85 percent correct responses in a single session, and Hanggi and colleagues (Hanggi, Ingersoll, and Waggoner 2007) 80 percent correct responses in two runs of twenty trials.

In addition to these differences, it is worth asking about the description of colors by the authors of the various studies. Is every red, green, blue, and yellow the same? For example, the reds used by Blackmore and colleagues (609, 611, and 615 nanometers) border on orange, whose spectral range is around 592 to 620 nanometers, rather than centering on red, which ranges from 620 to 700 nanometers. Moreover, as these authors themselves observe with respect to their green and yellow filters, the range of wavelengths allowed through the filters is not insignificant, and the use of broadband filters, each of which has its own spectral curve, may have a considerable influence or even explain apparent contradictions. For example, horses might plausibly have similar difficulties with red or yellow filters, which both have significant orange components in them.

How Well Do Horses Discriminate Color in Half-Light?
(Roth, Balkenius, and Kelber 2008)

As I already indicated in chapter 6, because of their very different sensitivity to light intensity, cones and rods are basically active either for day vision (photopic), in the former case, or night vision (scotopic), in the latter case. Under conditions of dusk, or moonlight, so-called mesoscopic vision, the two systems function simultaneously, but "visual accuracy is only mediocre because the light is insufficient for appropriate functioning of cones and too high for appropriate functioning of rods" (Bagot 2002, 127).

Nonetheless, these conditions are precisely the ones chosen by Lina Roth, Anna Balkenius, and Almut Kelber (2008) in investigating the extent to which horses can perceive color in half-light and whether they do it differently from humans.

The authors note that color represents a source of information that is equally valuable for evening and night, and that, at dawn and dusk, when the color and intensity of light are changing, color information is more reliable than that provided solely by achromatic contrast. In fact, some species of insects and reptiles are known to exploit this phenomenon for their benefit (Kelber and Roth 2006).

According to Roth, Balkenius, and Kelber, horses possess certain compelling characteristics in this respect. Aside from being vertebrates whose visual apparatus is based on cones and rods, horses are equally active day and night. They also have among the largest eyes of terrestrial species, with a pupil that dilates significantly to capture sparse photons in the dark. The equine tapetum lucidum also enables photons that have not been absorbed by photoreceptors a second opportunity to be trapped, which further increases the animal's response to light.

The experiment was first conducted with three horses and later reproduced with six humans (including the authors of the study). The experimental apparatus was the same as that used by the authors in their study on the neutral point (Roth, Balkenius, and Kelber 2007) and similar to that used at different times by Timney, with two swinging trapdoors onto which the stimuli were projected (Timney and Keil 1992; Timney and Keil 1996; Macuda and Timney 1999).

During the training phase, the horses learned to associate a specified color with a reward of carrots, namely, blue for Chap (a fourteen-year-old gelding) and Rosett (a thirty-three-year-old mare) and green for Rex (an eleven-year-old gelding).

In the test phase, the three horses were given two-choice tests between a blue and a green stimulus. The positive stimulus was green for Chap and Rosett and blue for Rex. The horses were generally given twenty choices per day, five days a week. However, for the three lowest intensities, from 0.08 cd/m^2, the horses were given five to fifteen minutes to adjust to the darkness prior to the sessions. Moreover, the horses were tested on no more than ten choices per day to maintain their motivation.

The authors used seven greens and five blues, selected to counter the effects of brightness. They began with a relatively high[14] brightness intensity of 50 cd/m^2, using a fluorescent tube present in the barn and dimming it gradually once the animals had achieved at least 75 percent correct choices in three days. To do so, they used two halogen spotlights with an intensity of 15 cd/m^2 and equipped them with neutral density filters to achieve even lower light intensities. Although the horses chose at significantly high frequencies up to 1.2 cd/m^2 (the intensity of light at sunset), Rex lost all motivation to such an extent that he even began producing random answers at intensities of 15 cd/m^2. His results were thus discounted. Rosett lost motivation at a brightness intensity of 0.2 cd/m^2, at which point she developed a marked right preference. After choosing the left side in only six presentations out of thirty, three days in a row, her tests were stopped.

In contrast, Chap continued to choose the blue stimulus in 31 presentations out of 40 ($p < 0.05$, binomial test) at an intensity of 0.02 cd/m^2, corresponding to moonlight, but did not succeed

---

14. Which corresponds approximately to a well-lit street at night. Recall that Kelber and Roth (2006) indicate light greater than 5 cd/m^2 for photopic vision, less than 0.005 cd/m^2 for scotopic vision, with the intermediate region corresponding to mesopic vision.

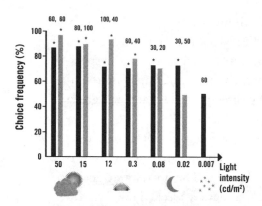
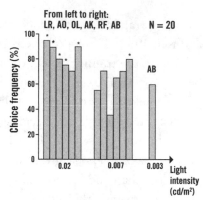

Figure 8.21 Choice frequencies at different light intensities for horses and humans (*: significance level, binominal tests, p < 0.05). Left: results for Chap (black) and Rosett (gray) from day 2, where they achieved at least 70 percent correct choices. The total number of choices is indicated above each bar. Right: results of twenty choices for the six human subjects. Source: Roth et al. 2008, 3.

beyond 30 presentations out of 60 at an intensity below 0.007 cd/m². Right afterward, he was again tested at a higher intensity, 15 cd/m², and chose correctly nine times out of ten, thereby showing that his motivation was intact. Thus, some horses do appear to be able to discriminate color by the light of the moon.

The human participants (the three female authors of the study and three men) were tested with only twenty presentations at brightness intensities. The experimental conditions were the same; however, they were tested only at intensities of 0.02 cd/m², for which five of them satisfied the success criterion, and 0.007 cd/m², where only Anna Balkenius met the criterion. Consequently, she was tested at an intensity of 0.003 cd/m², but in this case she did not satisfy the success criterion. These results show that horses and humans have roughly similar color vision thresholds in darkness and that both are able to discriminate colors by moonlight (see figure 8.21).

The authors note, however, that overall, color appears to be a less powerful stimulus for horses, although they do show a certain amount of interindividual variability.

In conditions of mesopic vision, the rods emit stronger signals while the cones are emitting weaker ones, which translates in humans to colors that are less and less saturated. Such a phenomenon might be reinforced in horses. In fact, unlike humans, who have no rods in the area around the fovea, horses have rods throughout their retina. Morever, rods predominate over cones in most of the equine visual streak. This arrangement appears to be compensated by the larger size of the equine pupil compared to that of humans.

Thus, the horse's large eyes would not appear to predispose it especially to night color vision, but rather, enabled by rods, good achromatic vision with no loss of spatial resolution in attenuated light. In this way, horses benefit from the advantage of color vision during periods of greatest change in illumination color, that is, sunset and twilight: "At dimmer light intensities the ability to detect movements from possible predators may be the most important visual task, and achromatic vision may therefore be favored." (Roth, Balkenius, and Kelber 2008, 4).

Colors That Can Be Fairly Well Discriminated across the Light Spectrum (Timney and Macuda 2009)

The ideas presented below are the product of an article in preparation[15] by Brian Timney and Todd Macuda. They are based on a working document provided me by the authors, which they kindly allowed me to use here.

Timney and Macuda observe that earlier behavioral studies showed that horses can make chromatic discriminations by distinguishing chromatic targets from gray ones in the context of experimental setups designed to make brightness cues irrelevant. The authors note a general consensus on the existence of color vision, with results suggesting dichromacy, but also certain inconsistencies with respect to the colors that can be discriminated. These differences

---

15. The conclusions from this research also appeared in a communication by Brian Timney (2009) presented at the International Equine Science Meeting of 2008, titled "Photopic Spectral Sensitivity and Wavelength Discrimination in the Horse (*Equus caballus*)."

may be due in part to the difficulty in determining the luminance of the gray target to be matched with any individual color to enable brightness to be adjusted appropriately. Moreover, almost all studies use broadband color filters or colored paper surfaces. Although for humans, brightness may be approximately matched using the standard human spectral luminosity function, the same cannot be assumed for nonhuman species.

The authors point out that anatomical and physiological studies have shown the existence of two classes of cones in the horse, S and M/L, with sensitivity peaks around 430 and 540 nanometers. They note, however, that aside from the presence of neural mechanisms required for dichromatic color vision, functional presence must also be shown through behavioral estimates: namely, an assessment of the extent of color vision and just how far the visual system can go in using chromatic information.

Accordingly, Timney and Macuda carried out a pair of experiments aimed at determining the photopic spectral sensitivity of horses and at establishing their wavelength discrimination function across the light spectrum.

The goal of the first experiment was to determine the spectral sensitivity of the two horses Timney had used in earlier studies on vision (Timney and Keil 1994; Timney and Keil 1996; Macuda and Timney 1999), with the same apparatus as before.

However, the new experiment involved detecting a circular patch 13 centimeters in diameter on one of two screens, each of which had a dim, uniform white background. The patch was projected by one of two projectors situated behind the apparatus and could be varied in luminance and wavelength using neutral density and narrow-band interference filters.

To derive the spectral sensitivity function, the authors tested the detection threshold of eleven wavelengths in the 430- to 700-nanometer region, in increments of 25 to 30 nanometers. Once the threshold points were determined, the absorption spectra of the photoreceptors were estimated by iteratively fitting the sensitivity data points to a template curve designed to approximate the spectral characteristics

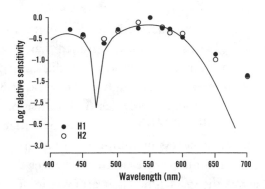

Figure 8.22 Increment-threshold spectral sensitivity function for horses H1 and H2, and the corresponding spectral sensitivity curve. The curve through these sensitivity values represents the best-fitting subtractive combination of the estimated photopigment absorption spectra. Source: Timney and Macuda, personal communication, 2009.

of any vitamin A$_1$-based photopigment[16] (Carroll et al. 2001). As shown in figure 8.22, the spectral sensitivity curves fit very well with experimental data, except for the two points corresponding to the longest wavelengths. The estimated spectral peaks are 429 and 545 nanometers, which (as the authors point out) is consistent with the estimates of Joseph Carroll and colleagues (2001), based on electro-retinographic measures of 428 and 539 nanometers, respectively, as well as that of Shozo Yokoyama and Bernhard Radlwimmer (1999), of 545 nanometers for the M/L photopigment, based on molecular methods.

The authors also note that, unlike other ungulates, such as pigs, cows, sheep, and goats, whose M/L photopigment resembles the human L cone photopigment, and whose S cone is shifted toward longer wavelengths by 30 to 40 nanometers, in the horse, the M/L cone is shifted toward the human M cone (that is, a position somewhere between the human M and L cones) and the S cone is closer to the human S cone by some 20 to 25 nanometers (that is, clearly quite close to it).

---

16. Recall that the different cones contain different varieties of iodopsin, which contains retinal, an aldehyde present in vitamin A.

The second experiment concerned the wavelength discrimination function across the light spectrum. It assumed that an animal may be able to tell a color from a gray, but still have difficulty distinguishing one color from another. Although the first situation supports the existence of color vision, it provides no information about how well horses see color. For this reason, Timney and Macuda turned to a measure of the acuteness of equine color vision that consisted of calculating the minimum wavelength differences that horses can discriminate across the spectrum. Their aim was to provide an indication of spectral regions where different wavelengths are confused. And because a behavioral measure was involved, it would be possible to deduce perceptual and cognitive aspects of color vision.

The shape of the wavelength discrimination function is linked to the degree of excitation of the photoreceptors. The best discrimination occurs where the excitation ratios change most rapidly, that is, where the photoreceptor response curves intersect. Thus, the number of minima in a wavelength discrimination function can be used to infer the dimensionality of the color vision system. Dichromats have a single minimum, trichromats have two minima, and tetrachromats three.

To do the second experiment, the authors used the same apparatus as for the first. They gave the horses a two-choice test between a specific wavelength and a second one that was measurably either longer or shorter, and then reduced the distances until they reached the discrimination threshold, while simultaneously varying the brightness of the stimuli to minimize its effect. They proceeded in this way, moving up and down the scale for five wavelengths—450, 480, 532, 568, and 590 nanometers—and averaged the results. Accuracy was nearly identical for the two horses participating in the experiment. The discrimination function was obtained by fitting sigmoidal functions to the data (see figure 8.23).

This function produces a single minimum at 480 nanometers, a wavelength at which the horses could discriminate around 12 nanometers. For the other reference wavelengths, discrimination was significantly poorer.

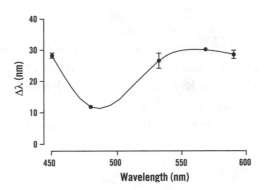

Figure 8.23 Wavelength discrimination function for the horse. Source: Timney and Macuda, personal communication, 2009.

The authors emphasize that their experiments show that horses can discriminate wavelengths in the absence of luminance cues, and that the fineness of discrimination varies across the spectrum. There is a single minimum in the discrimination function toward 480 nanometers, which is consistent with the existence of only two classes of photoreceptors as well as with all the data published elsewhere. These findings imply that horses recognize various wavelengths as perceptibly different and thus that they possess true color vision.

As with discrimination of the shortest wavelengths, the ability to discriminate medium and long wavelengths is relatively poor, particularly beyond 532 nanometers. Thus, horses' subjective color experiences may be similar in these regions, and it may be that green and yellow appear similar to them, whereas green and blue appear quite different. Yet, as Roth and colleagues showed (2007), horses do not appear to establish categories of colors, as humans do; rather, they perceive a continuum of hues.

Timney and Macuda also recall that previous studies generally agree on the ability of horses to make chromatic discriminations at two extremes of the spectrum, that is, to distinguish what we see as blue or red from gray. The discrepancies show up more for medium wavelengths, corresponding to green and yellow for us, which is also consistent with the weak discrimination of wavelengths between 532 and 590 nanometers (see plate 15).

They conclude that, aside from confirming that dichromacy in the horse is based on a combination of short- and medium- to long-wavelength photopigments, the demonstration of the horse's ability to discriminate on the basis of hue suggests that the equine visual system can compare levels of excitation produced by two distinct photoreceptors and that the animal also possesses the post-receptor neural architecture necessary to make these comparisons.

A Provisional Summing Up

First and foremost, both behavioral, and anatomo-physiological data suggest strongly that horses have some color perception, in the sense that they perceive certain hues as qualitatively different, independent of their brightness.

To the frequently asked question, "Do horses see colors?" the "gut" response is yes, insofar as (with the exception of the neutral point, which corresponds to an extremely narrow wavelength band) they perceive different wavelengths corresponding to our visible spectrum as colored, not achromatic.

Nonetheless, because horses are dichromats, their "palette" of colors is naturally comparatively less rich than ours, with an apparent reduction in the array of hues more or less analogous to what we observe in human dichromats. For the latter, as Valerie Bonnardel (2005) puts it: "The study of confusion in dichromats shows that their color vision is reduced to two tonalities: Hot colors are assimilated to yellow, and cold colors to blue" (22). Similarly, it appears that, although horses in fact perceive different spectral wavelengths as colored, except for the neutral point, for the horse the spectrum is reduced to a continuum of variations based on two main hues, with the "gray" of the neutral point constituting an intermediate shade (see plate 16).

In other words, taking into account its dichromacy, the "right" question should rather be: "How do horses discriminate between colors?" Horses' ability to distinguish colors that are close to each other varies across the spectrum. Thus, horses do a good job at discriminating hues close to the neutral point, particularly neighboring hues in the spectral range of what we perceive as blue or blue-green. But horses may, for example, less easily discriminate green from yellow or orange from red.

As noted by Jack Murphy and colleagues (Murphy, Hall, and Arkins 2009), although dichromacy presents drawbacks in evaluating colors, it also has the advantage of minimizing certain forms of variegated color-based "camouflage," as has been shown experimentally (Morgan, Adam, and Mollon 1992). Thus, in certain circumstances, dichromacy can help horses to better discern differences in shape or texture that are less visible to humans, which might explain the ease with which the animals navigate certain cross-country obstacles that we have trouble distinguishing in the landscape.

Yet, the subjective properties (qualia) of equine color perception remain beyond our comprehension. We can state that horses discriminate between two different hues, and thus that, to a certain extent, they perceive different colors. We can evaluate their chromatic discriminative capacities, but we cannot know what color their brain constructs based on a given wavelength.

9

HEARING IN HORSES

Due as much to the diffuse nature of sound wave propagation, which sends them around obstacles, as to the capacity of the auditory apparatus to detect sounds in all directions (both night and day), the auditory perceptual apparatus complements that of visual perception: It is a warning system that allows horses to detect, locate, and identify potential predators. Moreover, as is the case with many species, the horse is not only a "receiver" but also an "emitter" of sounds. In other words, auditory perception also plays a significant role in communication between individuals, and thus in their social life. I will return to these points.

Before embarking on a behavioral exploration of equine auditory perception, and how it relates to visual perception, I will first briefly review the nature of sound information, and how it is represented and measured. I will also touch on the anatomical and physiological characteristics of the horse's hearing system.

Nature, Representation, and Characterization of Acoustic Information

Sound is the product of vibrations; hearing is the "sensory modality that allows us remote access to object vibrations" (Bovet 2003, 129).

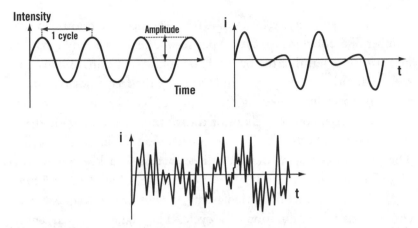

Figure 9.1 Pure sound (top left), musical sound (top right), and noise (bottom).

Vibrations translate into small pressure changes in the environment (generally, for our purposes, air), giving rise to acoustic (sound) waves that travel in all directions. In other words, unlike light, sound requires material "support" to propagate; it cannot do so in a vacuum.

Contrary to waves that move on the surface of water, for example, like those caused by skipping a stone, sound waves are longitudinal: The movements of matter occur (locally, around a point of equilibrium) not perpendicular but parallel to their direction of propagation. The result is a local series of compressions and depressions that cause, among other things, the ear's tympanic membrane to vibrate.

In a given environment, sound waves propagate at constant speed (around 340 meters per second in air, 1,500 meters per second in water, and 1,000 to 5,000 meters per second in solids).

The waves may give rise to pure sounds, musical sounds (or periodic complex sounds), and noise (nonperiodic complex sounds) (see figure 9.1).

Pure Sound

Pure sound corresponds to sound vibrations whose local pressure variation is periodic: It repeats itself at regular time intervals. Plotted as a function of time, its pressure variation is a sinusoidal curve whose amplitude waxes and wanes with the intensity of the sound. Pure sounds are characterized by a single frequency, corresponding to the number of oscillations (or cycles) per second (expressed in Hertz, or Hz), which determine the pitch of the sound and remain constant for the duration of propagation: "Studying the frequency of a pure sound is especially interesting, for frequency is a parameter that does not alter throughout the chain of transformations that leads from the sound vibrations of an object to excitation of the sensory cells. The perceptual system thus has direct access to a key dimension of a vibrating object: the frequency of the vibration" (Bovet 2003, 129). By definition, the period is the inverse of the frequency: It corresponds to the duration of a cycle. The higher the frequency, the more piercing the pitch. The lower the frequency, the lower the pitch. Generally speaking, low-pitch sounds are those with frequencies under 400 Hz; medium sounds have frequencies between 400 and 1,600 Hz; and high-pitch sounds frequencies over 1,600 Hz. The physical "height" of a sound wave (frequency) is actually not strictly proportional to its perceived amplitude (pitch). However we will ignore this distinction, which is immaterial for our purposes. A good illustration of pitch is music, where an octave corresponds to the interval separating two sounds one of whose frequencies is twice that of the other. The pure sounds perceivable by humans are approximately those comprised in a range from 20 to 20,000 Hz, which is about ten octaves. In music, seven of these octaves generally coincide with sounds between 30 and 4,000 Hz; however, an organ with a range of frequencies around 16 Hz to 16 KHz stretches ten octaves. More specifically, the lowest note on a modern 88-key piano (an A) has a frequency of 27.5 Hz, and the highest note (a C), a frequency of 4,186 Hz. The reference note A3 corresponds to a medium sound of 440 Hz. In fact, this A3 frequency is purely by convention: It was established by an international congress in London in 1953. Incidentally, the value given to A3 varied widely in the past, from 506 Hz for the A3 of the organ in Halberstadt Cathedral in 1495, to 1,750 to 390 Hz for the Dallery organ in Valloire Abbey. Before 1930, in the United Kingdom, a low A3 at 435 Hz was even distinguished from a high A3 at 452 Hz.

In contrast, the amplitude of pure sounds, which corresponds to their acoustic (or physical) intensity, is subject to variation due in particular to a natural process of attenuation during propagation in a given environment, or the inverse by virtue of amplifiers.

Thus, for example, the tuning fork normally used by musicians to "give the A" produces a pure sound—a 440 Hz A3 (at a temperature of 20 degrees centigrade)—that diminishes in intensity when moved away from the ear. But it can be amplified by placing the fork handle on a resonance box. In acoustics, the attenuation coefficient is a function of the sound frequency, low frequencies being less susceptible to attenuation than high frequencies.

The acoustic intensity of a sound is expressed by a logarithmic unit, the decibel (dB), whose reference value of 0 dB corresponds by convention to the value of the absolute threshold of human hearing for a pure sound of 2,000 Hz, or an amplitude pressure variation of 10^{12} watt/m². However, insofar as our hearing sensitivity is not maximum at 2,000 Hz, the absolute threshold of human hearing for more propitious frequencies (actually, the maximum sensitivity of the human ear is at about 3,700 Hz) is measured in negative dB. The logarithmic character of this unit of measure is due in part to the fact that the human auditory system is very sensitive and covers a very large range (which, as we will see, is nearly the same for the horse): At the perception threshold, the movements of the eardrum cover a distance less than the diameter of a hydrogen molecule; at the pain threshold (120 dB), the sounds correspond to an acoustic intensity 10^{12} times higher (1 watt/m²). The ratio between the two thresholds is 1,000,000,000,000. As a consequence of the logarithmic character of the unit of measure, sound sources must be amplified ten, a hundred, or a thousand times to obtain a 10, 20, or 30 dB increase in acoustic intensity. In contrast, doubling the sound power requires raising the acoustic intensity by only 3 dB. To paraphrase what I stated earlier with respect to pitch, the relationship between the physical intensity of a sound and its perceived intensity (loudness) may not be perfectly linear, but this distinction is not significant for our purposes.

Below are several specific examples of approximate sound levels (in dB) for a few fairly common situations:

- Ariane rocket taking off at 10 meters: 140 dB
- fighter jet taking off: 130 dB
- F1 racing car: 120 dB
- singer at top volume: 110 dB
- high-cylinder motor: 100 dB
- crying child: 90 dB
- congested street: 80 dB
- robot vacuum cleaner: 70 dB
- normal conversation: 60 dB
- quiet office: 50 dB
- running faucet: 40 dB
- murmur: 30 dB
- whisper: 20 dB
- normal breathing: 10 dB
- ant walking on sand!: 0 dB

The International Organization for Standardization (ISO) established the limit for harmful noise at 90 dB over eight hours, dividing this time by 2 for each 3 dB increase in noise. Beyond 110 dB, hearing damage can occur within minutes.

Music

Musical sounds—periodic complex sounds—are composed of different frequencies that form an overall pattern that repeats regularly. Thus, the sounds themselves are not sinusoidal; rather, as determined by the French mathematician and physician Joseph Fourier at the beginning of the nineteenth century, they can be decomposed into a group of sinusoidal sounds—harmonics—that are whole-number multiples of a base frequency. The lowest, called the fundamental frequency (or simply "fundamental"), is F0, with multiples denoted F1, F2, . . . Fn (see figure 9.2). In fact, it is the fundamental that determines the pitch of a sound; the harmonics contribute to its richness

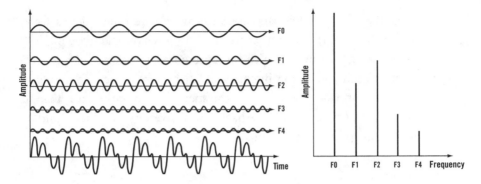

Figure 9.2 Left: the musical sound of the lower line can be decomposed into five pure sounds: the fundamental and four harmonics. Right: harmonic spectrum of the same musical wavelength.

and timbre. The fundamental is not necessarily the greatest intensity. Moreover, some harmonics may have a specific intensity and constitute what are called formants. For example, the sounds produced by vibration of the vocal cords, "voiced" sounds, have formants due to the reinforcement of specific harmonics by the resonators of the vocal apparatus (the supraglottal cavities). What is called the singing formant—particularly in the case of lyric artists, owing to the way they place their voices—corresponds to reinforcement of the harmonics between 2,000 and 4,000 Hz and enables their voices to carry far; indeed, human sensitivity is greatest around 3,000 Hz. The blast of a clarion or a trumpet, or the crowing of a rooster, all derive from the character of their formants.

In terms of survival value, knowledge of pitch is significant in the sense that it contributes much to the ability to identify sounds, and thus to the nature of their source. It may well be the reason that the fundamental enjoys a perceptually privileged status; it can even be suppressed without affecting the perceived pitch of the sound, a remarkable phenomenon known as the absent fundamental. And in fact, when the fundamental frequency of a musical sound is suppressed through filtering, the listener hears no change in the initial pitch of the sound. The same phenomenon is observed in playing a C on the piano, whose fundamental is 65.41 Hz: Played *pianis-*

simo, the note is below the human hearing threshold; played forte, and thus above the auditory threshold, it preserves the same pitch.

A musical sound can also be represented by its harmonic spectrum, which shows the intensity of all the frequency components of the sound (for a pure sound, the spectrum contains only a single component frequency). Some harmonics have zero amplitude and do not show up in the spectrum.

Noise

Noises are nonperiodic complex sounds. "The principle of Fourier decomposition still applies, but the number of constituents increases considerably, and their frequency is no longer a whole number multiple of the fundamental frequency. They are no longer harmonics, but are called 'partials.' . . . Often, the constituents are so numerous and so close together that one cannot be distinguished from the other. They form a continuous spectrum, and their graph is a spectral envelope" (Bagot 2002, 68) (see figure 9.3).

A special case of noise is white noise, where all the frequency constituents of the spectral envelope have the same amplitude. This can be obtained by summing many pure sounds having the same amplitude and frequencies that are very close to one another across a range of the sound spectrum (see figure 9.4).

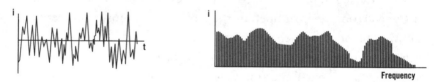

Figure 9.3 Noise (left) and spectral envelope of a noise (right). Source: Bagot 2002, 64.

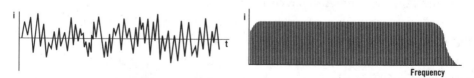

Figure 9.4 White noise (left) and spectral envelope of white noise (right). Source: Bagot 2002, 64.

Monaural and Binaural Information

Thus far, I have ignored the fact that sound perception relies on not one receptor, but (at least for mammals) two located on opposite sides of the head. Consequently, except in the case of pathology, the perception of sound waves is not monaural, but binaural (or two-eared). It was just such an observation that, at the beginning of the last century, inspired Lord Rayleigh's (Strutt 1907) duplex theory of binaural listening, according to which the head creates an "acoustic shadow" for the ear farthest away from the sound source.

In fact, except for sound waves emanating from a sound source located precisely in front of or behind the sagittal plane of the head, neither ear receives exactly the same information at the same time. Rather, each ear receives information at different times and of varying intensity.

Although the sound frequency remains constant throughout its propagation, the same is not true of intensity, which varies not only (and especially) as a function of distance from the sound source, but also the obstacles in its way. In fact, these may or may not inhibit passage of sound depending on its wavelength and the size of the obstacles: "Propagation of a vibration is perturbed only if its wavelength is shorter than the dimensions of the obstacle" (Bagot 2002, 102). Thus, as the wavelength is the inverse of the frequency, the head acts as a low-pass filter and allows passage of low frequencies around it because of their long wavelengths while blocking high frequencies (acoustic shadow) for the ear farthest from the sound source, where the complex sounds arrive with weakened intensity. These intensity differences between the ears only occur with high frequencies, at rates around 2,000 Hz and over in humans.

Moreover, the further an ear is from the sound source, the longer the sound takes to reach it. For a sound source located at an angle of 45 and 90 degrees with respect to the head axis, this time difference amounts to several tens of milliseconds. As a result of the time delay, the sound waves reaching one ear are no longer in phase with those reaching the other ear. This phase difference is all the greater when the sound source is situated off to the side. However, the difference is only processed by the nervous system for sounds lower than a certain frequency, around 1,500 Hz in humans.

These intensity and time difference cues enable sounds to be localized. In addition to these binaural cues, there are also monaural cues relating to the way in which the external ear modulates the intensity of the various frequency constituents of a sound based on its angle of incidence (Bovet 2003). Being deaf in one ear does not completely inhibit localization. In particular, moving the head makes it possible to resolve some ambiguities due to the fact that, for example, a sound signal located to the front or back may translate into similar differences between the ears. Remember, too, that horses can move their ears independently in practically any direction.

The Equine Auditory System: Anatomy and Physiology

The Peripheral Auditory System

As in other mammals, the peripheral auditory system of the horse includes the external ear, the middle ear, and the inner ear. As pointed out by Jonathan Levine and colleagues (2008): "Limited information is available on the physiology of the vestibular and auditory systems of the horse" (157). The overall anatomical and physiological organization is, however, common to mammals as a group (see figure 9.5).

External Ear

The external ear, which is the receptor organ for sound, includes the pinna and auditory canal, closed at its extremity by the tympanic membrane, also simply called the eardrum.

Composed of cartilage covered by skin, the pinna of the horse conducts sound toward the auditory canal especially well owing to its extreme mobility: The pinna can move in nearly any direction, its movements commanded by a group of sixteen auricular muscles (in comparison, humans have only three, which are atrophied and basically nonfunctional). The equine auricular muscles are innervated by

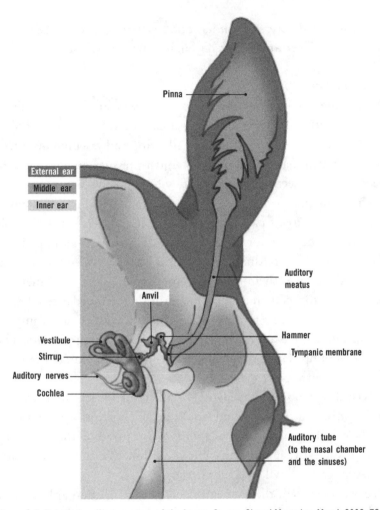

Figure 9.5 External auditory system of the horse. Source: Cheval Magazine, March 2008, 78. Illustration: Sylvianne Gangloff.

the auriculopalpebral branch of the facial nerve (Mair et al. 1998). The auditory canal is relatively long and fairly vertical in the horse, which makes examination difficult.

Given its shape, and in particular, its convolutions, the role of the mammalian pinna in localizing sounds is generally important for high frequencies. These are amplified in the auditory canal by resonance phenomena before the sounds reach the eardrum.

The skin that covers the pinna contains many glands that secrete a waxy material, cerumen, as well as hairs that catch dust and insects.

Middle Ear

The middle ear includes the tympanic cavity and the eustachian tube. The latter connects to the nasal cavities through the epipharynx. Normally closed, it opens in swallowing and yawning by letting air pass, which makes it possible to adjust pressure on either side of the eardrum as needed, allowing it to vibrate freely.

The tympanic cavity is an air-filled cavity within the temporal bone. It is the largest part of the middle ear and plays an active role in hearing. The basic function of the middle ear is as an impedance adaptor. It enables transmission of sound waves between the external ear, which receives them from the environment, and the inner ear, where they propagate in a liquid medium (here, impedance refers to resistance of a medium to the passage of sound). The acoustic impedance of a medium is a function of its density and compressibility. The greater the impedance difference between two mediums, the more the wave is reflected.

Accordingly, as air impedance is substantially less than that of liquid, only some of the sound would be able to cross it without the middle ear. This impedance adaptation is realized by the three tiny bones (ossicles) that transmit vibrations from the tympanic cavity: the hammer, the anvil, and the stirrup.

As we have seen, the eardrum constitutes the external boundary of the middle ear. Its internal boundary is a bony wall punctuated by two orifices: the vestibular window (or oval window) and the cochlear window (or round window). The eardrum is a thin, flexible membrane in the shape of a flat cone that protrudes into the middle ear through attachment to the "handle" of the hammer. Sounds that arrive at this membrane through the auditory canal cause the drum to vibrate, which in turn sets the ossicles to vibrating. The sounds undergo strong mechanical amplification as a result of the way ossicles interact as well as the size difference between the drum and the faceplate of the stirrup, which concentrates the vibrations.

Thus, the tympano-ossicular apparatus recuperates the roughly 30 dB of acoustic vibration lost to the impedance difference between the air and the labyrinthine liquid of the internal ear (Portmann 1992). However, an attenuating stapedius reflex (from *stapedius*, stirrup) that notably brings into the play the stirrup muscle protects the ear from overly loud sounds.

Inner Ear

The inner ear comprises the organs of balance, the vestibular apparatus (the vestibule and the semicircular canals), and the auditory organ that transforms sound vibrations into nervous impulses—the cochlea, or snail. The cochlea is a bony cavity within the temporal bone. It fulfills three functions: spectral analysis, amplification of sound stimulation, and mechano-electrical transduction. It consists of three tubes filled with liquid (perilymph and endolymph) that, in the horse, coil 2.5 times (Waring 2003) around a hollow axis that serves as a conduit for the fibers of the auditory nerve. The three tubes are the vestibular ramp, at whose base the oval window opens to communicate with the middle ear, and where the stirrup is located; the membranous cochlea, or cochlear canal, situated in the median portion, where the organ of Corti, the sensory element of hearing, is located; and finally the tympanic ramp, whose base opens on the round window that communicates with the middle ear. The vestibular and tympanic ramps are filled with peri-lymph and communicate at their tip (or apex) by means of a narrow channel called the helicotrema. The cochlear canal, which is closed at its apex and connects at its base with the membranous structures of the vestibular apparatus, is filled with endolymph. It is separated from the vestibular ramp by Reissner's membrane and from the tympanic ramp by a thick layer known as the basilar membrane. The organ of Corti rests on and is supported by the basilar membrane, which extends along the length of the cochlear canal. The organ of Corti contains structural support cells (the pillars of Corti and layers of Deiters' cells, which sit on the basilar membrane) and ciliary cells (sensory cells), and is covered by the stiff, rigid tectorial membrane into which the ciliary cells project by means of stereocilia at their tips (see figure 9.6).

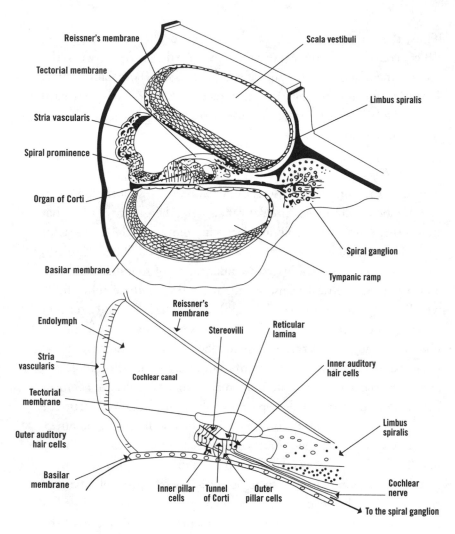

Figure 9.6 Transverse section of the cochlea and organ of Corti. Source: MacLeod and Sauvageot 1986.

The vibration of the stirrup, which translates into a piston-like movement from back to front within the oval window, transmits a periodic pressure variation synchronous with the sound waves to the vestibular ramp. Because the liquids contained within the cochlea are incompressible, the waves follow two pathways: On the one hand, they travel the tympanic ramp by means of the helicotrema and

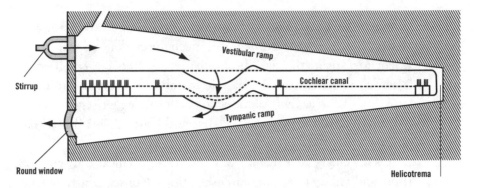

Figure 9.7 Schematic of the working principle of the cochlea. The cochlea, normally snail shaped, is here shown unfurled. Source: Bagot 2002, 88.

provoke compensatory movements of the flexible membrane of the round window, which relieves pressure; on the other hand, they take a shorter path to join the tympanic ramp by crossing the cochlear canal. In so doing, they lead to deformation of the vestibular and basilar membranes as a result of the pressure difference between the vestibular and tympanic ramps, which in turn affects the ciliary cells (see figure 9.7).

The basilar membrane is thick, narrow, and very stiff at its base, close to the stirrup. It gradually thins and widens to become very floppy at its apex. Consequently, for a given continuous pressure, the deformation of this membrane is much greater at the apex than at the base. As Georg von Békésy[1] showed (1947, 1956), at the level of the basilar membrane, the pressure fluctuations of the vestibular ramp translate into a transverse propagating wave, moving from the base to the apex with increasing amplitude toward a maximum whose position is a function of its frequency, then stopping abruptly. Consequently, the parts of the basilar membrane situated at the base of the cochlear canal, near the stirrup, become displaced by vibrating at high frequencies, and those located near the apex by low frequencies. In between, the location of the vibration from the base varies

---

1. Georg von Békésy received the Nobel Prize in physiology or medicine in 1961 for his work on the cochlea.

as an inverse function of the frequency of the sound wave. However, as Ichiji Tasaki (1954) showed, this mechanism alone appears insufficient to explain the fine discrimination of frequencies observed. "Passive" selectivity of the basilar membrane is accompanied by strong "active" selectivity of the ciliary cells (more precisely, the external ciliary cells) for frequency (Dallos and Harris 1978; Brownell et al. 1985).

It is the ciliary cells that are responsible for cochlear transduction, rising and falling in response to oscillations. Consequently, depending on their contact with the tectorial membrane the stereocilia pivot to the right or the left. This leads to the periodic release of neurotransmitters toward the dendrites of bipolar neurons with which their cells synapse, which happen to be spiral ganglion neurons in the organ of Corti. The axons of these neurons then emit an action potential that propagates within the cochlear nerve, which joins the vestibular nerve in the inner auditory canal to form the vestibulocochlear or auditory nerve (that is, pair VIII of the cranial nerves, as we saw in chapter 4).

Localization of the stimulated ciliary cells on the basilar membrane varies according to the frequency of the sound wave. These cells have strong frequency selectivity; the different fibers of the cochlear nerve are only activated by specific frequencies. For complex sounds, the group of fibers in tune with the different frequencies comprised by the sound are activated simultaneously. This is called spatial coding of frequency or "tonotopy." Moreover, as in the case of retinotopy for vision, tonotopy is conserved up to the primary auditory cortex of the brain.

However, this sound wave frequency is also encoded in another way. I have shown that the frequency of a sound wave can also be expressed by a temporal parameter, that is, its period, which is measured by the value of the inverse of the frequency, and which corresponds to the duration of the periodic acoustic pressure cycle. Now, the emission of action potentials is synchronized to this periodic regularity, but only up to a limit of around 5,000 Hz. It is therefore a temporal encoding of frequency, or frequency encoding:

"Spatial encoding is indispensable for capturing the high-frequency sound vibrations that the nervous system cannot follow owing to its slowness, whereas frequency encoding makes it possible to refine the perception of low frequencies by realizing a veritable chronometry of the sound signal. The existence of this double system shows very well the special role that identification of the sound frequency plays for the organism" (Bovet 2003, 141). Finally, the intensity of the sound wave is also encoded at the level of the inner ear. The basis of this encoding is a recruiting process among neighboring cochlear fibers in tune with the same frequency, but having staged release thresholds, the average rhythm of the action potentials emitted by each of these fibers also varying as a function of intensity, within the limits of small changes in intensity. "In short, the cochlear nerve transmits all encoded auditory information, on the one hand, in terms of sound intensity, by average rate of AP (action potential) emission and number of active fibers, and on the other hand, in terms of sound frequency, by temporal distribution of APs and localization of specific active fibers" (Bagot 2002, 97–98).

The Central Auditory System

A nonprimary, nonspecific auditory pathway, where sensory modalities converge, selects information for priority processing. This pathway passes through the reticular formation; it is linked to the centers that direct motivation and sleep-wake cycle, as well as the vegetative and emotional systems.

In contrast, a primary, specific auditory pathway, which I will touch on briefly, is entirely dedicated to decoding and to interpreting auditory information. The trajectory described includes just one side—left or right—but in fact, it is a paired pathway.

This primary auditory pathway, composed of large myelinated fibers, is fast and short; proceeding from the first neuron of the cochlear nerve, whose cellular body is situated in the spiral ganglion of the inner ear (where it is joined by the vestibular nerve to form the auditory nerve), it involves very few relays and sends information directly between the inner ear and the primary auditory cortex.

The first relay occurs in the brainstem, at the level of the rachidian bulb (myelencephalon), with the cochlear nucleus, which constitutes an integrating center that ensures recoding of sound (duration, intensity, frequency) and amplification contrasts detected at the level of the cochlea.

From this nucleus, there is partial crossing of the fibers (decussation), but with contralateral dominance (Barone and Bortolami 2004) such that each of the following structures still receives information from both ears.

Situated at the level of the pons (metencephalon), the superior olivary complex represents a second relay. It consists of nine nuclei, including three major ones: the lateral superior olivary nucleus, the medial superior olivary nucleus, and the medial nucleus of the trapezoid body. Most of the fibers coming from the cochlear nucleus relay there. Based on the information it receives from both ears, the superior olivary complex plays a role in the spatial localization of sound. However, as Rickye S. and Henry E. Heffner have shown (Heffner, R. S. 1986, 1997), the lateral superior olive of the horse is relatively small and limited in that it does not have the laminar arrangement of the bipolar cells characteristic of this structure in most mammals that is conducive to measuring the intensity difference of sound perceived by the two ears.

The next relay, also situated at the level of the pons, consists of the nucleus of the lateral lemniscus, which contains a small number of cochlear fibers that do not pass through the superior olivary complex. It responds especially to the commencement of sound and to its duration. The fibers leaving it converge on the inferior colliculus.

A fourth relay is found at the level of the mesencephalon, at the inferior colliculus, which receives afferences from nearly all the nuclei of the levels already mentioned. It is also involved in refining spatial localization (as mentioned above, in the horse this plays no role in intensity differences between the ears, but mainly is involved in phase differences). It may also respond to specific sounds (e.g., predators). Finally, as the inferior colliculus is linked to oculomotor nerves III and IV, it may be involved in orienting the head and eyes in the direction of sound (orienting reflex).

"The medial geniculate body is the last relay of the auditory pathway and represents the highest level of integration" (Barone and Bortolami 2004, 363). Also called the medial geniculate nucleus, it is situated in the thalamus (diencephalon) and filters all the information ascending toward the auditory cortex. It is the seat of the phenomenon of multisensory convergence (which enables connections between heterogeneous information) that is the basis for mechanisms of integration that underpin perceptual behaviors. In particular, these mechanisms of integration prepare for eventual motor response (vocal, for example).

At the level of the auditory cortex, with its primary, secondary, and associative areas, within the temporal regions of the telencephalon, the message is recognized, processed as conscious perception, memorized, and can then be integrated in a voluntary response.

Having concluded this brief outline of the primary auditory pathway, I note also that the connections are not organized in a rigidly hierarchical way, since levels may be skipped. Only the cochlear nucleus and the thalamus constitute a fully obligatory pathway, along with, though to a lesser degree, the inferior colliculus.

In addition to these ascending connections, from the ear to the brain, many descending links among different levels, from the auditory cortex to the cochlea, ensure feedback control at the lower levels and may contribute to selective processing of sounds, improving discrimination, and protecting against overstimulation. The stapedius (attenuation) reflex, mentioned earlier, is one example. Originating at the cochlear nucleus, its purpose is to protect the inner ear against overly loud sounds.

Behavioral Exploration of Equine Auditory Perception

Several decades ago, René-Guy Busnel (1963) suggested that the horse was endowed with very fine hearing since it was apparently capable of hearing sounds at a distance of up to 4.4 kilometers. But information regarding its auditory capacities remained incomplete, and essentially anecdotal, nearly up to the 1980s (Timney and Macuda 2001). In fact, it was only in 1978 that Frank Ödberg published his pioneering study on high-frequency hearing. And it was not until 1983 that Henry Heffner and Rickye Heffner—specialists in the comparative study of animal hearing—published systematic investigations of the horse's auditory register and its ability to localize sounds.

Referring to Bruce Masterton and Irving Diamond (1973), Heffner and Heffner (1992a, 1998) recapped the three main sources of selection pressure on the capacity of animals to perceive sound: detecting a sound, which permits the animal to determine the presence of the sound source, mostly other animals; localizing the source to decide whether to approach or avoid it; and finally, identifying the biological importance or relevance of the source to respond appropriately. However, the authors emphasized, little was known about this last aspect.

Detecting Sounds: Equine Hearing Acuity

The ability to detect sounds comes back to the issue of measuring hearing acuity. The measure of hearing acuity is based on pure sound. As we have seen, a pure sound is characterized by its frequency and intensity. These two values contribute to determining hearing acuity: In fact, the sensitivity of the auditory apparatus differs according to frequencies. For this reason, hearing acuity is not determined by a single value; it is represented by a graph, the audiogram, which indicates the perceptual thresholds (in dB) for the sound frequencies over the range of audible frequencies for the animal species considered. In humans, these limits are located at 20 Hz, just above infrasonic frequencies, and 20 kHz, just below ultrasonic frequencies. In an extreme case, two species might each be virtually deaf to the sound world of the other: This is true, for example, of the Asian elephant, which cannot hear sounds above 11.8 kHz, and the small brown bat (one of the species of bats living in North America), which can only perceive sounds above 10.3 kHz (Heffner, H. E., and R. S. Heffner

2003). Comparing the range of audible frequencies for sounds requiring good hearing acuity, that is, sounds with an intensity of 10 dB, also reveals significant variability. For example, under these conditions, the frequency band extends 6.6 octaves for the domestic cat, whereas it is only 0.4 octave for the mouse or the hamster (Heffner, H. E., and R. S. Heffner 2007).

Experimental Procedures

From the experimental point of view, measuring hearing acuity assumes that, for a given sound frequency, one is in a position to determine the threshold of intensity at which an animal detects the sound. Two types of experimental procedures make that possible.

First, one can assume that the spontaneous reaction of an animal hearing a sound in a silent environment will be either fright (translating, for example, into escape or avoidance) or, in the horse, a Preyer's reflex, consisting of pointing its ears toward the source of the sound. As I will show, this reflex has been used (Ödberg 1978), but it nonetheless poses a challenge in terms of reliability.

A second option is a more controlled procedure based on getting the animal to reproduce a response previously learned when detecting a sound. This makes it possible not only to reward the animal when it gives a correct response, but also to better control responses made in the absence of the sound (Heffner, H. E., and R. S. Heffner 2003). Such a procedure assumes a preliminary phase of discriminative learning before proceeding to the main testing phase. Insofar as the choice the animal must make is all or nothing (sound perceived or not), the "go/no go" procedure is appropriate. In this approach, the experimental subject must accomplish a motor act (like pressing a panel with its nose) when the stimulus is "positive" (sound detected) and only in this case. If the response is correct, it is followed by positive reinforcement (drink for a thirsty animal). In the case of a "wrong" response (reaction when there has been no sound), negative punishment (temporarily halting testing) or even positive punishment (light electric shock, for example) may also be used. The duration of a daily test session depends on the length of time until a horse is thirsty, say, around an hour. To determine the auditory threshold for a given frequency, its intensity is gradually diminished until the animal can only detect the sound 50 percent of the time (Heffner, H. E., and R. S. Heffner 1992a). In normal experimental conditions, the sound is emitted for several seconds. Under these conditions, the perceived sound intensity (loudness) is independent of the duration of emission; for a sound of constant intensity, the intensity increases as a function of duration until 180 milliseconds, after which it remains relatively constant (save for sounds lower than 30 dB, for which it only decreases for longer durations).

The Pioneering Work of Frank Ödberg (1978)

As I mentioned above, Frank Ödberg devised the first experimental approach to auditory perception in the horse (Ödberg 1978) to test the animal's capacity to detect high-frequency sounds. His interest in these sounds was sparked in particular by the observation of various authors (Waring 1971; Kiley 1972; Ödberg 1974) that horse vocalizations most often occur between 1 and 5 kHz and that, generally speaking, taking into account the total range of vocalized frequencies, including harmonics, they do not exceed 12 kHz. Ödberg noted, however, that in certain individuals, vocalizations were recorded at ultrasound frequencies, from 16 to 20 kHz, and might even have been higher, since the recording instrument used was itself limited to 20 kHz.

Ödberg's experiment involved 10 horses, 5 mares and 5 geldings, aged five to eighteen years, divided into adults for the 7 horses aged five to nine years and old for the 3 horses aged fifteen to eighteen years.

Using an ultrasound whistle (the Galton whistle), Ödberg made a preliminary test of a few horses to ensure that sounds produced by the whistle elicited the Preyer's reflex in the animals.

He then tested ten horses on the twelve frequencies, broadcast by a frequency generator connected to a loud speaker, and increasing every 1,000 Hz from 14 to 25 kHz. The apparatus used could not broadcast frequency sounds higher than 25 kHz. Nor was it possible to conserve a constant intensity according to the frequencies; thus, the intensity decreased from 65 dB for 14 kHz to 36 dB for 25 kHz. The horses were all tested in the same "loose" box located in an isolated building, and were observed through a hole in a cloth screen that hid the observer and the apparatus except for the loudspeaker. The horse's head was 1 to 3 meters from the speaker, depending on its position in the box.

For each horse, each of the frequencies was broadcast ten times, in a random sequence of 120 sounds, each being separated from the others by at least one minute. The broadcast was interrupted if the horse's attention strayed for any reason, such that the number of test sessions, total test time, and number of test days varied from one horse to the other.

Either of the two criteria, which consisted of spontaneous reactions by the horse—the Preyer's reflex or a fear (avoidance) reaction—was taken to indicate that the animal had heard the sound broadcast. Fright was recorded only twelve times out of all the reactions observed (see table 9.1).

As it turned out, all the horses perceived the sounds of each of the frequencies emitted. The animals generally had a harder time as the frequency increased, though given the concomitant diminution of intensity, it is difficult to draw firm conclusions about this last finding.

Although no significant difference in "performance" was noted between mares and geldings, these results nonetheless reveal a difference between "adult" and "old" horses. The difference appears above 22 kHz for mares ($\chi^2 = 5.64$; $p < 0.02$) and 23 kHz for geldings ($\chi^2 = 2.91$; $p < 0.05$). This experiment does appear to show that the general ability of horses to perceive the highest-frequency sounds decreases with age (which is also the case for humans). However, the authors also hypothesized that this phenomenon might be less marked in males, due to the role played by the stallion in protecting his family from potential danger.

Table 9.1	Average number of reactions to the ten broadcastings.						
Transmitting frequency (KHz)	Adult mares* (N = 4)	Adult geldings** (N = 3)	All adults (N = 7)	All old horses*** (N = 3)	All mares (N = 5)	All geldings (N = 3)	All horses (N = 10)
14	10	10	10	9.7	10	9.3	9.9
15	10	10	10	10	10	10	10
16	9.8	9.7	9.7	9.3	9.8	9.4	9.6
17	9.5	9.7	9.6	7.3	8.4	9.4	8.9
18	9.5	9.3	9.4	9	9.2	9.4	9.3
19	9.8	8.3	9.1	7.3	9.4	7.8	8.6
20	7.8	9	8.4	6	7	8.2	7.6
21	7	8.3	7.7	8	7	8.4	7.7
22	7.8	7.7	7.7	7	7	8	7.5
23	5.5	8	6.8	4.7	5.6	6.4	6
24	4	7	5.5	1.7	3.4	5	4.2
25	6.3	8.3	7.3	2.3	5	6.4	5.7
	t = 24.5 NS			t = 0.3 p < 0.01		t = 19 NS	

t: Wilcoxon test. NS: Not significant. *Aged 5, 6, 6, 7. **Aged 7, 9, 9. ***Aged 15, 18 (geldings), 18 (mare). Source: Ödberg 1978, 83.

Henry and Rickye Heffner's Audiogram (1983)

The exploration of the horse's field of hearing was first carried out systematically by Henry and Rickye Heffner (Heffner, H. E., and R. S. Heffner 1983b; Heffner, R. S., and H. E. Heffner 1983c).

They used three horses, aged eighteen to twenty months, that they initially trained by operant conditioning to place their nose on an observing plate. On hearing a tone emitted by a loudspeaker, the animal was to remove its nose within 3 seconds and place it on a reporting plate, below the observing plate. The observing plate served to position the head of the horse firmly in front of the loudspeaker, and thus to minimize variations in the listening conditions of both ears (see figure 9.8). The two supports were metal, and each was connected to a device enabling detection whether or not the horses were in contact with it. The device controlled the loudspeaker, which played tones from 16 Hz to 50 kHz, practically without distortion at

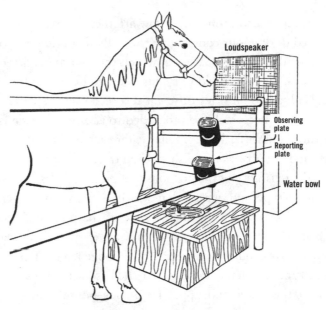

Figure 9.8 Apparatus for the auditory sensitivity test. Source: Heffner, H. E., and R. S. Heffner 1983b, 28.

an intensity of at least 70 dB. The tones, which lasted 3 seconds, were played at intervals that varied randomly from 3 to 27 seconds. The same interval could not repeat more than three times during a single run of twenty-seven trials.

The positive reinforcement, when the horses correctly detected a tone, consisted in the automatic delivering of 250 milliliters of water into a bowl located 10 centimeters below the reporting plate. Water was not given to the horses outside of the periods of experimentation. If the animals gave an "incorrect response" by pressing on the reporting plate when no tone was present, the experiment was halted for 15 seconds (mild negative punishment).

Once the horse had learned to do this procedure reliably, the authors proceeded to the phase of the test aimed at determining the auditory thresholds for all the frequencies included in the animal's hearing range in octave steps.

Accordingly, for a tone of a given frequency, the researchers gradually reduced the intensity following a correct response, and increased it after a nonresponse to a tone, until they attained an intensity that was not detected during four consecutive trials. By definition, the threshold corresponded to the intensity at which the sound was detected in half the cases. The threshold was considered to be reached when the thresholds obtained on three different test sessions were within 3 dB of each other. If not, the horse proceeded to additional sessions until a stable threshold was achieved. Finally, for each of the horses, after a complete audiogram was established, each frequency was tested again to assure the reliability of the previously obtained results.

The horse-type audiogram that resulted from this experiment was obtained by calculating the average of the measures recorded for each of the three horses for each frequency. Taken together, the results turned out to be generally homogeneous. They have a U-shaped function that is fairly flat at the base with thresholds that are high at the lower and upper frequency boundaries compared with intermediate frequencies where thresholds are lowest, as is true for most mammals, and more generally for most vertebrates (Moss and Carr 2002).

The authors draw several conclusions from the audiogram. Beginning with the lower frequencies, the horses' hearing sensitivity increases gradually as frequency is increased to about 500 Hz, and then levels off up to 16 kHz (with a slight dip at 4 kHz). Above 16 kHz, sensitivity decreases rapidly to the upper limit of audibility. At an intensity of 60 dB (corresponding to the sound intensity of normal conversation), the horses' lower and upper limits of hearing are found at 55 Hz and 33.5 kHz, respectively. Thus, they are respectively worse and better than the limits of human hearing, for which the corresponding values are 29 Hz and 19 kHz (that is, a gap of around an octave). Consequently, horses are less sensitive than humans to lower frequencies, but their hearing acuity enables them to perceive ultrasound. In this, horses are not unique. Broadly speaking, mammals that are endowed with good hearing at high frequencies have limited hearing at low frequencies (Heffner R. S., and H. E. Heffner 1983c; Heffner, H. E., and R. S. Heffner 2003; Heffner, H. E., and Masterton 1980). Moreover, at mid-range frequencies of 500 Hz to 8 kHz, equine hearing sensitivity

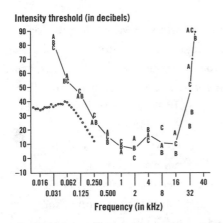
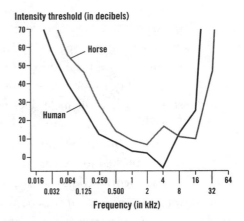

Figure 9.9 Left: audiograms for horses A, B, and C. The points correspond to the background noise. Right: comparative audiograms for humans and horses. Sources: (left) Heffner, R. S., and H. E. Heffner 1983c, 302; (right) Heffner, H. E., and R. S. Heffner 1983b, 30.

(7 dB at 2 kHz) is markedly less acute than that of humans (–4 dB at 4 kHz). Above 8 dB, however, the hearing sensitivity of horses is clearly better than ours (see figure 9.9).

In sum, the hearing capacities of horses are roughly comparable to those of humans.

Horses' Ability to Localize Sounds

Sound localization, that is, the ability to determine their source, is critical to survival. In predators, it contributes not only to locating prey but also to figuring out how to advance without being seen. In prey, localization provides information about the direction of approach of predators, even before they can be seen, and thus also indicates possible avenues of escape. In this respect, hearing and sight appear to play complementary roles. I will return to the relationship between these two perceptual modalities, including their anatomical and physiological aspects.

Localizing a sound source requires determining its position based on three parameters: whether approximately right or left, and front or back on the horizontal plane, called the azimuth; up or down in the vertical plane, which we know as altitude; and, finally, near or far, termed distance. Most of the experimental work done to date focuses on azimuth. As with hearing acuity, substantial variability has been observed between species, around 1 degree, for example, in elephants (similar to humans), 33 degrees in the mouse, and a near absence of an ability to localize sounds in certain subterranean mammalian species, such as the mole (Heffner, R. S. 1997; Heffner, R. S., and H. E. Heffner 1992b; Heffner, H. E., and R. S. Heffner 2003).

For the horse, too, studies have focused on azimuth, and the only ones published to date are those of Heffner and Heffner.

As I mentioned previously, binaural cues relating to differences of intensity and time are essentially what permits localization of sounds. Intensity differences are associated with higher frequencies, which are the only ones blocked by the acoustic shadow produced by the head. Time differences apply to lower frequencies, as the nervous system cannot interpret the phase difference of sound waves reaching both ears above a certain frequency.

Experimental Procedures and Main Results
Although the ability of an animal to localize a sound could possibly be assessed from its natural tendency to direct its head to the source of the sound, that approach is rather crude.

Heffner, H. E., and Heffner, R. S. (1984) employed two more reliable and precise methods for determining the sensitivity of localization: one, a two-choice test, and the other, conditioned avoidance, which they used in parallel for confirmation. Moreover, applied to different species, these methods appear to have produced consistent results (Heffner, R. S. 1997). For the two-choice test, following an initial learning phase, the horse was led to touch its head to a metal plate ("observing" plate) situated in front of it, centered on the horse's midline with respect to a row of symmetrically positioned loudspeakers for broadcasting sounds to the right or left of the animal. The loudspeakers could be moved laterally to vary their angular position vis-à-vis the horse's head. The angular separation between the loudspeakers was gradually reduced until the animal began making incorrect choices. Two other metal ("response") plates were situated to the left and right of the observation plate. The horse was trained

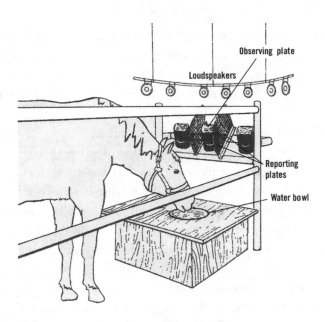

Figure 9.10 Procedure for the two-choice test for localizing sounds. Source: Heffner, H. E., and R. S. Heffner 1984, 543.

to touch them when a sound was emitted to its left or right. Each of the plates was equipped with a sensing switch that detected when the animal touched it. When the horse, which had been deprived of water, touched the "correct" response plate (on the same side as the sound), a click alerted the animal that water (35 milliliters, dispensed by a valve under electrical control) was being delivered to a bowl located in front of it under the observing and response plates. If, however, the horse touched the "incorrect" response plate, it had to wait 3 to 15 seconds, during which the light in the experimental room was dimmed, before the next trial. In a two-choice test, random responses are correct responses 50 percent of the time. The detection threshold is generally 75 percent correct responses, or midway between random responses and 100 percent correct responses. One can also take into account the rate of correct responses statistically superior to chance, based on a given probability. This the authors did in choosing a significance level of $p < 0.01$, which is equivalent to a rate of correct responses on the order of 63 percent (see figure 9.10).

The two-choice test was performed with three young horses (two geldings and one mare, aged twenty-one to twenty-three months). The tests generally lasted around an hour, during which two hundred to six hundred trials took place, and 6 to 20 liters of water were consumed.

In the conditioned avoidance procedure, the horse was also deprived of water beforehand. Here, too, loudspeakers that could be moved sideways were placed to the left and right of the animal. Following an initial learning phase, the horse kept its head centered while drinking from a bowl positioned at its midline and supplied by a thin but constant stream of water (140 mL/min). A touch-sensing switch, connected at one end to the bowl and at the other to the horse's flank, detected when the animal made contact with the bowl. When a sound was emitted, it was followed 4 seconds later by a slight electric charge in the bowl if and only if it came from the loudspeaker on the left. Nothing happened if the sound came from the loudspeaker on the right. Thus, the horse was conditioned to stop drinking (to avoid the electric charge) only when it perceived the sound as coming from the left. In this situation, random correct responses did not account for 50 percent of correct responses, but rather nearly zero percent. Accordingly, the detection threshold was generally set at a score of 50 percent correct responses (and not 75 percent), midway between the random responses and a 100 percent rate of correct responses. As before, the rate of statistically significant correct responses can simultaneously be taken into account, based on a given probability. Choosing a significance level of $p < 0.01$ resulted in a rate of correct choices between the two stimuli of 25 percent (see figure 9.11).

The conditioned avoidance experimental sessions were carried out with two other young horses (a male and a female of three years). The sessions generally lasted sixty to ninety minutes, during which 275 to 375 trials took place, and 6 to 9 liters of water were consumed. Both experiments involved periodic complex sounds: broadband white noise lasting 100 milliseconds and brief, 1-millisecond clicks. Both sounds had an intensity of 40 dB.

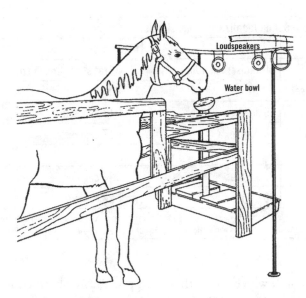

Figure 9.11 Avoidance procedure for localizing sounds. Source: Heffner, H. E., and R. S. Heffner 1984, 546.

The results of the experiments are presented in table 9.2. In the table, the thresholds shown correspond to the smallest angular difference between a sound source perceptible to the horse and the sagittal plane of its head taking into account the detection thresholds indicated above. For the two-choice test, these thresholds are those of the best-performing horse (i.e., having the smallest angular difference) in response to the clicks, and the average of the two best-performing horses (which were nearly at the same level) in response to broadband noise. For the conditioned avoidance experiment, detection thresholds are the average thresholds obtained for each of the two horses, as the results showed close agreement.

Although the results of the two experiments do reveal some interindividual variability, they are still generally consistent, indicating an angular acuity for azimuthal sound localization in the horse averaging around 22 degrees for broadband noise and 30 degrees for clicks, based on commonly accepted detection thresholds.

In particular, better localization acuity was observed to be obtained with broadband noise lasting 120 milliseconds than with a click of

Table 9.2	Sound localization thresholds.	
	Threshold	
Task	50% discrimination[a]	Statistical threshold[b]
White noise		
Two choice[c]	19 degrees	13 degrees
Avoidance	25 degrees	17 degrees
Click		
Two choice[d]	31 degrees	20 degrees
Avoidance	30 degrees	22 degrees
[a]75 percent correct for the two-choice task, and 0.50 level of performance for the avoidance task. [b]0.01 chance discrimination. [c]Average of horses A and B. [d]Horse A only. Source: Heffner, H. E., and R. S. Heffner 1984, 549.		

1 millisecond, which the authors had previously observed in a study on the Asian elephant (Heffner, H. E., and R. S. Heffner 1982), and which may be due to the longer duration and broader frequency spectrum of the white noise.

The authors note in addition that localization in the horse is markedly poorer than that of other mammals, whereas its good hearing at high frequencies, and the size of its head, in particular, provide it with intensity difference cues between the two ears that help to interpret complex sounds. Accordingly, the authors were curious about the nature of the horse's binaural cues and what it does with them, and conducted a third experiment on the magnitude of binaural time (Δt) and spectral (Δfi)[2] cues for sound direction angles from 0 to 90 degrees.

To do so, the researchers placed a microphone at the entry of a recently deceased horse's ear, positioning its head in the test stall in the same way as for the horses in the behavioral experiments. They measured the time difference in the arrival of the sound at both ears, using clicks limited to 10 microseconds to simplify measurement, and varying the placement of the loudspeaker used to emit the sounds.

For purposes of comparison, the researchers performed similar

2. Δfi: Frequency-related intensity difference for the sound spectrum of a complex sound.

measurements with a live human subject, whose localization acuity in response to clicks was 0.8 degrees, and with an anesthetized cat, whose acuity based on noise had previously been shown to be 5 degrees (Casseday and Neff 1973).

To measure the difference in intensity, the researchers inserted a microphone at the base of the horse's left ear, directly in front of the auditory meatus. The loudspeaker was initially positioned at 0 degrees, that is, in front of the horse's head. Measurements of broadband noise were taken at eight frequencies from 125 Hz to 16 kHz in octave steps, using high-pass and low-pass filters set to the same frequency to obtain a narrow-band pass. These measurements were then repeated with the loudspeaker positioned at different angles to the left and right of midline. The Δfi for a specific frequency band was calculated by subtracting the measured sound level for a particular angle to the left from that of the corresponding angle to the right.

Comparing the measurements for horse, human, and cat, the authors observe that horses have Δt and Δfi cues largely sufficient for good localization and that, moreover, the mobility of their ears may facilitate their detection of intensity difference. The authors (Heffner, H. E., and R. S. Heffner 1983a, 1984) conclude: "The poor localization acuity of the horse cannot be due to an insufficiency of binaural cues. Instead, it follows that the horse has not developed the neural capacity to take advantage of the binaural cues available to it" (Heffner, H. E., and R. S. Heffner 1984, 552).

High Frequencies and Monaural Pinna Cues

In their later comparative studies, these same authors note that the ability of a species to use binaural cues can be assessed by determining the ability of the animal to localize pure tones (Heffner, H. E., and R. S. Heffner 2008). If animals can localize low-frequency pure sounds that pass around the head with little or no intensity difference between the ears, that shows they can use interaural time-difference cues resulting from interaural phase-difference cues; if they can localize pure sounds of sufficiently high frequency to be attenuated by the acoustic shadow formed by the head (according to their angle of incidence), that shows they can use interaural intensity-difference cues (Heffner, H. E., and R. S. Heffner 2003, 2008). Such experiments show that most mammals, for example, humans, monkeys, and cats, use these two binaural cues; they also show that, although species such as hedgehogs, mice, and some bats do not use time difference cues, others (Heffner, R. S. 1986; Heffner, R. S., and H. E. Heffner 1989; Heffner, H. E., and R. S. Heffner 2003), such as horses, domestic pigs, and cows, have the characteristic that they cannot localize high-frequency sounds, whereas they hear such sounds very well (up to 35 kHz in the case of the horse). Similarly, the researchers observe that removing the high frequencies from broadband noise degrades localization of the sound by these species (Brown et al. 1982; Butler 1986; Heffner, H. E., and R. S. Heffner 2008).

A question that arises is how to explain, from an evolutionary perspective, the apparent paradox between the good hearing acuity for high frequencies in species that otherwise are unable to use the same high frequencies to localize sounds. A first factor to consider is that detection of high frequencies and their use for localization does not necessarily involve the same mechanisms or, more precisely, the same anatomical and physiological components. That being the case, these species' inability to localize is linked (as I mentioned earlier) to a neural deficiency that prevents interaural comparison of high-frequency intensities, in other words, the absence of a neural device for instantaneously comparing stimuli arriving at both ears. With respect to the ability to hear these high frequencies, it is worth referring to a simple detection device related to what I mentioned previously in the first part of this chapter, that is, the existence of monaural cues for localizing sounds linked to the way in which the external ear modulates the intensity of the different frequency constituents of a sound based on the angle of incidence, in particular the role of the pinna in focusing high-frequency sounds by virtue of its convolutions, and their amplification in the auditory canal due to resonance phenomena. Thus, unconsciously moving the head to better localize a sound translates into directional changes of the pinna, which in turn enables the nervous system to compare high-frequency intensities. This mechanism helps, for example, in eliminating ambiguity between a sound coming from the front and one coming from the back that present the same binaural localization cues. Similarly, because the pinna of the species mentioned, especially horses, is not only directional but extremely mobile, its mobility enables the animals to avoid front-back confusions (Heffner, H. E., and R. S. Heffner 1983a, 2008). This aspect is critical for orienting, even approximately, the direction of looking in locating potential danger.

These considerations naturally invite a more specific examination of the functional relationship that unites hearing and vision.

The Coevolution of Hearing and Vision: Localization of Sounds and the Field of Best Vision

As I mentioned, sound localization acuity varies dramatically according to species, even within the same class of mammals; indeed, its accuracy varies from around 1 degree in the elephant to 33 degrees in the mouse.

Different factors may explain this variability. Their relative significance is shown by calculating their correlation to localization acuity for broadband sounds given a sufficiently large sample of species. Such calculations were made initially by analyzing thirteen species of mammals (Heffner, R. S., and H. E. Heffner 1992c), work that produced the results shown in table 9.3, and later by increasing the sample to thirty-six species once the data were available (Heffner, R. S. 1997). These studies are the basis for the following commentary.

Interaural distance has the advantage that it provides intensity and time difference cues that increase with increasing distance. However, it appears to be only weakly related to the accuracy of sound localization, as the correlation coefficient is small (−0.404). When it is calculated taking into account thirty-three nonsubterranean species,[3] the result is little changed (−0.425, $p = 0.014$).

---

3. The vision and hearing of subterranean species are severely limited: In fact, these animals have practically no need to orient themselves in the dark, and sound does not reach well underground.

Table 9.3	First-order correlation coefficients.					
	Sound localization threshold	Interaural distance	Binocular visual field	Field of best vision	Visual acuity	Trophic level[1]
Sound localization	1.000					
Interaural distance	−0.404	1.000				
Binocular visual field	−0.732	0.103	1.000			
Field of best vision	0.911	−0.542	−0.611	1.000		
Visual acuity	−0.532	0.762	0.336	−0.747	1.000	
Trophic level	0.606	−0.026	−0.536	0.729	−0.240	1.000

[1]All factors logarithmic except trophic level. Source: Heffner, R.S., and H.E. Heffner 1992c, 227.

The orienting reflex, which turns the head and the eyes in the direction of the source of an unexpected sound, brings into play the *width of the binocular visual field*, whose correlation coefficient appears promising (−0.732). However, when it is calculated based on a larger sample, thirty-one species, the value falls to −0.596; it falls farther when removing the three subterranean species from the sample, in addition to a number of more typical species deviate strongly from the regression line.

However, in turning one's gaze toward a sound source, it is not only the binocular visual field that is oriented, as the authors point out, but the part of the eye with the best visual acuity. In humans, and generally also for primates, this part of the eye is linked to movements whose purpose is to center a stimulus on the portion of the retina with the highest resolution, the fovea. However, most animals have no fovea, but a larger area called the area centralis. For many of these animals, their optimal field of vision extends to a large "visual streak." This is the case for the horse in particular, and an extreme case at that: With its eyes positioned to the side and its large visual streak, the horse's field of vision encompasses almost the entire surroundings, as I showed in chapter 6. Having thus examined the relationship between the size of the *width of the field of best vision* (whose estimate is based on measuring the area of greatest density of ganglion cells in the retina) and the impreciseness of localizing sounds, the authors effectively obtained a strong correlation coefficient (0.911) for sound localization thresholds. When the sample was extended to twenty-four species for which the latter data were available, the correlation coefficient was somewhat lower (0.825). But it increased to 0.922 (p < 0.0001) when discounting the three subterranean species included in the sample. Consequently, there appears to be a strong relationship between sound localization acuity and the width of the field of best vision.

In addition, two supplementary observations support the finding that interaural distance does not in itself determine sound localization acuity. Removing the influence of field of best vision from the correlation between sound localization acuity and interaural distance reduces the residual correlation almost to zero (0.04). In contrast, removing the influence of interaural distance from the correlation between sound localization acuity and the width of the field of best vision has little effect on the residual correlation, which remains at 0.893.

However, a question arises regarding the possible relationship between sound localization acuity and visual acuity. Not only does the initially calculated correlation coefficient of −0.532 not appear significant when it is calculated for thirty species (−0.306, p = 0.106), but many of these species then appear to deviate strongly from the regression line. A final factor that may influence visual characteristics and, thus, their correlation with a species' hearing acuity is life style. An example is trophic level, that is, the degree to which a species is predator or prey, with 1 being strictly predatory and 5 being strictly prey (as is the horse). The initial correlation coefficient of 0.606 increases slightly (0.643) when the calculation is done over thirty-six species. However, predators generally focus their field of best vision forward; thus, eliminating the influence of the width of the field of best vision causes the residual correlation to fall to 0.132, and it is no longer significant.

What emerges from this group of observations is reassertion of the strong correlation between sound localization acuity and width of the field of best vision, which leads Rickye Heffner (1997) to conclude: "A primary function of audition is to direct visual attention for scrutiny of sound sources. This function seems to have been a major source of selective pressure affecting the evolution of sound localization among mammals" (51).

Identifying Sounds and Vocal Communication in the Horse
As mentioned earlier, Henry and Rickye Heffner (1998) highlighted a decade or so ago how little was known about the way animals identify the meaning or biological relevance of sound sources. However, Henry Heffner also noted that vocal communication between conspecifics is widespread throughout the animal kingdom, in particular in the search for sexual partners. A variety of studies have also shown that, among different species of farm animals, a mother (cow, goat, sow, or hen) and her offspring recognize each other by their vocalizations (Walser, Hague, and Walters 1981; Walser 1986; Kent 1987; Barfield, Tang-Martinez, and Trainer 1994; Weary, Lawson, and Thompson 1996).

In the case of the horse, it is only recently that several studies (Basile et al. 2009; Lemasson et al. 2009; Proops, McComb, and Reby 2009; Lampe and Andre 2012; Proops and McComb 2012) heralded a new era in the understanding of vocal communication in social and individual recognition.

Before presenting these findings, allow me a rapid overview of the variety of equine vocalizations.

Vocalizations and Auditory Acuity
A social animal like the horse naturally uses its hearing capacities for intraspecific communication. Thus, it is hardly immaterial to touch on their relevance, however briefly, to the sounds that the animals make.

In briefly summarizing George Waring's (2003) and Seong Yeon's (2012) work on the subject, I would like to highlight seven main sounds, the first four of which, strictly speaking, are voiced emissions only, made through the larynx; the last three consist of nonvoiced exhalations and inhalations.

Squeals are high-pitched cries, uttered as a defense warning or threat ("Don't provoke me") in agonistic situations such as between stallions in sexual competition, or when a mare protests a stallion's advances. When the conflict is more intense, squeals are stronger and prolonged, almost a scream. The fundamental of a squeal is generally close to 1 kHz, with many harmonics, whereas the maximum sound energy is concentrated above 4 kHz.

Nickers fall into three different modalities, but are always characterized by low-pitched, broadband vocalizations with a guttural pulsated quality. The sound energy is typically below 2 kHz.

- Nickers frequently accompany the distribution of food by horse handlers and are audible at a distance of at least 30 meters; they are less broken into syllables than in the other two cases.
- These sounds are also produced by stallions in connection with sexual behavior, particularly when they are led to a potentially receptive mare. The nicker shows the horse's sexual interest and is audible at a distance of at least 30 meters. It consists of repeated sounds and is unique to each individual.
- Nickering may also be directed by a mare toward her young foal on the approach of potential danger, or for any other reason that attracts the mare's interest to her young. The sound causes the foal to draw near to its mother. It consists of many pulsations, but it is produced so softly, like a breath, that it is only audible up close ("sigh nicker").

Whinnies (neighs) facilitate social contact at a distance. They can be heard up to 1 kilometer. Whinnies occur especially during separation from group members at the moment contact is broken. They are also heard among stable horses as their morning feeding approaches (Pond et al. 2010). A whinny generally comprises three phases, beginning at a high pitch, like a squeal, continuing with a more or less rhythmic vocalization, and ending with lower-frequency sounds similar to nickers. Whinnies are the longest vocalizations, lasting an average of 1.5 seconds.

These three types of vocalizations also appear to involve individual recognition cues (Tyler 1972; Ödberg 1974; Wolski, Houpt, and Aronson 1980; Lemasson et al. 2009; Proops, McComb, and Reby 2009).

Groans appear to be the expression of mental conflict, suffering, or physical effort (for example, for a birthing mare in the lateral decubitus position who has not yet evacuated the placenta). With a bandwidth under 300 Hz, and very rapid pulsations, a groan may be followed immediately by a broadband, nonvoiced but audible supplementary exhalation. Groans are audible only within a few meters.

A **blow** is an expression of alarm that reflects a state of anxiety or suspicion. It alerts other nearby horses and informs a potential intruder that it has been detected. It consists of a forceful expulsion of air through the nostrils, which is audible within 30 meters. The sounds it produces may reach 8 kHz, although the maximum sound energy is below 3 kHz.

Snorts also consist of a forceful expulsion of air through the nostrils that produces broadband sounds that are audible up to some 50 meters. But they are emitted in the form of pulsations accompanied by vibrations. They serve to relieve irritation of nasal passages, especially by dust, and as a displacement activity by expressing restlessness in response to an obstacle, constraint, or frustration (for example, when the horse has been pushed to its limit or is prevented from galloping). The maximum sound energy is under 3 kHz.

Snores are broadband, raspy sounds that accompany inhalation. They occur in two types of situations, only one of which is related to communication and serves an alarm function: Here, the horse briefly makes such sounds prior to emitting an alarm blow. Otherwise, horses may snore as a sign of difficult breathing, when they are lying down. The maximum sound energy is under 500 Hz.

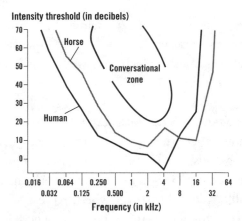

Figure 9.12 Position of the conversational zone vis-à-vis human and horse radiograms.

On the basis of these few calls, it is easy to see that, aside from groans, these sound emissions, which can also serve to communicate, fall within a range compatible with the best conditions for hearing. If, now, turning more specifically to the domestic horse, and to our interactions with it, we ask how well it can hear us, we note that its auditory acuity enables it to hear without difficulty all the sounds that fall within what is called the conversational zone (see figure 9.12).

This zone approximately circumscribes all the sounds used to communicate in spoken language and represents the zone of usefulness for daily interactions. It is characterized in particular by the fact that the speech we produce in the context of normal conversation ranges between 250 Hz and 4 kHz, with most speech falling in the range of 1,000 to 2,000 Hz. The intensity of these sounds is on the order of 20 to 70 dB, depending on how loudly a person is speaking. Although for humans this intensity corresponds basically to the maximum audible sensitivity frequencies, the situation for the horse is not very different, though with two minor qualifications that are easily shown on the graph of the conversational zone.

One general consequence of aging for mammals is substantial loss of sensitivity to high frequencies, which for humans is a social handicap. The horse, on the other hand, with the same loss of acuity, nonetheless hears us just as well as before. In contrast, horses are clearly more sensitive than we are to a decrease in sensitivity, however modest, to low frequencies, which may well contribute to a certain mutual incomprehension!

Whinnies: A Source of Social Information

Everybody knows that when you separate horses living together, they whinny vigorously, effectively maintaining contact for a time even when they are out of sight of each other. Alban Lemasson and colleagues (2009) asked themselves whether these whinnies constitute acoustic cues unique to the caller, and whether other horses within earshot could, based on this possible acoustic "signature," recognize whether the caller was a familiar animal.

To answer the question, they first asked whether it was possible to identify among whinnies parameters relating to the sex, age, size, or identity of the caller. To this end, they recorded whinnies from thirty horses either put out to pasture or living in stalls, alone or in a group. The horses included ten geldings, ten stallions, and ten mares, each category covering a large range of ages and different breeds. For each horse, sixteen whinnies of sufficiently good quality were recorded for analysis, across multiple days, following brief isolation from its stall or pasture mates. The analysis of the whinnies covered thirteen parameters presenting significant variability among individuals. Parameters with the weakest intraindividual variability, and thus the most stable, were the frequencies and especially the fundamental frequency of the first of the three whinny phases called the "introduction."

Sex appeared to be an important source of variation. The stallions showed significantly lower frequencies than the mares and the geldings, which exhibited no significant difference between them. However, this observation held only for eight of the ten stallions, which proved to be the ones living alone or having dominant social status. In fact, the two other stallions, which lived in a group and had a subordinate status, produced the highest frequencies (given their small number; however, no comparison was made with the mare frequencies) (see figure 9.13).

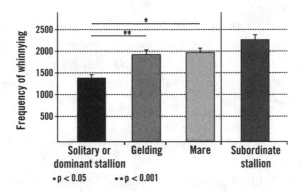

Figure 9.13 Dominant frequency of the introductory phase of whinnying according to sex and social status. *p < 0.05 and **p < 0.01. Source: Lemasson et al. 2009, 696.

The authors note that for geldings hormonal influence on the pitch of the whinnies was more important than a mere sexual dimorphism and that the status of the subordinate stallions might have induced strong physiological changes. Although they were reluctant to draw any firm conclusions given the low number of individuals in the study, they pointed out that work on feral horses showed that they can distinguish calls of dominant stallions from those of subordinates (Rubenstein and Hack 1992).

Sex aside, the size and weight of individuals appear to be another source of interindividual variation in the call pitch. Larger, heavier individuals emit lower-frequency whinnies, which may be linked to concomitant fluctuations of the vocal apparatus, although the data to support such a hypothesis are currently insufficient. In contrast to findings for many other species, no relationship was found between the emitter's age and the call parameters.

The authors also investigated whether, based on prolonged whinnies, horses could discriminate individuals varying in their degree of familiarity. This experiment consisted in individually testing twelve horses, used to living in groups of two to four individuals, with loudspeaker emissions of whinnies recorded beforehand in natural conditions. The subject, held by a long rope, was visually isolated at a distance of around 200 meters from its social group and positioned so that its group mates were behind it. The loudspeaker was placed 10 meters behind the horse. Two tests, separated by a few minutes' walk, were performed during each period of isolation.

Each horse underwent a total of sixteen tests, structured in the following way:

- four "group whinnies," produced by members of their social group;
- four "familiar whinnies," produced by horses living in their environment and with whom they had permanent vocal contact and occasional visual contact;
- four "unfamiliar whinnies"; and
- four control white noises, lasting 1.5 seconds.

The test began only when the subject displayed a straight posture, with its two ears oriented forward. The test sessions were video-recorded.

In analyzing the data, the authors distinguished immediate reactions (turning of the horse's ears or head toward the loudspeaker), for which they measured latencies and angles of orientation, from short-term reactions (vigilance and observation posture, contact with the experimenter, grazing, chewing, whinnying, kicking, defecating) during the two minutes preceding and following playbacks, and scored the animals for each behavioral item by subtracting the former from the latter.

The researchers found that the first immediate reaction consisted in turning the ears toward the loudspeaker, 1 second on average after the playback, then turning the head, 2.8 seconds on average at the end of the playback, though with differences depending on the categories of the sounds emitted:

- All three categories of whinnies induced a significantly high proportion of immediate reactions (70 to 80 percent), whereas reactions to white noise did not differ significantly from chance.

- Latencies to react did not differ significantly among the four stimuli categories.
- Ear orientations did not differ significantly among the four stimuli categories.
- In contrast, head turning fell along a gradient according to the familiarity of the caller; the less familiar the caller, the more the subject turned its head. White noise, however, elicited only a weak reaction.

Similarly, white noise induced no particular reaction in the short term, whereas subjects whinnied in response to all the categories of whinnies heard. Familiarity with the emitter proved again to play an important role; subjects oriented their body toward the loudspeaker or pulled on the rope when they heard group member or familiar whinnies. They adopted a posture of alert only when whinnies were not those of the group, stopped grazing at the sound of unfamiliar whinnies, and only relaxed their posture of observation on hearing familiar whinnies (see figure 9.14).

Since the horses were able to distinguish whinnies from other horses based on familiarity, they categorized these individuals according to the degree of familiarity, which assumes (as I mentioned

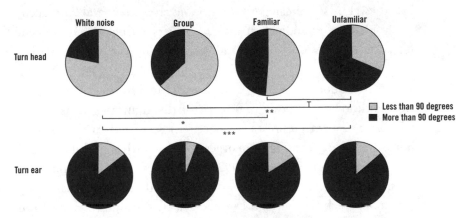

Figure 9.14 Orientation of the head and ears in response to different categories of sounds. *p < 0.05; **p < 0.01; ***p < 0.001. Source: Lemasson et al. 2009, 699.

in chapter 2 in discussing animal representation) representation in memory. Thus, as the authors make clear, the ability of horses to distinguish among the calls of other horses "suggests a process of social learning of vocal signatures and long-term memorization of the different degrees of social bonds maintained among horses" (Lemasson et al., 701).

This mental capacity, which is matched by a behavioral adaptation the animals exhibit (more or less vigilant, for example) depending on the category of individual detected, also suggests "abilities in planning/prediction of the issue of dyadic encounters" (ibid., 702).

Contribution of Hearing to Intraspecific Individual Recognition in Horses

In the words of René Zayan (1992): "In the light of comparative psychology founded on the complexity and development of the neurobiological structures of knowledge, it is most often acknowledged that the process of perceptual recognition implies the existence of mental representations previously formed with respect to objects that have become familiar and reactivated when these objects are discriminated from novel objects" (143–44).

For individuals of a given species, recognizing other individuals of the same species as conspecifics appears to correspond to a fairly broad level of social categorization. Conversely, being able to accurately identify an individual meets the very stringent requirement for accuracy in social categorization and, hence, in discrimination (Zayan 1994).

It is not a question simply of differentiating, for example, known from unknown conspecifics, or even of distinguishing conspecifics within one's own group, as we have just seen with whinnies that horses perceive as coming from other horses, which is straightforward social recognition. As Zayan (1992) emphasizes: "This recognition of a familiar individual must not be confused with the individual recognition by which two familiar conspecifics are distinguished one from the other as unique individuals—that is, identified on the basis of idiosyncratic properties" (147).

Indeed, individual recognition requires a much finer categorization than does social recognition: "Individual recognition . . . refers to a subset of recognition that occurs when one organism identifies another according to its individually distinctive characteristics" (Tibbets and Dale 2007, 529).

As I indicated in chapter 1, the reality of such recognition in horses is strongly suggested by various ethological considerations, such as the existence of hierarchies in family groups (Tyler 1972) or bachelors (Feist and McCullough 1976), as well as between stallions from different harems (Miller and Denniston 1979), preferential interindividual relationships within groups (Feist and McCullough 1976), or the selective bond of mares with respect to their foal (Tyler 1972). And yet, except for three early studies, no longer current (Ödberg 1974; Wolski, Houpt, and Aronson 1980; Leblanc and Bouissou 1981), no experiment to date has incontrovertibly demonstrated intraspecific individual recognition in horses nor explored its various mechanisms.

Recalling that recognition of individuals is widespread among animal species (Tibbets and Dale 2007), Leanne Proops, Karen McComb, and David Reby (2009) emphasize that, in humans, it is fundamentally multimodal, that is, founded on the increased information to be had from at least two different perceptual modalities, by matching immediate sensory cues of identity and stored information to specific individuals.

To determine whether horses are similarly capable of individual recognition, on the basis of integrating information from different perceptual modalities, these authors carried out an experiment that consisted of providing their subjects with a visual identity cue from one of their familiar conspecifics, that is, by showing the subject a herd mate and then, once the mate was out of sight, playing a "whinny" either from that associate or from another familiar herd mate.

The authors used two stable groups of sixteen horses each. Within each of these groups, among the individuals that had been part of a herd for at least six months and thus familiar to each other in terms of physical appearance and whinnies, the authors selected:

- on the one hand, twelve horses divided into two groups of six "horse subjects," that is, a total of twenty-four subjects (thirteen geldings and eleven mares aged three to twenty-nine years); and,
- on the other hand, four horses chosen at random (among those for which high-quality whinnies had been recorded) to form two groups of stimulus horses, that is, a total of eight stimulus horses.

The whinnies of the stimulus horses were recorded, and six good-quality recordings of each were chosen (a total of forty-eight recordings) to avoid using the same stimulus horse call for two different subjects.

Thus, the authors had available a total of four groups of six horse subjects, each of which would be tested, first visually, then aurally, using one of the four pairs of stimulus horses. The subjects and the stimulus horses to which they were exposed were not related. The experiment consisted of bringing a stimulus horse a few meters in front of the subject, and holding it there for around a minute and then leading it behind a barrier, where it could no longer be seen. After a delay of about 10 seconds, two whinnies were played by a loudspeaker behind the barrier with a 15-second interval between each call.

The ten-second delay between the disappearance of the stimulus horse and the playing of the call derived from the work of Andrew McLean (2004), who concluded that short-term spatial memory in the horse lasts less than 10 seconds;[4] accordingly, the subject is forced to rely on stored memory. The playback procedure was repeated four times for each subject with the participation of one of the pairs of stimulus horses from the same group as the subject horse, according to four different combinations:

---

4. An early experiment by Bernhard Grzimek (1949) left open the possibility that equine short-term memory might be as long as 60 seconds. Since then, Evelyn Hanggi (2010a) and Paolo Baragli and colleagues (2011c) have shown durations of 30 seconds.

- visual presentation of stimulus horse 1, then calls of stimulus horse 1 (congruent trial);
- visual presentation of stimulus horse 1, then calls from stimulus horse 2 (noncongruent trial);
- visual presentation of stimulus horse 2, then calls of stimulus horse 2 (congruent trial); and
- visual presentation of stimulus horse 2, then calls of stimulus horse 1 (noncongruent trial).

The order of the trials varied according to the subject for overall balance. An intertrial interval of at least four days was imposed to avoid habituation. All the experimental sessions were video-recorded and analyzed by two different observers, and the results were checked for consistency.

The researchers hypothesized that, if horses are capable of cross-model representation of known individuals, presenting a cue to the visual identity of such an individual should lead them to expect that subsequent auditory information should match the visual cue.

Violation of this expectation should, accordingly, translate into reactions of surprise similar to those one observes in infants using the preferential looking method (Spelke 1985; Baillargeon 2004), namely, a shorter response time and a longer looking time toward the direction of the call.

Indeed, the authors observed that subjects responded more quickly to incongruent calls than to congruent calls, and they also looked in the direction of the stimulus horse more often and for a longer time in the incongruent trials.

In a commentary on this experiment, Robert Seyfarth and Dorothy Cheney (2009) stress that it was the first study to show cross-modal integration of information on identity in animals. Although research on golden hamsters (which possess at least five different odors that are individually distinctive) had shown that the animals integrate several cues relating to the same perceptual modality in forming a representation of an individual (Johnston and Bullock 2001), Proops and her colleagues had, for the first time, shown that a nonhuman animal recognizes individual members of its own species across sensory modalities.

Proops and colleagues themselves point out that research had already shown that certain species spontaneously integrate visual and auditory identity queues for familiar humans. One example is the dog (Adachi, Kuwahata, and Fujita 2007), which associates the fact of its owner to the sound of his voice. However, the contribution of Proops and her colleagues was to demonstrate that "horses possess cross-modal representations that are precise enough to enable discrimination between, and recognition of, two higher familiar associates" (Proops, McComb, and Reby 2009, 949).

Contribution of Hearing to Equine Interspecific Recognition of Individuals

Two recent studies published at roughly the same time aimed to show the cross-modal basis of recognition of humans by horses.

The first (Lampe and Andre 2012) was inspired by the experimental expectancy violation paradigm described above (Proops, McComb, and Reby 2009).

The experiment involved twelve subject horses (eight geldings and four mares) who were exposed to the sight and smell of a stimulus person who was either familiar (that is, with whom the horse interacted daily and thus knew well by sight, sound, and smell) or unknown. Twelve seconds after the presentation of the human, a voice recording either of the same human or another, speaking the same words, was played. The same two familiar and unknown humans were used with all twelve horses.

Consistent with their working hypothesis, the authors observed that, when the auditory, visual, and olfactory cues were incongruent, the horses showed more interest, as evidenced by looking more quickly (lower response latency) toward the source of the sound, more frequently, and for a longer time (duration of the first look and total looking time).

Nonetheless, although this study confirmed the capacity of horses for cross-modal recognition of familiar humans, it does not, strictly speaking, prove that horses can recognize individual humans. As the authors themselves suggest, complementary studies are needed, for example, to test the equine ability to distinguish between humans that are equally familiar to them. Moreover, it would be desirable to differentiate visual and olfactory stimuli.

Proops and McComb (2012) have reported two experiments, one in social recognition and a second in individual recognition.

Starting from the premise that horses can discriminate between different human faces as well as between familiar and unknown persons (Stone 2010; Krueger et al. 2011), Proops and McComb sought to determine whether a correlation exists between the sound of a previously recorded human voice and the sight of the speaker, that is, whether horses are capable of cross-modal recognition of individual humans. The task for the animals was to choose between two humans after hearing the voice of one of them.

For the experiments, the authors used a preferential looking paradigm (Fantz 1958; Spelke 1979, 1985). In such a paradigm, two stimuli are presented simultaneously to the subject (and not in succession, as in an expectancy violation paradigm). The horse exhibits cross-modal matching ability if it looks more rapidly and preferentially toward the congruent stimulus, in this case, toward the person whose voice the animal has just heard.

In the first experiment (social recognition), the human pair presented to the horse comprised one person the horse was familiar with (its owner), and one person that the horse did not know. In the second experiment (individual recognition), both people were familiar to the animal.

During the experiments, the authors also tested for laterality response as evidence of hemispheric specialization in cross-modal processing of recognition.

Both experiments used the same apparatus (see figure 9.15). The horse was positioned in front of a loudspeaker hidden from view, with two humans on opposite sides of it facing the horse.

For the first experiment, the voice of one of the two humans was heard over the loudspeaker pronouncing the name of the horse in an

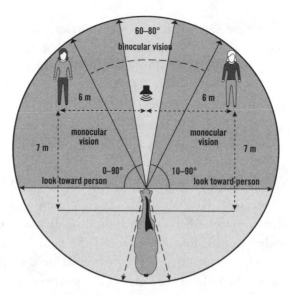

Figure 9.15 Diagram showing the experimental setup. The behavioral coding of the subject head orientation defined as looks toward each person (10 to 90 degrees from center), the speaker (less than 10 degrees from center), and elsewhere (greater than 90 degrees from center) are also shown in relation to the binocular (60 to 80 degrees) and monocular visual fields of the horse. Source: Proops and McComb 2012, 3132.

authoritative tone. The playback was repeated twice, with an interval of one second. Fifteen seconds of silence followed during which the results were noted. Then the procedure was repeated with the voice of the other person, followed by a fifteen-second period of silence to note the results.

In the second experiment, the procedure was the same except that the playback of the second person's voice was done at least a week after the first playback.

For each emission, the person who made the recording was situated either to the right or left of the horse.

The usual methodological precautions were taken regarding the order of the playbacks, and the left- or right-side placement of the congruent stimulus, and so forth. In addition, the humans acting as stimuli were equipped with earphones emitting white noise to mask the playbacks directed at the horses to minimize the risk of unintentional cues on their part.

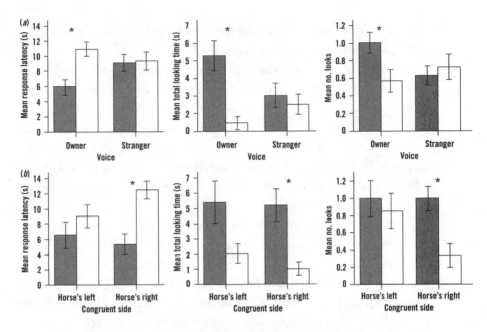

Figure 9.16 Experiment 1. Cross-modal discrimination of familiar human handlers and strangers. (a) Mean ± s.e.m. for the behavioral responses of subjects to the matched (congruent person: gray bars) versus the mismatched (incongruent person: white bars) person during the trials in which the owner and the stranger were heard (*p < 0.05). (b) Mean ± s.e.m. for the behavioral responses of subjects to the owner's voice during the trials in which the owner was on the right side of the horse versus when they were on the left (*p < 0.05). Source: Proops and McComb 2012, 3133.

The first experiment (social recognition), during which the horses were exposed to their owner and an unknown person, involved thirty-two horses (fourteen geldings and eight mares), aged 1.6 to 31 years.

In fact, the findings show (see figure 9.16) that horses responded more rapidly to the congruent person than the noncongruent person, and spent more time overall looking at the former. However, these response speed and time differences were apparent only when the playback was that of the owner, not that of a stranger. The authors caution that, at present, the question whether the failure of matches in the latter case are due to the horse's inability to infer that an unknown voice belongs to a stranger or whether it indicates a lack of motivation to respond to the call of a stranger remains open. Failed matches were omitted from the results relating to strangers' voices in the analysis of lateralized responses.

Consequently, when the voice was that of the owner, the horse looked more quickly toward the person, for a longer time, and more often when the person was on the right side of the horse. In contrast, when the person was on the left side of the horse, there was no significant difference save that the difference in total time of looking only approached significance (p = 0.061). In any event, it is clear that the horses had much more difficulty making a match when the owner was not on the right side.

No other significant differences were evident in either the direction of first looking or in total number of looks in each direction.

The second experiment (individual recognition), during which the horses were shown two human handlers who were both familiar to them, involved 40 horses. However, the results of only 39 of them (1 stallion, 23 geldings, and 15 mares, aged two to twenty-five years) were included, as one horse was sold before the end of the study.

The humans acting as stimuli in the second experiment were grouped into ten pairs of handlers. Each pair of handlers shared the care of the four horses presented to them, and thus were equally familiar with them.

When the horses were exposed to both of their handlers (see figure 9.17), they were quicker to look at the handler whose voice they had heard; they also looked more often and longer at that handler. "Crucially," write Proops and McComb, "in the second experiment, subjects proved able to match a specific familiar voice with a specific familiar human handler. This indicates that the sight of the handler activated a multimodal memory of that specific individual, allowing subjects to match the sight of that particular person with the sound of their voice. Furthermore, the ecologically valid methodology and the large number of handlers pairs presented suggest that horses use this recognition strategy naturally to identify numerous individual people in their day-to-day lives" (3135). Recognition was also more rapid when the speaker was to the right or the left of the horse. It thus clearly seems that good correlation of voice and vision occurs more

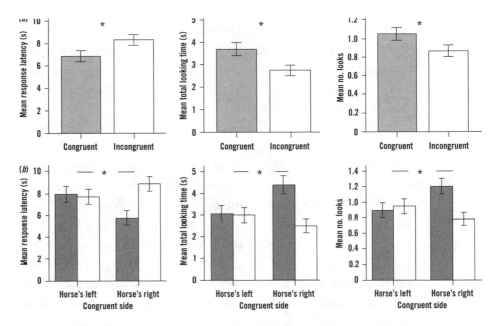

Figure 9.17 Experiment 2. Cross-modal individual recognition of familiar human handlers. (a) Mean ± s.e.m. for the behavioral responses of subjects to the matched (congruent person: gray bars) versus the mismatched (incongruent person: white bars) person (*p < 0.05). (b) Mean ± s.e.m. for the behavioral responses of subjects during trials in which the congruent person was on the right side of the horse versus when they were on the left (*p < 0.05). Source: Proops and McComb 2012, 3134.

readily with the right eye in each of these two experiments, and the result is even more striking when two familiar human handlers are involved.

On the basis of this observation, in particular, and analysis of the results of complementary unimodal controls subsequently carried out with the same experimental apparatus (either visually, presenting two humans without voice playback, or aurally, using playback without humans present), the authors conclude that, apparently, "the left hemisphere dominance observed in the experimental trials is due to specific features of cross-modal recognition rather than of the presentation of the visual or auditory stimuli per se" (3135).

Auditory Laterality and Social Familiarity
Muriel Basile and her colleagues (Basile et al. 2009) wished to show the existence of auditory laterality in the horse in response to calls from other horses. The question they asked was whether the level of social familiarity influences this laterality.

As previously mentioned, the acoustic fibers of the auditory pathway run from the inner ear and partially cross at the level of the rachidian bulb, but with contralateral dominance: Although each hemisphere receives information from both ears, it is nonetheless processed predominantly by the opposite hemisphere. Recall also that the ears of the horse, whose pinnae are controlled by sixteen muscles, are extremely mobile and can rotate 180 degrees to the back. Thus, the orienting reflex (or orientating-investigating reflex), which in mammals serves to turn the head toward the sound source, in the horse acts to rotate one or both ears (Preyer's reflex), with or without movement of the head.

The experiment conducted by these authors was to individually expose twelve horses (six geldings, four stallions, and two mares) to previously recorded whinnies played by a loudspeaker located 10 meters behind the subject. Two good-quality recordings (very little background noise or echo) were made of the calls of nineteen horses (the twelve horses tested and seven other horses), and equalized for intensity.

These horses were from different populations and lived in groups of two to four individuals. The whinnies they listened to came from other horses selected from three categories of social familiarity:

- group members, thus horses with whom they were used to having physical, visual, and auditory contact;
- neighbors, horses with whom the subjects had only visual or auditory contact, but no physical interaction; and
- strangers, or horses unknown to them.

For each test session, the subject was isolated and removed about 200 meters from its social group. Attached to a long rope, it was allowed to graze and walk freely for two minutes. Then, with the horse standing straight with its head and ears facing forward, the stimuli were played by a loudspeaker. Each session lasted around a quarter of an hour and comprised two tests. In between tests, the horse was walked. The tests were video-recorded.

Each subject was tested sixteen times, two to seven trials per day over a maximum of six days, with four whinnies from group members, four from neighbors, four from strangers, and four instances of white noise, played in random order. The stimuli within categories consisted of two different samples from two different individuals. The white noise used as a control was in fact pink noise, that is, composed of all the frequencies without any weighting for human hearing.

The authors measured the latency between the call and the initial response, and established three categories of lateralized responses:

• backward rotation of the first ear to move;
• backward rotation of the first ear to move (as above) or, in the case of both ears moving simultaneously and the return of only one to its initial position, the other ear still facing backward (this category includes the preceding category); and
• turning the head either left or right.

However, with regard to the distribution of recorded latencies, turning the ear toward the loudspeaker was considered a valid response only when it occurred within a delay of 1.8 seconds following the playback, and that of the head within a delay of 4.8 seconds, based on the general distribution of latencies observed. On the other hand, the first of these three types of responses was observed infrequently, and was not statistically significant. Consequently, it was dropped from the analysis.

The general response rate of subjects to whinnies, whether from group members, neighbors, or strangers, was high: an average of around 80 percent for head turning and 72 percent for ear turning. In contrast, the animals only reacted around 44 percent of the time to white noise (see figure 9.18).

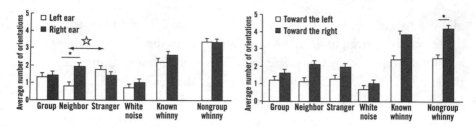

Figure 9.18 Laterality of ear (left) and head (right) orientations based on familiarity with the horse making the calls*: Level of significance p < 0.05, binomial test; open star: level of significance p < 0.05, Fisher exact test. Source: Basile et al. 2009, 616.

Horses turned their right ear (left hemisphere predominance) significantly more (p < 0.05) in response to neighbor whinnies, but not to whinnies of group members and strangers. However, the authors also note a significant difference in laterality between rotation of the ears in response to whinnies of neighbors (right ear) and strangers (left ear).

For all categories of whinnies, subjects mainly turned their head to the right. However, this preference was significant only for whinnies from the group composed of neighbors and strangers (the "not group members").

Rotations of the right ear and head to the right were positively correlated (r = 0.767, p = 0.01) only in response to whinnies from known horses, that is, group members and neighbors.

These results indicate auditory laterality in the horse, related to social factors. Although the animals often respond by turning their ears or head to the sound of whinnies coming from other horses, the same is not true of white noise; moreover, these responses differ according to the degree of familiarity of the caller.

The authors note that the fact that their subjects responded more often, but not as quickly, in turning their head than their ears, no matter the category of whinny, suggests a sequence of behavioral responses that enables them to orient their sensory receptors toward the perceived sound source, in order to process information in a unimodal fashion (auditory processing), then in a multimodal fashion (enrichment of information by visuo-auditory processing).

Given the right ear preference, the left hemisphere does appear to be more involved in processing calls from neighbors. This involvement in processing sound signals from known conspecifics (group members and neighbors) in general is further indicated by the correlation between ear and head responses. In contrast, the ear side preference patterns for neighbors and strangers are opposite. Thus, there appears to be some involvement of the right hemisphere in processing calls of strangers. Furthermore, note the authors, as laterality does not appear to be independent of social familiarity among horses, this observation would support sharing of interhemispheric tasks.

Finally, the authors note that their data support a potential influence of the subject's level of arousal on the expression of its laterality, namely, the fact that the social stimuli induce more lateralized responses than white noise, and particularly the calls of neighbors. They authors cite unpublished data indicating that the angle of the horse's head rotation in the direction of a whinny emitter is inversely proportional to the level of familiarity between them, and that horses display more vigilance and observation postures in response to nongroup-member calls. Thus, horses present a particularly high level of arousal when hearing the call of a nongroup member, especially when ecological conditions are such that the groups are naturally separated and hearing a familiar nongroup member in proximity is unusual. This could be related to the emotional content of calls and the emotional state of subjects, as well as the expression of lateralized responses, all of which, however, merit further investigation.

In any event, they conclude, their study confirms a socially dependent auditory laterality in horses and highlights the importance of taking into account the perceptive modality of response and the subject's attention.

As Rickye Heffner has remarked elsewhere (1997): "Animals do not use each of their senses in isolation, but rather in coordination to obtain information about their environment" (52).

Let us continue our exploration of equine perception by other sensory modalities.

10

EQUINE CHEMICAL PERCEPTION: ODORS, PHEROMONES, TASTES, AND FLAVORS

Chemoreception, which enables living organisms to respond to chemical stimuli, is universal in the animal world: "From the amoeba to man, all living creatures are endowed with chemical sensitivity" (Lefèvre-Balleydier, MacLeod, and Holey 2006, 91).

Actually, in mammals, chemoreception corresponds to different sensory modalities that make it possible to detect odors and pheromones[1] (primary system and accessory olfactory or vomeronasal system), tastes (gustatory system), and irritating or nociceptive substances (trigeminal system).

Chemoreception has the particular characteristic that, in comparison with vision and hearing, it relies on information that depends on the physicochemical configuration of volatile, low-molecular-weight chemical components and not physical parameters, such as wavelength for vision or pressure variations for hearing: "As vibrational systems, light and sound are easily described and characterized by

---

1. Recall that pheromones are chemical substances secreted by individuals and function for intraspecific communication: Received by other individuals of the same species, they provoke a specific reaction, behavior, or biological modification (Karlson and Luscher 1959).

simple physical parameters (frequency, intensity, and polarization). The same is not true of a fragrance, which may contain many chemical substances at what may be very weak concentrations. Even in pure compounds, it is not as easy to characterize a molecule using a few parameters" (Meierhenrich et al. 2005, 29). Here, what goes for olfaction also goes for all of chemoreception.

As I will show, each of these perceptual modalities is distinct. For example, although smell and taste share certain characteristics, they are two separate perceptual systems with two fundamentally different modes of functioning. Smell derives from the perception of chemical substances borne on the air or water, far from their source, whereas taste requires physical contact with the source of the substance detected.

Olfactory Perception in the Horse

Structure of the Equine Nose

The horse's nose extends the length of its forehead. It consists of a pair of left and right nasal cavities (or nasal fossa), divided by a flat wall, the nasal septum, which is bony at the back and cartilaginous and then fibrous toward the front (see figure 10.1). At the front, these cavities open to the outside through the nostrils, which are widely separated and oriented in divergent directions. At the back, on the ventral side, the nasal cavities reach toward the nasopharynx (upper part of the pharynx) through the choanae (posterior nares).

At the entrance to the nasal cavities, level with the mobile portion of the septum and at a depth of several centimeters, is the nasal vestibule, which controls the admission of air into the cavities. Outside the vestibule, the nostrils and the tip of the nose are covered with short, fine hairs, as well as long, strong tactile hairs; inside the nostrils, fine, short hairs act to filter the incoming air.

The nasal cavities themselves are lined with nasal mucosa (pituitary membrane) and covered with a ciliated respiratory epithelium: "The secreted mucus catches dust that the cilia then carry to the nostrils

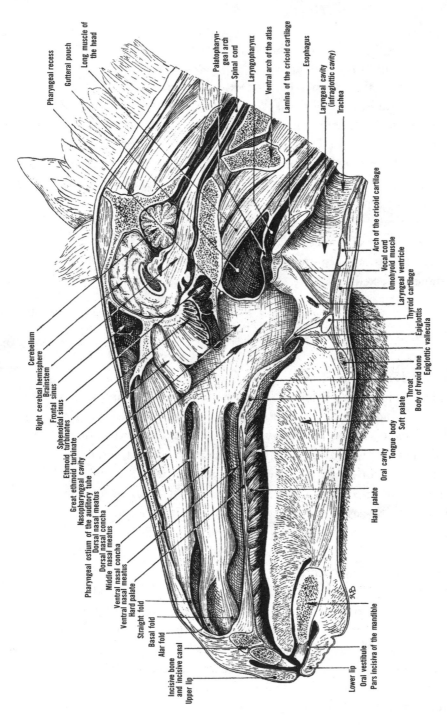

Pharyngeal recess
Gutteral pouch
Long muscle of the head
Palatopharyn-geal arch
Spinal cord
Laryngopharynx
Ventral arch of the atlas
Lamina of the cricoid cartilage
Esophagus
Laryngeal cavity (infraglottic cavity)
Trachea

Arch of the cricoid cartilage
Vocal cord
Omohyoid muscle
Laryngeal ventricle
Thyroid cartilage
Epiglottis
Epiglottic vallecula
Throat
Body of hyoid bone
Soft palate
Tongue body
Oral cavity
Hard palate

Cerebellum
Right cerebral hemisphere
Brainstem
Frontal sinus
Sphenoidal sinus
Ethmoid turbinates
Great ethmoid turbinate
Nasopharyngeal cavity
Pharyngeal ostium of the auditory tube
Dorsal nasal meatus
Dorsal nasal concha
Middle nasal meatus
Ventral nasal concha
Ventral nasal meatus
Hard palate
Straight fold
Basal fold
Alar fold
Incisive bone and incisive canal
Upper lip

Lower lip
Oral vestibule
Pars incisiva of the mandible

Figure 10.1 Sagittal section of the horse head (clockwise from upper right). Source: Barone 1984, 68.

or pharynx for elimination" (Barone 1984, 625). This mucosa also includes the serous glands, whose aqueous secretions moisturize and heat the incoming air, as well as a dense vascular network that adjusts the temperature to that of the body. Note that equids breathe only through the nose; owing to the shape of their soft palate, its free end fits tightly against the base of the epiglottis. As a result, they cannot breathe through the mouth (Barone 1984).

The lateral walls of the nasal cavities consist mostly of bony structures and at the very back the lateral cartilage of the nose. These bony structures exhibit "strong protruberances formed by the turbinate bones, which border the longitudinal depressions, or meatus" (Barone 1984, 613). Each nasal cavity has two of these turbinates (conchae), which look like elongated seashells and are also lined with nasal mucosa. The turbinates force air to circulate in the nasal cavities over the largest possible mucosal surface area.

Dorsally to the posterior end of the nasal cavities are the cavities of the ethmoid bone, which form the labyrinth of ethmoid. The caudal portion of the labyrinth is lined with olfactory epithelium (Kumar et al. 2000). Unlike the nasal epithelium, which is red, the olfactory epithelium is characterized by its yellowish-brown color.

The base of the nasal cavities "is made up of the bones of the hard palate, and its caudal portion is continuous with the soft palate" (Barone 1984, 617). It hosts the vomeronasal organ, or Jacobson's organ, named for the biologist who described it at the beginning of the nineteenth century. It refers to a pair of organs divided on either side of the ventral edge of the front of the nasal septum, between the posterior edge of the septal cartilage and the vomer (one of the bony components of the septum). The equine vomeronasal organ consists of an elongated vomeronasal tube protected by a cartilaginous capsule, 12 to 15 centimeters long, with a 3-millimeter-long lumen along its length, closed at its end and opening at the front into an incisive canal. In the horse, this incisive canal has the characteristic that, unlike in other mammals such as cows and pigs, it communicates only with the nasal cavity and not with the buccal cavity. This lumen is lined with a sort of respiratory epithelium, except for the medial surface, where it gives way to an olfactory epithelium (Kostov 2007; Okano, Fujimaki, and Onishi 1998), as in other mammals (Døving and Trotier 1998).

Knowledge of olfaction has progressed substantially in the last two decades, in particular with respect to the nature of olfactory receptors (both those of the vomeronasal epithelium and the olfactory epithelium), identification of protein families with which they are coupled, and the transduction phenomena that they induce. However, many issues regarding this perceptual modality, which for a long time aroused little interest, still remain unsettled, to the extent that knowledge believed to have been gained is once again being called into question.

The same is true of the respective roles of the main olfactory system, including the olfactory epithelium, and the accessory olfactory system, including the vomeronasal organ, which we formerly thought to be exclusively concerned with perception of odors and pheromones, respectively. It now appears that this functional division of labor is far from clear-cut and that a fair amount of overlap exists between the two systems, in terms both of the chemical signals that they detect and the physiological and behavioral reactions that they provoke (Baxi, Dorries, and Eisthen 2006; Brennan and Zufall 2006; Keller et al. 2009; Lévy, Keller, and Poindron 2004; Liberles and Buck 2006; Martínez-García et al. 2009; Restrepo et al. 2004). As Fernando Martínez-García and colleagues point out (2009): "There is an emerging consensus that the two olfactory systems do not have independent functions (e.g., odors vs. pheromones) but play an integrated role in mediating the behavioral responses to chemosensory cues . . . In fact, some molecules are detected by both the olfactory and vomeronasal systems . . . However, there is no consensus on how exactly this olfactory-vomeronasal interplay takes place" (283).

Through the intermediary of its basilar membrane, the olfactory epithelium rests on chorionic connective tissue that includes, among other things, the serous glands (Bowman's glands) whose excretory channels penetrate into the olfactory epithelium and secrete a watery, proteinaceous liquid that aids in producing the mucus lining the olfactory epithelium. This mucus is regenerated roughly every ten minutes and contains substances such as odor-binding proteins.

As in other mammals, the olfactory epithelium of the horse comprises three types of cells: basal cells, supporting cells, and olfactory cells (Kumar et al. 2000). The basal cells, near the basement membrane, constitute a reservoir of stem cells that replenishes the neuronal population of the olfactory epithelium. The supporting cells are distributed throughout the surface of the basal membrane and surround the olfactory cells, nourishing them and protecting them by degrading potentially dangerous substances.

The olfactory cells—olfactory neurosensory cells—are bipolar neurons. The single dendritic extremity of these olfactory receptor neurons has a blub at its end (an olfactory bouton) that carries olfactory cilia, embedded in the mucus layer. These neurons are the seat of olfactory transduction: They contain the receptors that generate action potential in response to binding of odorant molecules. The axons of the olfactory cells cross the basement membrane and form olfactory nerve bundles surrounded by glial cells (Schwann cells). They then cross the skull by way of the cribriform plate of the ethmoid bone, before reaching the lower portion of the olfactory bulb. Olfactory cells are constantly degenerating and are replaced within four to six weeks with new cells generated by basal cells.

The vomeronasal epithelium, which lines the central portion of the vomeronasal organ lumen on its medial surface, has a structure much like that of the olfactory epithelium, consisting of basal, supporting, and receptor cells. However, the receptor cells are not ciliated, like normal olfactory cells, but rather have microvilli. The axons of these cells come together to form the vomeronasal nerve, which joins the accessory olfactory bulb (beside the olfactory bulb). The accessory olfactory bulb sends information to the hypothalamus and the amygdala, which are involved in reproductive and feeding behavior (Kostov 2007).

Detection of Odors by the Olfactory Epithelium

"An odor is defined as the specific impression that certain bodies produce on the smell organ through their volatile emanations. These consist of molecules that are diffusible through air and that have the ability to interact with the nervous system to produce a pattern: smell. The idea of pattern is retained in the meaning employed by psychologists (gestalt), that is, a structure whose elements cannot be separated without compromising the identity of the whole . . . The meaning of an odor is thus not, contrary to popular opinion, vague and rudimentary, but precise; its complexity corresponds to that of the pattern that it recognizes" (Lledo and Vincent 1999, 1211). Indeed, write Uwe Meierhenrich and his colleagues (2005): "Nearly all volatile substances are odorant, and it is remarkable to note that no one has reported two different molecules having exactly the same smell" (31).

Linda Buck and Richard Axel (1991) showed that the receptor organ, located at the level of the olfactory cilia, consists of membrane receptors belonging to the G protein coupled-receptor (GPCR) superfamily, that is, receptors coupled to one of the G-protein family intramembrane proteins, which are proteins that play an important role in cellular communication.

Briefly, when a smell molecule—generally hydrophobic to some extent—comes into contact with the olfactory mucosa, it is captured by an odorant-binding protein in the mucus, which enables it to cross the aqueous barrier of the mucus and to travel to an olfactory receptor capable of recognizing it. This recognition is stereochemical, that is, it is based both on the molecule's chemical composition and its three-dimensional structure. Thus, in particular, two isomeric mirror-image molecules (called chiral molecules), one levorotary and the other dextrorotary, will not be recognized as identical and will not give rise to the same odor; this is the case, for example, with carvone, whose L form smells like cumin and whose D form smells like mint (Lledo and Vincent 1999).

However, the correspondence between a smell molecule (designated by the term *odorant*) and a receptor is not one to one: "After having discovered the first olfactory receptors in 1991, Americans Linda Buck and Richard Axel, who received the Nobel Prize in Physiology or Medicine in 2004, showed that each olfactory neuron possesses a single type of receptor, that each receptor responds to several molecules, and that each molecule is recognized by several receptors" (Lefèvre-Balleydier, MacLeod, and Holley 2006, 92). Thus, olfaction is based on a combinatory process. Accordingly, Jean-Claude Pernollet and his colleagues, on the subject of human olfaction, noted: "This capacity of olfactory receptors to interact with several smells is well adapted to encoding a vast olfactory space: with only 350 different receptors and a system in which 10 receptors on average are required to discriminate a smell, there are more than 6.7 $\times$ 10^{18} combinations of receptors, more than the number of possible chemical compounds" (684). This combinatory process is even more complex, as the diversity of the repertoire of active olfactory receptors has also been shown to increase with increasing concentration of the odorant substance, resulting in different smells based on the concentration. Moreover, the same odorant substance can be both an activator for certain olfactory receptors and an inhibitor for others (Zarzo 2007). In fact, the complexity of this system is such that it "enables discrimination of molecules that have not yet been found in Nature" (Pernollet, Sanz, and Briand 2006, 685), thus also making possible the perception of odors from molecules produced by synthetic chemistry. As Meierhenrich and colleagues point out (2005): "We still have a way to go before we reach a comprehensive understanding of the sense of smell sufficient to explain olfactory perception based on the molecular properties of odorant substances" (34).

In all mammals, around a thousand genes code for olfactory receptors, distributed throughout nearly all the chromosomes. However, not all of them are functional. As suggested earlier, only some 350 of the 950 human genes are functional, that is, around 40 percent. These genes give rise to as many proteins and thus different olfactory receptors. In contrast, in the dog, for example, 80 percent of its 1,122 genes are active, which would seem to suggest a much more acute sense of smell, especially given that, whereas the human olfactory epithelium comprises ten million receptors (Engen 1982), that of the dog has twenty times more (Marshall and Moulton 1981). However, as I will show, in this respect things are not so simple.

The axons of all the neurons carrying the same type of olfactory receptor converge in the same region (and sometimes two) within the olfactory bulb, called the glomerulus, where they contact the mitral cells, which transmit information to the neurons of the primary olfactory cortex. A single glomerulus thus receives action potentials from several thousand olfactory neurons. The mouse, for example, has twenty-five thousand neurons per glomerulus, each glomerulus connecting with twenty-five mitral cells, and each of the two olfactory bulbs having two thousand glomeruli (Lefèvre-Balleydier, MacLeod, and Holley 2006). "If we make an analogy with photography, the olfactory neurons would be innumerable small grains (or pixels), whereas the mitral cells would form a larger-grain image (large pixels) of the sensitivity of the olfactory organ" (Holley 2007, 52–53). From the olfactory cortex, which acts as a hub, the olfactory information is redirected toward different brain structures (hypothalamus, thalamus, amygdala, hippocampus, orbitofrontal cortex) involved in memory, learning, and emotions. Finally, the vomeronasal organ comprises two distinct categories of sensory receptors (VR1 and VR2). These receptors are different from those of the olfactory epithelium, but they belong to the same GPCR superfamily (Kostov 2007).

Controversial Relationship between the Neuroanatomical Characteristics of Olfactory Structures and Their Performance

Pascale Quignon and colleagues (2003) emphasize, among other things, that "several elements such as the size of the olfactory epithelium, the density of neuronal cells and the number of Ors [olfactory receptors] that they express on their surface, as well as the size of the olfactory bulb, have to be taken into consideration when comparing the sensory capacities of different mammals. The dog olfactory epithelium, although variable in size between different breeds, can express up to 20 times more ORs than that of humans. Undoubtedly, this contributes to the ability of dogs to detect odorant molecules at much lower concentrations" (7).[2]

If one looks for similar research on the horse, the pickings are very slim; but just suppose, taking the same point of reference, that its olfactory capacities are closer to those of the dog than of humans. A search of a database created by Chiquito Crasto and colleagues (2002), which includes gene sequences identified for olfactory receptors, turns up information on, for example, dogs, cats, cows, chickens, pigs, rats, and mice, but, admittedly, not horses. It is, however, plausible, based on phylogenetic considerations and the nature of the selective pressures at play, that the horse benefits from a much larger proportion of functional genes than do humans, and in this respect, the horse is closer to the dog. Furthermore, primates, in particular Old World primates, and especially humans, appear in fact to constitute an exception among mammals; their large proportion of nonfunctional genes (pseudo-genes) linked to olfaction appears to be related to the acquisition of full trichromatic vision: "It has not been definitively proven that improved vision caused the decline of the sense of smell, but the coincidence of the two evolutionary changes is striking . . . The hypothesis of the lessening of selective pressure on the olfactory system due to improvement in vision is thus certainly plausible" (Holley 2006a, 117).

The size of the olfactory epithelium in the horse does not appear to have been evaluated. Henri Pihlström and colleagues (2005) nonetheless showed that the surface of the olfactory epithelium of mammals is connected to a part of the surface of the cribriform plate of the ethmoid bone, which, as I mentioned earlier, is traversed by the axons of olfactory cells, and to the skull. Among the measurements of these surfaces that they made, based on a sample of 150 mammals

---

2. In the same spirit, Carol Saslow (2002) writes (though somewhat speculatively): "While in primates, and especially humans, olfactory structures have greatly diminished, the horse brain has extremely large olfactory bulbs with a convoluted surface. Since densities of olfactory receptor cells remain constant per unit surface area, the extent of olfactory epithelium determines the total quantity of receptors. The extensive size of the smoothed-out epithelium of the horse olfactory bulb implies that volatile odors should form a much more significant part of their Umwelt than is the case for humans" (212).

from 130 different species, are those of another equid, the plains zebra (*Equus burchelli*), as well as those relating to humans and dogs.

The cribriform plate of humans measures 132 mm², that of the dog 578 mm², and that of the zebra 1,408 mm². When these absolute values are compared to those of the skull, the following ratios are obtained: 0.008 for humans, 0.041 for dogs, and 0.020 for zebras. Thus, here again, the horse comes out closer to the dog, whose sense of smell is proverbial, than to humans, who are traditionally believed to have a relatively poorly developed sense of smell and who are usually classified as microsmatic.[3] Moreover, as the authors of this study remark: "Large ungulates have impressive olfactory organs" (Pihlström et al., 957). Finally, comparing the size of the olfactory bulb, Barone and Bortolami (2004) note that the bulb of humans is miniscule (10 to 12 millimeters long and 3 to 4 millimeters wide), whereas that of dogs and horses is huge. In the horse it is oval in shape and measures 30 to 35 millimeters long and 20 to 25 millimeters wide.

Nevertheless, other voices have weighed in, casting doubt on the validity of what appears to be a strong correlation between the neuro-anatomical data reported and olfactory performance. For example, Gordon Shepherd (2004) not only contests (as have others, such as Schaal and Porter 1990) the meager olfactory acuity of humans but also the absolute supremacy of dogs over humans in this respect, based on a certain number of observations. In particular, he points out that if, as Stephanie Bisulco and Burton Slotnick (2003) have shown, 20 percent of the glomeruli of the olfactory bulb of the rat (after ablation of 80 percent), and thus the 20 percent of the olfactory receptor genes they represent, suffices to ensure the functions normally carried out by 1,100 genes, "350 genes in the human are more than enough to smell as well as a mouse" (573). He notes, too, that Matthias Laska and colleagues (Laska, Seibt, and Weber 2000), among others, have shown that primates, including humans,

---

3. It is Paul Broca (1888), the French anatomist who determined "the language area" of the human brain, which bears his last name, and who made the distinction between mammalian macrosmatics (rodents, carnivores, ungulates) and microsmatics—in particular primates, including humans (cited by Bonnet 2003).

have a surprisingly good sense of smell and that, depending on the substances tested, either dogs or humans perform better. For the latter authors: "Between-species comparisons of neuroanatomical features are a poor predictor of olfactory performance" (473). In the same vein, Jess Porter and colleagues (2007) conducted an experiment in which they found that not only are humans capable of tracking a scent trail to its source by using their nose to follow the scent along the ground, but that they deviate less from the trail in doing so than do dogs. Moreover, limiting training to three trials per day over three days within a two-week period enabled experimental participants to accomplish the trajectory twice as fast, while making even fewer deviations from the trail, leading the researchers to conclude that longer training might improve human performance even more.

Finally, as a matter of interest with respect to human olfactory discrimination acuity, Meierhenrich and colleagues (2005) point out: "The number of different smells that can be discriminated is subject to discussion, but can reach over ten thousand for a trained professional, and at concentrations that may be extremely weak" (34). In this respect, they touch in particular on a chemical compound whose odor can be detected by a human at a concentration corresponding to one microgram in an Olympic swimming pool.

In short, as Timothy Smith and Kunwar Bhatnagar (2004) put it at the conclusion of a study on primates, which are assumed to be microsmatic compared with macrosmatic mammals and in which they specifically consider different anatomical parameters relating to the olfactory epithelium and the olfactory bulb: "The question remains: is there a quantitative difference in olfactory abilities, or is this only an impression based on the relative importance of vision in the two groups?" (30). Thus, it must be noted that, for now (whatever the comparative olfactory performance of humans and dogs!), the anatomical data that are the basis for assessments of the quality of the equine sense of smell have on balance shown themselves to be fairly uncertain measures.

There is also consensus that the sense of smell itself is still rather mysterious: "We cannot define a certain number of classes of stimuli that correspond to a limited number of detectors. It is also impossible to associate possible classes of odors to olfactory messages. And today, we still do not know how to predict the odor of a molecule based on its structure" (Holley 2007, 52). Under these circumstances, it is easy to appreciate the difficulty of establishing procedures amenable to the behavioral exploration of olfactory perception in animal species, analogous to those available for the other senses. It is thus not very surprising that the situation has hardly changed since the recent past, where Carol Saslow (2002) suggested that, so far as the olfactory perception of the horse goes, very few studies have been published on the subject. However, some data can be extracted from field observations and the few existing experimental efforts, despite their specialized orientation.

Olfactory Perception in the Horse under Natural Conditions

The fact that the current state of our knowledge does not point to a clear correlation between the neuroanatomical characteristics of the olfactory system of a mammalian species and its olfactory acuity does not obscure the fact that the horse, in any case, and mammals in general possesses an extremely effective olfactory system and that the variety of situations in which it makes that obvious attests to the importance of this perceptual modality for the animal, especially in its relationships. Moreover, although, as I mentioned previously, it is currently hard to establish any clear correlation between olfaction *sensu stricto* and perception of odors, on the one hand, and an accessory olfactory system and perception of pheromones, on the other, it is still the case that the horse is endowed with a sophisticated vomeronasal system whose role should not be underestimated.

Activation of the vomeronasal organ appears to be facilitated by the characteristic expressive gesture known as a flehmen (or grimace), a reaction that the horse shares in particular with other ungulates and felines. After having vigorously sniffed the source of the odor (a trace of urine, for example), the horse holds its breath, then raises its head while extending its neck, ears generally to the side, and pulls back its upper lip to reveal its teeth and gums. This posture is usually most associated with the functioning of the vomeronasal organ (König et al. 2005). It may be observed in a variety of circumstances (not necessarily, as is often believed, in a sexual context) and independent of the sex and age of its perpetrator (Weeks, Crowell-Davis, and Heusner 2002). Moreover, it is not strictly limited to analyzing pheromones in a context of intraspecific communication; for example, Saslow (2002) reports the case of a mare that, while nursing, reacted with a flehmen to the smell of dog food that remained on Saslow's fingers when she had washed her hands too hastily. It would also be speculative to draw a distinction a priori between perception of odors and perception of pheromones in the observational situations that I am about to describe.

The question of the olfactory acuity of the horse is sometimes illustrated by anecdotes referring to the distance at which an animal in an unfamiliar setting is capable of perceiving the smell of its stable. However, experiments carried out long ago by Moyra Williams (1976) highlight the difficulty of drawing precise conclusions on the subject based on simple anecdotal correlations. Although, in fact, she had managed to observe several times that the horses with whom she was working had spontaneously found the way back to their stable, after having been transported to places they had never been and at distances as great as thirty kilometers, she also noted that what had really been observed corresponded to a tendency on the part of the horses to walk upwind each time that it blew hard. In addition, she realized that this behavior manifested when the horses were in relatively familiar environments, whereas they tended to go downwind when they were in strange surroundings. It is tempting to see this behavior as the tendency of various animals to move upwind when they are hunting and downwind when they are hunted. Consequently, writes Williams, "one must be careful not to read into the results of an experiment more than the observations really warrant" (101), reasoning that her attempt to discover how horses find their way home was inconclusive.

In this same vein, it must be acknowledged that the oft-cited observation of Daniel Janzen (1978) remains an anecdote unsubstantiated by other observations. In short, Janzen had the occasion to observe a horse that was following tracks that its stable companion had left in the sand an hour earlier along the beach use both its vision and its sense of smell. When the tracks of the horseshoes were visible in the sand, the horse kept its head at normal height for several hundred meters, after having sniffed the ground. But in places where the sea had wiped out the tracks, the animal kept its nose close to the ground . . . like a dog . . . until it picked up the tracks again!

If, now, we turn to the various situations in which horses use their sense of smell spontaneously and systematically, the observations made by researchers in the field show that it is in the animals' relationships that olfactory perception plays a significant role, which obviously complements the other perceptual modalities, especially vision. In fact, the likelihood of such complementarity is precisely what the authors of a recent experimental study (Hothersall et al. 2010) concluded. The study was an exploratory-type investigation of the ability of horses to discriminate conspecifics based on samples of smells from their excretions (urine, feces, body odor).

Without trying to be exhaustive, and without going into the details of the particular situations, which have been reviewed elsewhere (McDonnell 2003; Waring 2003; Leblanc, Bouissou, and Chéhu 2004), as a matter of interest let me note that quasi-systematic sniffing has been observed in the following contexts:

- Social investigation: Two individuals meet and engage in nose-to-nose mutual sniffing of the partner's respiratory breath (Keiper 1985; Feh 2005).
- Olfactory marking: Horses sniff dung piles, especially along trails and near watering holes, and stallions that compete to be the last to add a deposit to a dung pile deposit new feces, the order of defecation among stallions within a group being a function of the dominance hierarchy (McCort 1984). A harem stallion pays highly ritualized olfactory attention to the recently released urine and feces of the members of his harem, particularly mares, including covering urine and feces (Feist

and McCullough 1976; Tyler 1972; Miller 1981; Turner, Perkins, and Kirkpatrick 1981; Salter and Hudson 1982; Keiper 1985). Similarly, in places where horses frequently roll, and which consequently are dusty and devoid of vegetation, one of the members of the group rolls, preceded by sniffing the ground, which induces others to roll, who also imbue the ground with their odor (Keiper 1985).

- Fighting between stallions: Stallions suspend the fight to defecate in turn in a ritualized way on a dung pile, smell it, cover it with new dung, and so forth (Keiper 1985), as fights frequently take place near dung piles (Welsh 1975; Miller 1981).

- Reproductive behavior, during the precopulatory phase: Stallions sniff the shoulder, flank, rump, or perineal area of the mare and also sniff traces of the mare's urine on the ground, frequently accompanied by flehmens (Feist 1971; Keiper 1985). Note, however, in this respect, that it does appear that olfaction is not absolutely necessary for a stallion to detect an estrous mare; moreover, olfaction alone would not be enough in any event. Indeed, "sight of the mare and interactions between mare and stallion are essential" (Briant et al. 2010, S121);

- Parturition and mother-offspring relations: The mare sniffs the fluids and membranes of parturition, frequently followed by flehmens, contributing to impregnation of the odor of the newborn (Tyler 1971; Keiper 1985). The mare also sniffs the newborn (Tyler 1972; Keiper 1985; Crowell-Davis and Weeks 2005) and the base of the tail of the nursing foal (Leblanc and Bouissou 1981; Crowell-Davis and Weeks 2005). The foal ingests dung (coprophagia) that it recognizes by smell as being that of its mother (Tyler 1972; Francis-Smith and Wood-Gush 1977; Crowell-Davis and Houpt 1985; Crowell-Davis and Caudle 1989; Marinier and Alexander 1995).

Sniffing is also very often involved in all young foals exploring the world, in the context of investigation, whether of members of other animal species, potential food, or intriguing objects (Waring 2003).

Olfactory Perception of the Horse in Experimental Situations
Smells of More or Less Appealing Plants

A number of studies have begun to explore the olfactory universe of the horse. Almost all, however, are concerned with perception of flavors, that is, retro-olfactory, gustatory, and trigeminal sensations simultaneously perceived when ingesting food. I will come back to this point. Very few studies appear to focus only on olfaction to the exclusion of taste.

Accordingly, Bonde and Goodwin (1999) investigated the way in which forty horses (twenty-six geldings, twelve mares, and two stallions of varying breeds and ages) react to the smell of eleven plant or animal substances: aniseed oil, ground coriander seeds, whole cumin seeds, dried curryplant, ginger, dried lemon balm, orange oil, and dried sage, as well as dried feces from a potbellied pig, a tiger, and a maned wolf. The palette of these substances was supplemented by a control consisting of odorless cotton wool pleat.

The horses were tested with a maximum of six different odors per test day. Two odors were presented simultaneously (in randomized order according to a Latin square design), with a minimum period of five minutes between presentations of successive pairs. The test horses were in a box and held at the back of the box while two cotton bags containing the test substances were positioned on each side of the door. The horses were then released, and their reactions to the substances were video-recorded for 90 seconds. The authors scored the horses as showing interest when they directed the tip of their nose toward a bag and looked it at from a maximum distance of 40 centimeters. The authors also counted investigative and manipulative behaviors, such as sniffing, touching, licking, or biting one of the bags.

Without going into details of the results, substance by substance (except that the preferred substances appear to have been coriander and ginger), the most striking finding from this experiment is that the horses showed only limited interest in the control bag. This suggests that they were, in fact, interested in the smells presented and that they did more than simply respond to the bag as a visual stimulus. Unlike the bags containing plant substances, which were licked and even bitten, the bags containing feces induced no manipulative behaviors. Moreover, the horses took longer to make contact with the bags of feces than those containing plant substances, and contact lasted only for a short time.

Soothing Plant Odors?

Aromatherapy has been practiced in many civilizations for thousands of years. It has always enjoyed great popularity, and anecdotes attesting to its benefits are legion.

Although scientific studies of its biological effects are still incomplete, a recent review of the subject, which used lavender essential oil among others, suggests the possibility of a short-term reduction in anxiety in humans (Cooke and Ernst 2000). In animals, studies of mice (Buchbauer 1991), dogs (Komyia et al. 2009), and sheep (Hawken, Fiol, and Blache 2011) in particular show some evidence for sedative effects of lavender essential oil and highlight the need for additional experiments.

As regards the horse, to my knowledge, only one study (just recently published) exists (Ferguson, Kleinman, and Browning 2013). This study, also based on lavender essential oil, involved seven adult horses. The results show a reduced heart rate after fifteen minutes of aromatherapy, following stress induced by subjecting the horses to two successive blasts from an air horn at close range. The authors also note that the horses they used were rodeo horses who were accustomed to hearing loud sounds during rodeo opening ceremonies and that, consequently, more striking results might be obtained with free-ranging horses or foals. At any rate, although aromatherapy based on lavender essential oil might be used to calm nervous show horses during a precompetition examination or aid in restoring calm after the competition, the authors note that, as with studies of humans, only short-term results can reasonably be expected.

Olfactory Recognition of Traces of Odors from Conspecifics

As mentioned above, many field observations refer to the interest shown by horses in the olfactory traces left by conspecifics, especially feces. The few experimental studies done in the past to determine to what extent horses are capable of discriminating based on social (sex, status, familiarity, and so forth) or individual characteristics, in situations where only smell is involved and not connected to other perceptual modalities, report inconclusive results, not least because of the small number of individuals tested (Houpt and Guida 1984; Marinier, Alexander, and Waring 1988; Stahlbaum and Houpt 1989).

More recently, two teams have revisited the subject. The first of these (Hothersall et al. 2010) managed in particular to show that, in the presence of the feces and body odor of unknown females and geldings, mares and their foals were able to discriminate sex based on the smell of the unfamiliar urine, but could not discriminate individuals.

The second study (Krueger and Flauger 2011) involved mares and geldings some of which were familiar to each other, and some not. In an initial experiment, the horses were presented with samples of feces that might have been their own, with feces from other members of their group, with feces from unfamiliar mares, or with feces from unfamiliar geldings. Although the individuals tested were more interested in the feces of others than their own, the researchers noted no significant difference in the performance either of geldings or mares based on familiarity with the emitter of the feces. In contrast, individuals of both sexes showed more interest in feces emitted by a conspecific of the same sex than in feces of the other sex. In addition, although the animals were systematically exposed to feces of members of their own group, they paid significantly more attention to the feces of the individuals who had behaved most aggressively with them, thus showing that they could recognize individual competitors within their group based on the odor of their feces. It is worth noting in this respect that the discrimination was not based on social rank, as no such discrimination has been shown.

Although the objective of their study was different since it basically involved recognizing the odor of estrous in stallions, Christine Briant and her colleagues (2012) also showed the capacity of stallions to discriminate based on the sex of emitters of feces that they were led to sniff. Moreover, endocrine data collected "suggest that testosterone could be an indicator of the recognition of the odor of male feces and prolactin an indicator of the recognition of the odor of female feces. . . . In contrast, none of the results obtained suggest that stallions differentiate the odors of estrous and diestrous" (54).

The Odor of Predators

Janne Christensen and Margareta Rungren (2008) tackled the question of the reactions of domesticated horses to predator smells by evaluating the behavioral and physiological influence of the smells and comparing it with the response provoked by the sudden emission of an unknown auditory stimulus, based on whether the stimulus occurred in the presence of a predator odor or not. Heart rate was used as a reliable physiological measure of emotional response in horses exposed to a stress agent (Visser et al. 2002; Christensen, Keeling, and Nielsen 2005; McCall et al. 2006).

An earlier experiment (Christensen, Keeling, and Hielsen 2005) had shown that horses respond more strikingly to the presentation of an unknown stimulus when the stimulus is visual or auditory than when it is olfactory and plant based (eucalyptus oil, in that case). In particular, although in the latter case an increase in vigilance was found, the researchers observed no increase in heart rate, as they had in the first two experiments.

In the later work, the researchers carried out three experiments each of which used an essentially identical experimental apparatus: in a circular station (10 meters diameter), situated outdoors, or rectangular (8 by 10 meters) and enclosed, horses were individually offered food (wheat mixed with molasses) in a bucket placed in a crib the bottom of which was covered with litter (sand or wood chips) impregnated (or not) with different odors. Each test lasted two minutes.

The researchers used forty-five horses aged two years, pasture raised with their mother until weaning at around six months. The horses then lived in the meadow in large groups of the same age and sex during the summer and returned to their stalls in winter. These horses had been handled very little, except occasionally for other experiments, during which they were familiarized with the same experimental conditions as described here.

Of these horses, thirty-three participated in the first experiment, randomly assigned to three groups:

- For eleven of the horses, the litter in the crib was impregnated with wolf urine.
- For eleven other horses, the litter was impregnated with lion urine.
- For the final eleven horses, as a control, the litter was impregnated with the urine of an unknown horse.

The horses exposed to the predator odor spent significantly more time sniffing the food ($p = 0.004$) and interrupted their eating more often ($p = 0.068$) when the litter was impregnated with the odor of predator urine than when the urine odor was from a conspecific. In contrast, there was no significant difference in heart rate.

The second experiment, in which twelve horses participated, lasted four days. As a control, no odor was used in the apparatus for the first and third days. The second day, the walls of the bucket containing the food were coated with blood from a very stressed conspecific, at an abbatoir, and possibly contained warning pheromones (Terlouw, Boissy, and Blinet 1998). This blood did not come into contact with the food contained in the bucket. Finally, on the fourth day, fur-derived wolf odor was rubbed on the walls of the bucket, and the litter was also impregnated with the odor.

On exposure to a predator odor and that of a stressed conspecific, the horses showed significant changes in behavior compared with the control situation (increased frequency and duration of sniffing, as well as time spent looking around, more frequently interrupted eating, less time spent eating). In contrast, there was no significant increase in heart rate.

In the third experiment, which involved the participation of the same twelve horses as the preceding experiment, the animals were exposed to the sudden emission of the unusual sound of a plastic bag, out of sight, being dragged on the ground, as they began eating. For six of the horses, this occurred when no odor had been added to the litter or the bucket, whereas for the other six horses, the fur-derived wolf odor was present. In both cases, the horses reacted with a flight response; the horses of the second group looked longer ($p = 0.066$) before turning back to the crib and also showed a significant increase in heart rate ($p = 0.001$).

Based on these results, it appears that, despite a fairly widespread belief among riders, horses are not frightened by the smell of a predator and react to it simply by increasing their level of vigilance, which allows them to respond more easily to the appearance of other danger signs. For the researchers, this finding is consistent with the fact that, in the wild, horses have practically always lived in an open environment, within which they frequently found themselves in proximity to predators but could keep an eye out for them. Fear reactions are energetically costly. According to the authors, this behavior amounts to an adaptive strategy geared to living in the wild where equids share habitats with predators and must trade off time and energy that they devote to antipredator responses against time allocated to other activities, such as eating and reproducing (Lima and Dill 1990). Nonetheless, the researchers also note that these experiments were carried out with young domestic horses with no previous experience of predators, and it is possible that, following unpleasant experiments, domestic riding horses might react with fear to certain odors after having thus learned to associate them with imminent danger.

Olfactory Lateralization
Paul McGreevy and Lesley Rogers (2005) wished to investigate the possible existence of motor and sensory lateralization in the horse.

In addition, they wished to know if these preferences are found together. Note that olfactory lateralization, which has been shown in different animals, including humans (Brand et al. 2002; Royet and Plailly 2004), may be facilitated in the horse, whose nostrils, as I indicated earlier, are wide apart and oriented in quite divergent directions. Note, too, that unlike other sensory systems, the projection of information from the olfactory receptors toward the central structures is essentially ipsilateral, without decussation and (as we saw earlier) with short, basically direct pathways.

For their olfactory lateralization experiments, the authors used 157 freely moving English thoroughbreds and presented them with stallion feces contained in open plastic bags, held at nose height at an equal distance from the nostrils. When the horse flared, indicating thereby that it was going to inhale, the researchers noted the nostril used. They proceeded in this way, recording for each horse the number of inhalations and the nostril used for each until the horse lost interest in the feces. For each horse, the researchers calculated a laterality index of nostril preference based on the number of times inhaled with the left (IL) or right nostril (IR), using the following formula: $(IL - IR)/(IL + IR) \times 100$.

Based on the results thus obtained, they first considered the left or right laterality of the first nostril used to sniff, which showed a significant right nostril preference ($p = 0.04$) for the whole population of 157 horses, particularly for young horses. When the results of 32 males and 125 females were considered separately, there was no significant sex bias in nostril preference. Moreover, they found a good correlation ($r = 0.69$, $p = 0.01$) between the laterality of the first nostril used for sniffing and the index of nasal laterality preference, which suggests that horses make more inhalations with the nostril used first than with the other nostril, and thus that the first nostril used is a good indicator of olfactory laterality.

Analysis of the results for laterality index revealed no effect of age or sex on the bias of nostril use. Males, however, were more strongly lateralized than females. Horses two years old or younger inhaled more than did older horses, and males inhaled more than females;

among the latter, pregnant mares made more inhalations than nonpregnant mares.

A second test was made with seventy-eight of the mares the day following their first exposure to the bag of feces, as the novelty had worn off. The number of inhalations was then less in 81 percent of the cases, and the researchers observed only a mild (but significant) correlation between the results of the two days, for both the nostril first used for sniffing and for the laterality index and individual significance of the laterality bias. In other words, this test became less reliable as novelty waned.

In sum, olfactory lateralization in the horse does appear to exist at the population level, as noted in the earlier chapter on the brain, namely, in this case, the majority of horses tested appear to have the same laterality preference. In this earlier chapter, I also referred to the right hemispheric dominance observed in many species for detecting novelty (Rogers, Zucca, and Vallortigara 2004). The predominance of the right nostril use (in connection with the right brain hemisphere) observed for the entire population in the experiment described here (and particularly in the young), and its sharp decrease in a second presentation the following day when the novelty had worn off, is consistent with this general observation. Recall too that, in the chapter on auditory perception, I pointed out that Muriel Basile and colleagues (2009) had also noted that the horses used their right ear (processed by the left hemisphere) significantly more at the sound of a whinny from a familiar horse, but that the left ear (processed by the right hemisphere) turned significantly more than the right in response to whinnies emitted by unknown horses, and thus novel "auditory signatures."

Finally, consider that the study just described was concerned with motor and olfactory laterality and on their possible connection. In this regard, not having observed a significant correlation between the two, the authors conclude that "lateralization of the equine brain occurs on at least two levels of neural organization—sensory and motor" (Basile et al., 350), which would be consistent with conclusions drawn from observations made in other species.

Before ending this discussion of olfactory lateralization in horses, let us return quickly to the research done by Alice de Boyer des Roches and colleagues (2008), which I mentioned in the chapter on visual perception of shape and movement. This research not only involved lateralization during visual exploration by a horse of an object based on the emotional value of the object, but also on the investigation of a possible olfactory asymmetry during such exploration. I will not go into the details of the study except for the animals' olfactory response to novelty. In this study, the authors observed a slight tendency to use the right nostril rather than the left to make initial contact and also to use the right nostril more often to sniff a novel object, in contrast to the case of a known object (with negative emotional value). These findings were not significant, but they are consistent with observations described above, and their lack of significance might possibly suggest that, in the study by de Boyer des Roches and colleagues, olfactory exploration occurred only after an initial visual exploration.

From Taste to Flavor

In the strict sense of the word, taste or gustation is what enables perception of the five tastes (salty, sweet, sour, bitter, umami). By extension, when we speak of how a thing tastes, we are generally referring to a perception that combines olfactory, gustatory, and somesthetic sensations associated with ingesting food—more precisely, what we call its flavor. In particular, retronasal olfaction, which occurs naturally during chewing and especially swallowing, sends aromatic constituents of the food toward the olfactory system and is alone responsible for around 80 percent of what is commonly called taste (Leclerc, MacLeod, and Schaal 2002).

Tastes: Gustation *Sensu Stricto*
Morphology of the Equine Tongue
The taste receptors are essentially incorporated into the papillary mucosa that covers the tongue; accordingly, it is generally considered

to be the organ of taste (a few receptors are also scattered in the soft palate, as well as the epiglottis, the pharynx, and the inner surface of the cheeks). These receptors only function in a liquid medium. To be sapid (that is, perceptible to the sense of taste) a substance must be able to dissolve in water or in fat: "The job of the gustatory system is to signal to the brain the chemical properties of food that are not accessible to smell, either because the constituent substances are too heavy and not sufficiently volatile, or because they are trapped in water and fat. Consequently, the receptor elements of taste, the taste cells, are distributed over the tongue and the palate in areas reached directly by food that has been chewed and impregnated with saliva" (Holley 2006b, 69).

Generally speaking, herbivores are endowed with a long, narrow tongue. This is especially true of the horse: The equine tongue is around 40 centimeters long and 6 to 7 centimeters wide (and weighs around 1,200 grams!), though individual tongues vary widely (Barone 1984). The tongue involves a rostrally positioned, freely moving, very mobile portion, an apex, and a fixed portion whose "body" is situated in the middle and whose root is situated caudally and sloped toward the epiglottis (see figure 10.2). The back of the tongue is entirely lined with thick mucosa, whose surface epithelium is scattered with many small granulosities, called papillae. Some of these do not have a taste function but a mechanical one (keeping ingested substances at the surface of the tongue) or a tactile one, in particular filiform papillae, which are the most abundant.

In addition, this mucosa also contains taste buds, which themselves contain the sensory taste cells. Although some taste buds are found in the epithelium, most are grouped within the taste papillae: foliate papillae, in the form of sheets "that form on each side an oblong mass two to three centimeters long, just in front of the palato-glossal arch" (Barone 1984, 71), each of which may contain several dozen taste buds; two (and sometimes three) large circumvallate papillae (or calciform papillae) around 1 centimeter in diameter (ibid.), situated slightly forward vis-à-vis the foliate papillae, near the caudal end of the tongue, in a cup-shaped mucosal depression

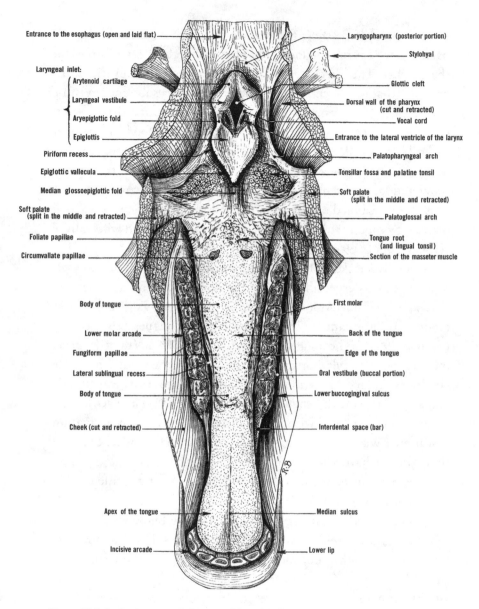

Entrance to the esophagus (open and laid flat)

Laryngeal inlet:
- Arytenoid cartilage
- Laryngeal vestibule
- Aryepiglottic fold
- Epiglottis

Piriform recess

Epiglottic vallecula

Median glossoepiglottic fold

Soft palate
(split in the middle and retracted)

Foliate papillae

Circumvallate papillae

Body of tongue

Lower molar arcade

Fungiform papillae

Lateral sublingual recess

Body of tongue

Cheek (cut and retracted)

Apex of the tongue

Incisive arcade

Laryngopharynx (posterior portion)

Stylohyal

Glottic cleft

Dorsal wall of the pharynx
(cut and retracted)

Vocal cord

Entrance to the lateral ventricle of the larynx

Palatopharyngeal arch

Tonsillar fossa and palatine tonsil

Soft palate
(split in the middle and retracted)

Palatoglossal arch

Tongue root
(and lingual tonsil)

Section of the masseter muscle

First molar

Back of the tongue

Edge of the tongue

Oral vestibule (buccal portion)

Lower buccogingival sulcus

Interdental space (bar)

Median sulcus

Lower lip

Figure 10.2 Equine tongue and pharyngeal floor (clockwise from upper right). Source: Barone 1984, 70.

that contains several hundred taste buds; and the fungiform papillae (shaped like mushrooms), essentially located near the edges and at the apex of the tongue, though some are found at the ventral surface of the tongue, which contain only the rare taste bud.

The taste buds are ovoid. They are generally said to be onion shaped. The cells they comprise are epithelial cells (or epithelio-cytes), arranged lengthwise side by side and linked through chemical synapses to the sensory neurons to which the buds communicate their excitation.

Embedded in the epithelium of the lingual mucosa, the taste buds are 80 to 100 microns long and 50 microns wide. Apart from basal and supporting cells, each contains fifty to a hundred sensory taste cells that regenerate every ten days. The taste buds are extended at their tip by microscopic outgrowths called microvilli, whose membrane contains molecular taste receptors; these microvilli protrude across the gustatory pore, which opens to the surface of the lingual epithelium.

Tastes and Their Detection

Tastes traditionally fall into four categories: salty, sweet, sour, and bitter, to which a taste called umami, meaning "delicious" in Japanese, has been added. Isolated at the beginning of the twentieth century by a Japanese researcher, Kikunae Ikeda, the latter corresponds to the taste of glutamate, which is found in many Asian dishes.

Throughout recent decades, many debates have taken place in an attempt to determine whether the repertoire of tastes was limited to these five, collectively known as basic tastes or primary taste qualities, and their combination, or whether there might be a continuum of intermediate tastes. As André Holley (2006b) puts it: "Perception of the taste qualities of foods depends on the activation of molecular receptors carried by the taste cells. It is a research subject that, after having stagnated for a long time, has experienced sudden and considerable progress over the course of the last few years thanks to the powerful methods of molecular biology" (70). However, he cautions (as will be shown below) that "the new knowledge is still uncertain" (ibid.). In this context, questions such as the identity of taste receptors and their response selectivity (or lack of it) to different tastes is obviously crucial.

Without preempting receptors that might be discovered in future, notably a potential fat receptor (Gaillard et al. 2006), it is generally possible to discriminate two broad categories of tastes. The first consists of what are called ion channels, which allow the passage of sodium and potassium ions for salty tastes and hydrogen ions for sour ones,[4] across the membrane of presynaptic cells that then release neurotransmitters that bind to sensory neurons that they synapse with (Holley 2006b). The second category consists of neurosensory cells that carry membrane receptors coupled to intramembrane proteins of the GPCR superfamily of proteins for sweet, umami, and bitter tastes.

Two families of GPCR-associated proteins have been identified. One is a small family of three proteins, T1R1, T1R2, and T1R3,[5] the first two of which combine with the third to form a receptor complex (a dimer): T1R1 + T1R3 for umami and T1R2 + T1R3 for sweet (natural or artificial). T1R3 by itself, however, is a protein that reacts equally to very strong concentrations of natural and artificial sugars. The second, larger family of more than thirty proteins is called T2Rn and detects a variety of bitter tastes (Scott 2004).

Two theories have been formulated regarding the encoding of information to explain the perception of a particular taste. According to the labeled line theory, each type of receptor is associated with a distinct type of cell that specifically detects one of the five taste categories. The information relating to a particular taste is then transmitted to the cortex along afferent fibers unique to this taste (Chandrashekar et al. 2006). According to the across-fiber pattern code theory, each taste cell responds differentially to different tastes. The information relating to a particular taste is then transmitted to

---

4. However, a specific sour receptor has now been identified: a membrane protein, PKD2L1, which belongs to the TRP family of cationic transmembrane channels and without which we would be unable to perceive acidic taste (Huang et al. 2006).

5. T for taste and R for receptor.

the cortex along afferent fibers whose response spectra overlap, and the recognition of a given taste is achieved in a combinatory fashion by bringing into play a range of different receptors, each of which is differentially sensitive to different tastes (Erickson 2008). In addition, the hypothesis was advanced that neurosensory cells carrying GPCRs might respond selectively to the three categories of sweet, umami, and bitter, while modulating the responses of the presynaptic cells corresponding to salty and sour tastes, thus in effect responding to all taste categories (Tomchik et al. 2007). In short, it does appear that the current state of knowledge is insufficient to definitively resolve the question (Breslin and Spector 2008).

In any event, very recent advances do not fundamentally call into question the traditional categorization of tastes, each of which appears to have a certain adaptive function, as explained by Paul Breslin and Alan Spector (2008). Many animals, with the notable exception of carnivores, actively seek out sweet-tasting foods whose components (glucose, fructose, sucrose) constitute an important supply of energy to the brain. As Roger Wolter (1975) notes: "The horse is consequently very sensitive to the presence of soluble carbohydrates" (141). Umami taste derives from amino acids and peptides, which are good indicators of the presence of proteins. For herbivores and some omnivores, which are constantly eliminating sodium without storing enough of it to satisfy their vital needs, having a detector of salty taste is critical. It seems in particular to be horses, whose alimentary regime is deficient in salt, who actively look for foods rich in this substance (Salter and Pluth 1980). Although sour taste (along with sweet taste) may be an indicator of fermentation, including the edible state of foods such as fruits, strong acidity is generally avoided, thus protecting the individual from injury to tissues and teeth. Finally, it appears that bitterness, with its strongly disagreeable taste, basically serves as a warning system against ingestion of toxins.

Nonetheless, I must also note that there is very strong interindividual variability in the way in which tastes are perceived. As Annick Faurion (1993) observed on the subject of humans: "From one individual to the other, the sensitivity to a product may vary by a factor of 10. If one of two individuals puts two sugars in his coffee and the other half of one, that does not necessarily mean that one likes sugar more than the other; this may result from the difference in their peripheral receptors for sugar . . . Such sensory differences, due to different chemoreceptor makeup, itself genetically determined, explains the lack of consensus in general, and the lack of words . . . The same individual differences translate into qualitative variants that are sometimes very large (and thus observable) due to the different proportions of receptors involved in the detection of the same product in different subjects . . . Some products are described as bitter by some subjects and sweet by others" (80–81). This intraspecific individual variability, which I will show later with respect to the horse, is further complicated by interspecific variability. For example, citing Morely Kare (1971), George Waring (2003) suggests that the horse does not discriminate between pure water and an aqueous solution of a certain saccharose acetate in concentrations that would be repulsively bitter to humans.

I would also add that, despite a fairly widespread belief, strictly speaking there is no exclusive localization of different tastes on the tongue. All tastes are found in different regions of the tongue, though some have response thresholds that are lower for certain tastes. For example, the tip of the tongue is especially responsive to sweet and salty tastes, the back of the tongue to sour tastes, and the soft palate to bitter tastes (Collings 1974).

The gustatory information detected at the level of the taste buds travels to the cortex in an ipsilateral fashion, without decussation, along three nervous pathways: The facial nerve (cranial nerve VII), whose salivary and gustatory fibers constitute the chorda tympani, gathers taste afferences from the front two thirds of the tongue, that is, cells from the fungiform papillae; the glossopharyngeal nerve (cranial nerve IX) gathers those of the back third, and thus from the follate and calciform papillae; and the vagus nerve (cranial nerve X) collects information from the taste buds scattered in the epiglottis in particular.

The taste fibers of these nervous pathways converge in the solitary tract along the rachidian bulb, and synapse in the nucleus of the solitary tract. They then travel partway along a thalamocortical pathway that ends at a second relay within the thalamus, the ventral postero-medial nucleus (somesthetic thalamus), where they project to the insula and the cortical taste area. It is here that the cognitive process-ing of taste information (analysis of its nature, duration, intensity, connection to mnemonic processes, integration) is thought to take place. The other taste fibers follow a hypothalamic-limbic pathway toward the hypothalamus and the amygdala, apparently giving rise to the affective, hedonic quality of taste information.

Equine Responses to Basic Tastes

As Breslin and Spector (2008) point out, the experimental procedure most commonly used to evaluate the taste function in mammals consists of putting the animal in a situation where it must choose between two bottles filled with liquids, one of which contains the test solution and the other water (or some other solution), then measur-ing the relative intake from the two solutions. Although interpreta-tion of the results must take into account in particular the potential influence of repeated ingestion on preferences and thus on the quan-tity of liquid drunk, as well as possible interference with olfactory perception, this method has the advantage of simplicity and may be useful as a first approach. That is exactly the purpose it served many years back in evaluating the responses of horses to sweet, salty, sour, and bitter solutions (Randall, Schurg, and Church 1978).

The authors used five weaning foals, fed on hay and pelleted feed concentrate according to their food needs, and maintained in indi-vidual box stalls for the duration of the experiment, except when they were taken for exercise. Each of the test sessions lasted eleven hours, during which the foals had the choice of drinking from one of two ten-liter plastic containers, one with water and the other one of four test solutions, after which the liquid taken from each of the containers was measured. To prevent visual or positional bias, each of the test sessions was repeated, also for a duration of eleven hours, by reversing the position of the two containers, and taking the average of the results.

For each taste, the concentration of the solution was gradually increased, by doubling it from one session to another. For the sweet taste, the researchers used saccharose solutions, with successive concentrations of 0.01 to 20 g/100 mL, for two times twelve sessions; for the salty taste, solutions of sodium chloride, with successive concentrations of 0.01 to 5 g/100 mL, two times ten sessions; for sour taste, solutions of acetic acid with successive concentrations of 0.00063 to 1.25 mL/100 mL, two times twelve sessions; and, finally, for bitter taste, quinine solutions with successive concentrations of 0.16 to 80 mg/100 mL, two times ten sessions. In addition, ten control sessions for each foal, with water in both containers, permitted evaluation of normal variations in the level of liquid in each recipient, with an average that never significantly exceeded 50 percent.

The preference thresholds and rejects (in other words, the preferences for water) were determined based on confidence intervals of 95 percent around the theoretical mean of 50 percent. Thus, for 40 to 60 percent of the liquid in a recipient for the averaged results of five foals, for a given taste and concentration, there was neither preference nor rejection. Up to 70 percent, the preference (or rejection) was considered weak; from 70 to 80 percent, moderate; and beyond,

Table 10.1 Variation in individual foal taste reactions to moderate to high concentrations of sucrose.

Foal number	Sucrose concentration g/100 mL		
	5	10	20
	(% intake[a])		
1	97	100	84
2	68	69	40
3	65	92	81
4	46	0	0
5	79	95	64
Average	71	71	54

[a]Test solution as percent of total intake. Source: Randall et al. 1978, 53.

strong. A weak preference for saccharose manifested only for concentrations between 1.25 and 10 g/100 mL, and indifference below or above; for sodium chloride, indifference below 0.63 g/100 mL, and above that, rejection became increasingly strong at 5 g/100 mL; for acetic acid, indifference up to 0.16 mL/100 mL, then increasingly stronger rejection, as well as for quinine from a concentration of 20 mg/100 mL.

Aside from the indications given by these values, what is most striking (as shown in table 10.1) is that they confirm for horses a strong interindividual variability of responses to tastes, previously mentioned in humans (Faurion 1993) and also found in many other species of mammals, for example, pigs (Kare, Pond, and Campbell 1965).

Flavors: How Things Taste

As I have said, perception of "how things taste," as the expression goes, cannot be reduced to a single taste perception. It is the product of interactions between different perceptual modalities, including taste, of course, but also, as I previously noted, smell through the retronasal olfactory pathway, as well as somesthesia via the trigeminal nerve pathway (cranial nerve V).

In fact, the trigeminal nerve, which (among other things) possesses free ends in the nasal cavity as well as in the buccal cavity, generates chemical, tactile, and thermal sensations that provide information about different qualities of substances absorbed: sharpness, astringency, texture, and temperature. Stimulation of the trigeminal nerve causes, among other things, sensations of pain or irritation in response to different chemical compounds, which may have neither smell nor taste, for example, chili peppers, black pepper, mustard, and ammonia. Accordingly, this sensory modality also plays a role in an organism's defense mechanisms. It also seems that, in interacting with the olfactory system, the trigeminal nerve plays a warning role, for trigeminal stimulation may induce lowering of olfactory perceptual thresholds (Jacquot, Monnin, and Brand 2004). However, no specific study of the horse appears to have been done on the contribution of the trigeminal system to flavor perception.

In any event, it is obvious that what we refer to as the taste of things, which corresponds more precisely to their flavor, is basically multi-modal: "Taste, in the everyday sense of the word, is an ensemble of gustatory, olfactory, retronasal, somesthetic (mechanical and thermal, trigeminal chemical) sensations. The stimulus presented in the mouth is always multimodal. Moreover, the convergence of different sensory pathways at the cellular level is responsible for multimodal, nonspecific responses in the gustatory sensory chain" (Faurion 1993, 79).

Flavors and Equine Alimentary Behavior
As noted by Patrick Duncan (1992): "The use herbivores make of available plant matter is rarely random—large herbivores are generally selective in their feeding, preferring some and avoiding other compo-nents of the available plant matter" (75). Deborah Goodwin, Helen Davidson, and Pat Harris (2005) point out in addition that free-rang-ing horses consume more than fifty varieties of food resources (Hansen 1976; Putnam et al. 1987) and that domestic horses show the same interest in diversity (Archer 1971, 1973). Moreover, it appears that stable horses offered access to several different foods may become sated with a given food whereas they become motivated to consume others (Goodwin, Davidson, and Harris 2002), which suggests a sensitivity to alimentary monotony or a satiety specifically linked to sensory char-acteristics with respect to certain foods. This observation is not limited to the horse, and in fact constitutes a fairly general principle (Meuret, Viaux, and Chadoeuf 1994; Baumont et al. 2000; Ginane et al. 2002).

Consequently, the authors identified fifteen flavors used in concen-trated food for horses, commercialized then as now all over the world, to determine how flavors are accepted by horses, what choices they make, and what their relative preferences are. The flavors used were the following: apple, banana, carrot, cherry, coriander, cumin, Echinacea, fenugreek, garlic, ginger, nutmeg, oregano, peppermint, rosemary, and turmeric, all presented in the form of powder.

The experiment was carried out with eight horses (six mares and two geldings) of different ages (from two to twenty-one years old) and took place in three phases.

The first phase consisted of evaluating the acceptability of the fifteen flavors. The horses were fed with lots of hay, and every morning were given 100 grams of low-energy cereal by-product meals to which one or another of the flavors was added in the ratio of 1 gram of powder to 100 grams, or 1 percent, except for Echinacea, which was mixed at 10 percent in accordance with the recommendations of the manufacturer. The researchers measured the quantity ingested (in grams), the time taken to eat everything, the partial rejections (disinterest in the food lasting for two minutes), and refusal (not eating by two minutes after presentation of the food). At the end of this phase, some horses were refusing three flavors: Echinacea, nutmeg, and coriander.

For the second phase, which aimed to evaluate preferences between flavors, the eight best-accepted flavors (those that had required the least time to be entirely consumed)—cherry, oregano, peppermint, cumin, banana, carrot, fenugreek, and rosemary—were presented in pairs, according to various possible combinations. The pairs included 3 grams of flavor for 300 grams of food, and the test was stopped when around half of the total quantity had been consumed. The order of preference was calculated on the basis of the ratio ingestion of A/(ingestion of A + ingestion of B), the measurements having been made in grams, followed by normalization of the data by arcsine transformation and a test of significance. The order obtained was the following: fenugreek, banana, cherry, rosemary, cumin, carrot, peppermint, and oregano.

In the third phase, the first two flavors were used to evaluate whether flavoring mineral pellets improved their acceptability. At the outset, the consumption time of unflavored pellets was measured.

Two days later, the horses were separated into two groups. For five days, one of the groups was offered fenugreek-flavored pellets, and the second group banana-flavored pellets. For the next five days, the flavors were switched for the two groups. From the very beginning, the banana-flavored pellets were consumed more quickly ($p < 0.05$) than the unflavored pellets. For the fenugreek-flavored pellets, although the response was only weakly significant on the first day (p < 0.075), it was significant (p < 0.05) for the next four days.

The researchers conclude that, in this study of short-term effects, aside from the preference shown for some flavors over others (the preferred flavors resulted in increased acceptability of the food, with a decrease in the time taken to consume them), additional studies are nonetheless needed to evaluate longer-term effects and to investigate the determinants of choice at the sensory level.

Contributions from Experience in Equine Chemical Perception

Thus, although horses clearly show preferences, avoidance, and even refusal characteristic for different flavors, that does not mean that the likes and dislikes shown by individuals are irrevocably set for life.

In discussing behavioral strategies developed by herbivores to protect themselves from toxic plants, James Pfister (1999) notes that, if indeed they exist, innate avoidance mechanisms based on taste and smell are not particularly effective. However, different forms of learning may contribute to protect individuals against the dangers they encounter. Inexperienced animals are the most vulnerable to poisoning.

Among these strategies, aversive conditioning has been the subject of several studies in the horse. Let me say briefly that this form of learning, long known and studied in the rat (Garcia and Koelling 1966; Rzoska 1953), is basically Pavlovian conditioning where the conditioned stimulus is a food or liquid with a particular smell or a taste (or flavor!), and the unconditioned stimulus is the negative biological effects of the food consumed (discomfort, nausea, vomiting).[6] Consequently, the animal subsequently avoids any substance with the same flavor. A flavor that was once appealing becomes disgusting.

Compared with conditioning where the unconditioned stimulus is an "external" stimulus (a light electric shock, for example, as

---

6. Horses cannot vomit, except in cases of stomach rupture, whence the mandatory precautions taken with the use of known emetic substances.

an unconditioned aversive stimulus in the laboratory) and the time interval between the conditioned and unconditioned stimuli must be very short, generally on the order of a half second, in conditioned aversion the time interval may be much longer: several minutes or even a few hours.

The existence of such conditioning in horses was shown with Shetland ponies by Katherine Houpt and Donna Zahorik (1981, 1990), who combined different foods and injections of apomorphine, which quickly induces nausea. Conditioned aversion was indeed observed, except for very appealing foods, and when the apomorphine was administered a half-hour at most after consuming the food.

Poisoning in certain species by oxytrope and astralagus, which contain swainsonine, an indolizidine alkaloid that causes nerve cell damage, possibly leading to death, is a worrisome problem in the western United States where these plants—in particular white loco-weed (*Oxytropis cericea*)—are widespread and favored by herbivores. Horses are particularly sensitive to these plants and more likely than other species to show signs of intoxication due to them.

With the aim of developing ways of protecting horses from these types of poisoning, Pfister and colleagues (2002) became interested in the possibility of inducing in horses a conditioned aversion to this oxytrope species, and began by investigating the most appropriate nausea-inducing substance for this purpose. After several trials, they substituted lithium chloride for apomorphine, which in the doses at which it can be used without side effects (extreme nervousness and exitability) proved to induce only partial, short-term conditioned aversion. In fact, apomorphine, which works for only a limited time, is not suited to creating strong conditioned aversion to foods, whereas lithium chloride induces nausea that may last several hours after ingestion. The experiments carried out by these researchers on horses showed the possibility of using lithium chloride to induce a strong, persistent conditioned aversion not only to oxytrope but also to other appealing foods. This aversion has been observed in horses put out to pasture as well as those maintained in stables.

The same authors (Pfister et al. 2007) also observed that, in general, horses (and sheep) that have previously experienced food poisoning either by oxytrope species (*Oxytropis cericea*) or astralagus species (*Astralagus lentiginosus* var. *diphysis*) were still able to form strong, lasting conditioned aversions to other food substances. However, these aversions were observed to be less strong in some individuals than for animals that had never been poisoned.

Finally, note that, although it is quite possible for horses to "reject" flavors based on repeated negative biological effects after eating foods containing the flavors, it does also appear possible for them to acquire preferences for foods with high energetic values (Cairns et al. 2002), as has been shown in other species.

11

TACTILE PERCEPTION IN THE HORSE

The field of tactile perception in animals, especially the horse, has produced very few psychophysical investigations (Saslow 2002). In the words of Paul McGreevy, this state of affairs is surprising "given the importance of tactile stimulation for communication both within human-horse dyads and between horses" (51).

Nonetheless, the histological structure of the skin has been described in detail not only in humans (Prost-Squarcioni 2006) but also in horses (Scott and Miller 2003; Wakuri et al. 1995; Wong, Buechner-Maxwell, and Manning 2005). All the same, substantial uncertainty exists regarding the molecular mechanisms underpinning different tactile modalities.

Structure and Function of Horse Skin

Epidermis, Dermis, and Hypodermis

Thick and not very elastic, the skin of the horse consists of the three common main layers: the epidermis, the dermis, and the hypodermis. Compared with other domestic mammals, however (including humans), equine skin has the peculiarity that the base of the skin at certain parts of the body has an additional compact fibrous layer that clearly separates the dermis from the hypodermis (see figure 11.1).

The thickness of the skin varies from around 1 to 7 millimeters (with an average of 3.8 millimeters), depending on the region of the body, and can even reach 1 centimeter at the mane and the base of the tail. The thickest skin is concentrated in the dorsal surfaces (forehead, mane, back, lumbar region, croup, base of the tail, and the limb extremities) and is thinner in the ventral parts (teats, axilla and interaxilla, and external portions of the genital organs). The skin constitutes an anatomical and physiological barrier between the external environment and the inside of the horse; it aids thermoregulation; and it is endowed with receptors for heat, cold, pain, itching, contact, and pressure.

The equine epidermis is a keratinized stratified squamous epithelium. With an average thickness of 0.053 millimeters, it is the most superficial layer of the skin. Containing neither blood nor lymphatic vessels, it nonetheless possesses many free nerve endings, as well as several types of cells: keratinocytes, melanocytes, cells of Langerhans, and Merkel cells.

The keratinocytes, which are the most numerous by far (around 85 percent) ensure epidermal cohesion, act as a protective barrier (mechanical and chemical) between the external and internal milieux, and help protect against light radiation. Migrating gradually from the base to the surface of the skin (the average length of the equine keratinocyte replacement cycle is around seventeen days), they are distributed in four layers: basal (which renews the epidermis by mitosis), spiny, granular, and corneal (whence the term *corneocytes*), with the structure becoming increasingly flatter. At the skin surface, corneocytes gradually detach from the epidermis (desquamation).

Melanocytes, which constitute around 5 percent of the epidermal cell population, are situated mainly in the basal layer. They are responsible in particular for the synthesis of melanin, which acts as a photoprotector and is the basis for the color of the skin, coat, and mane of horses.

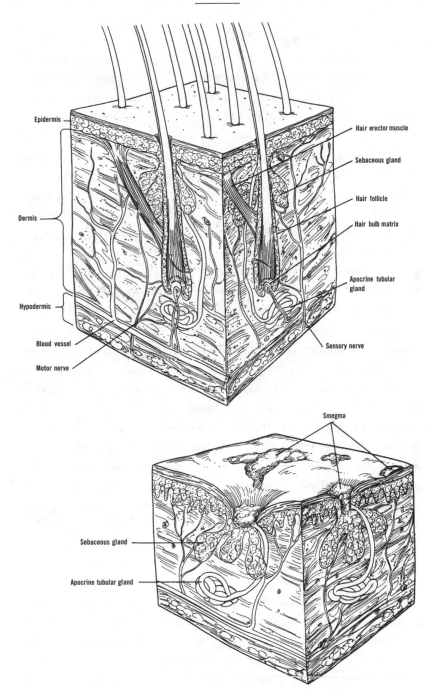

Figure 11.1 Top: hairy skin of the neck. Bottom: smooth skin of the prepuce. Source: Kainer and McCracken 1998, 5.

The cells of Langerhans are generally dispersed among the keratinocytes of the spiny layer and account for 3 to 8 percent of epidermal cells. They initiate and propagate immune responses to antigens (bacterial, viral, parasitic antigens) that come in contact with the skin.

The Merkel cells, which are scant (around 2 percent) and dispersed in the basal layer in contact with free nerve endings, are mechanoreceptors that react slowly to pressure. They also have inductive and nutritional functions vis-à-vis the nerve endings of the epidermis and the skin appendages (pilosebaceous follicules, sweat glands, sebaceous glands).

The dermis of the horse is much thicker than the epidermis and varies from 1 to 6 millimeters, depending on where it is on the body. Blood vessels and nerves run through the dermis and hypodermis.

The connective tissues of the dermis and hypodermis are characterized by an abundant extracellular matrix that comprises elastic and collagenous fibers (the latter provide thickness and resistance to the forces of traction), as well as an amorphous ground substance that is rich in water and soluble components and that supplies both structural support and nutrients to the other tissue components. The whole connective tissue structure forms a partly compressible gel.

The dermis contains many blood and lymphatic vessels, "the sensory receptors and nerves, the nerve endings of vessels and adnexa and, sometimes, smooth muscle tissue (hair, mammary aerolas, penis, perineum, scrotum) or striated skeletal muscle (extensions of the skin muscles of the face)" (Prost-Squarcioni 2006, 135). The horse has a superficial, papillary dermis that is relatively thin and consists of a rich population of fibers that protrude into a fine, loose connective tissue (containing the pilosebaceous follicules, sweat, and sebaceous glands) and a deep, thicker, reticular dermis, consisting of bundles of collagen fibers and a few elastic fibers. In horses, unlike small carnivores like cats and dogs, or laboratory animals (rats, mice, guinea pigs), the hair bulbs do not penetrate deep into the dermis nor, a fortiori, into the hypodermis. Moreover, the separation between the dermis and hypodermis in the horse is clearly easily demarcated and uniform. In addi-

Figure 11.2 Distribution of the additional fibrous layer separating the dermis from the epidermis of the horse. Source: Wakuri et al. 1995, 180.

tion, as I mentioned above, on the dorsal part of the body and at the extremity of the limbs, the equine dermis has a third supplementary layer, situated under the reticular layer, made of a dense tissue of fine collagen, elastic, and reticular fibers (see figure 11.2).

Smooth muscle fibers are also located in the dermis of the scrotum, the teats, and the penis, and a skeletal muscle connects the cutaneous muscle of the trunk to the whiskers of the facial region.

The hypodermis connects the dermis with deeper muscular and bone structures. Depending on the skin region, it contains more or fewer adipose cells, which constitute a reservoir of energy and provide protection and support as well as thermal isolation.

Skin Appendages

The skin appendages consist of the sweat glands, pilosebaceous follicles, and hooves.

The pilosebaceous follicules consist of a tubular invagination of the epidermis (with its four types of cells) that penetrates into the dermis and forms the epitheleal sheath of a hair, with an attached sebaceous gland. At its deepest end, it folds into a bulb, where the keratogenous cells of the matrix proliferate, giving rise to a hair, which in turn consists of a group of keratinized cells tightly linked together that migrate to the skin surface. The epithelial sheath is surrounded by a connective sheath, formed by the dermis, which is vascularized and carries sensory nerve endings. A smooth muscle, the arrector pili muscle, or horripilate muscle, is attached to the hair follicle just below the sebaceous gland. Its contraction, especially in response to fear or to cold, results in the hair standing up and release of a small quantity of sebum by the skin.

The long, fine, tapering hairs at the end of a horse's nose, on its lips and nostrils, around its eyes, on its cheeks, and above the zygomatic arch are tactile hairs, or vibrissae, rooted in tissues bearing many sensory nerves (the vibrissae at the front of the upper lip—light and wavy—are called a moustache).

Receptors: Equine Sensory Pathways and Skin Sensitivity

Skin Receptors

Tactile information is collected by the rich reservoir of nerves and sensory receptors in the skin. The sensory nerves may have free endings, or be linked to hair follicles or to receptors. The latter may be encapsulated in an envelope of connective tissue (Meissner, Ruffini, or Pacini corpuscles) or not (Merkel corpuscles).

Most of these receptors are mechanoreceptors, sensitive to stretching or to pressure. They may adapt slowly, responding continually for the entire duration of stimulation, as is the case with Meissner and Pacini corpuscles, or adapt rapidly, responding only at the beginning or possibly at the end of the stimulation, as is the case with Merkel and Ruffini corpuscles.

Only Merkel corpuscles, which are formed by a Merkel cell and a free nerve ending, are situated in the epidermis. As receptor surfaces, they are sensitive to light pressure at low frequency, which they recognize as distinct. They are very good at detecting the texture of objects. Merkel corpuscles are linked to horse bristles.

Meissner corpuscles are situated at the dermis where it meets the epidermis. They react to rapid pressure and to shaking. Ruffini corpuscles, which are more deeply rooted in the dermis by means of collagen fibers, are sensitive to stretching as well as to intense, prolonged pressure. Finally, Pacini corpuscles react in a diffuse way to rapid vibrations. They are found in subcutaneous adipose tissue, in the dermis, in connective tissue, in tendons, in joints, and in fascia muscles.

In addition to these mechanoreceptors, there are free endings of nerve fibers that respond to pain (nociceptive) signals from the skin. Some of these nociceptors, or nocireceptors, respond to intense mechanical stimulation (pinching, pricking), and others to high temperatures; still others are multimodal and react to all types of painful stimuli, whether mechanical, thermal, or chemical. Other free nerve endings constitute different receptors, thermoreceptors, some of which are sensitive to low temperature and others to high temperature.

Sensory Pathways

Except for the face, which is innervated by the trigeminal nerve (cranial nerve V), the nerves of the tactile receptors of many parts of the body have their neurons in the spinal ganglia at different levels of the spinal cord. From there, after having synapsed within the cord or in the rachidian bulb, they ascend to the thalamus, where they meet after crisscrossing of the fibers (decussation) via one or the other of three pathways distinguished notably by the nature of their sensory traffic (light or rough touch, heat, and so forth). There, they synapse again before projecting to the sensory areas of the cerebral cortex, which are located at the back of the central sulcus of the cerebral hemispheres.

The three branches of the trigeminal nerve, which innervate the different parts of the face, have their neurons in the Gasser ganglion, situated at the base of the skull. They then synapse within the pons (annular protruberance) of the brainstem, then join the thalamus, where they also synapse, before proceeding to the cortical sensory areas.

Skin Sensitivity

The tactile sensitivity of the horse, which varies according to the part of the body, is especially strong around the lips, nostrils, and eyes, given both the high concentration of receptors and the presence of vibrissae, which (as we have seen) are rooted in many nerve endings. Among other advantages, these vibrissae provide a horse with a "second view," which enables him to precisely analyze, by touch, what is in front of him in the blind spot of his visual field and to calculate how far he is from it. McGreevy (2004) reports that, in addition," horses are said to test electric fences with these whiskers before touching them" (50).

Consider, too, the presence of many filiform papillae with a tactile function within the buccal mucosa of the horse, which enhance its capacity for discrimination by allowing it to finely sort what it ingests, which explains the rarity of foreign bodies in horse intestines, compared with, say, cows (McGreevy 2004). This buccal sensitivity also comes into play in the different effects of the bits used on horses.

The sensitivity of the equine body is especially evident in its response to very light stimulation produced by the presence of flies, which causes a reflex twitch of the skin, the cutaneus trunci muscle reflex (Theriault and Diamond 1988), named for the broad, thick subcutaneous muscle that extends under the skin of the back and flanks. Saslow (2002) carried out a study of the tactile sensitivity of the horse that unfortunately remains unpublished. She measured the variation of the tactile sensitivity thresholds of different parts of the trunk, using the same methods as those used to test human tactile sensitivity.[1] A propos of this subject, she notes: "We were surprised to find that horse sensitivity on the parts of the body which would be in contact with the rider's legs is greater than what has been found for the adult human calf or even the more sensitive human fingertip. Horses can react to pressures that are too light for the human to feel" (215). This might, she continues, explain the negative consequences (failed training, "dead-sided" horses) of human instability in the saddle, and the rider unconsciously giving his horse irrelevant cues, as well as a trained animal's apparent "extrasensory perception" of the intensions of its rider, which may simply be a response to light movements or muscle contractions that the rider makes unawares.

In addition, there are also mechanoreceptors, thermoreceptors, and nociceptors in the hooves, with Ruffini and Pacini corpuscles, as well as free nerve endings (Bowker et al. 1993; Floyd and Mansmann 2007).

Skin Sensitivity and Comfort Behavior

As suggested by David McFarland (2006), the term *comfort behavior* applies to a group of heterogeneous activities that "have to do with body care, including bathing, dusting, grooming, preening, scratching, shaking, stretching, sunning, and yawning" (34). The attention that horses pay to the care of their skin manifests in many ways; it is not a matter of simple cutaneous reflexes, such as the skin twitch, which I just mentioned, in response to contact with flies, for example, but voluntary motor behaviors, such as chewing, rubbing, scratching, rolling, and so forth that contribute to the hygiene of the skin and to relieving itching due especially to parasites or diverse foreign bodies that manage to become entrenched in the horse's coat.

That is why, when (among other things) turning their head and stretching their neck, horses chew their rumps or flanks, front as well as back. Different parts of the face (eyes, forehead, cheeks, nostrils) can also be attended to by rubbing them along their foreleg, which they may extend for this purpose. They may also rub the parts of their bodies that they cannot reach, like the rump or buttock against the branches or trunk of a tree, or any other available surface. Horses also scratch certain parts of the head or the neck, or the elbow area, using a hind limb. When horses roll, they do it on one side, and then the other, then shake themselves vigorously, which jiggles their skin all over their body, thus freeing it of debris that may have accumulated in the coat.

However, comfort behaviors are not necessarily individual; for example, it is not rare for horses to pair up, one against the other,

---

1. Most likely von Frey filaments.

head to tail, swinging their tails as mutual flyswatters. A propos of which, there is one behavior that merits particular attention, as we will see, that is, mutual grooming.

Mutual Grooming and Neurophysiological Response

Mutual grooming, or allogrooming, during which horses stand head to tail and gently chew at each other's neck, mane, withers, back, rump, and even the base of the tail, is an activity that manifests more frequently during the month of April, around the time a horse's winter coat molts, which attests to its hygienic function. However, mutual grooming is also practiced regularly throughout the year between members of a social group (Feist 1971; Tyler 1971; Keiper 1985) and especially during summer when horses gather closely to share shady areas. In primates it is thought not only to help eliminate ectoparasites but also to have a social function (Schino et al. 1988) by calming the social tension caused by greater proximity.

Claudia Feh and Jeanne de Mazières (1993) were the first to explore this aspect of allogrooming in horses through an experiment aimed at determining, on the one hand, whether a particular area of the body constitutes a privileged target for grooming and, on the other hand, whether the activity translates into decreased heart rate for the groomed individual. The researchers first recorded body sites that were groomed during two months, in May and June, during which period the incidence of ectoparasites on horse skin is at its lowest. They observed that one clearly preferred site was the base of the neck, surrounded by a site of least preference, aside from which no grooming was done on various parts of the body, in particular at shoulder level, some 40 centimeters below the preferred area, called the "nonpreferred grooming" area. They also chose to stimulate this area for purposes of comparison (see figure 11.3, top).

They then investigated the influence of scratching on eight horses aged three to twenty years old and eight foals six to eight months old. One of the researchers simulated grooming by manual scratch-

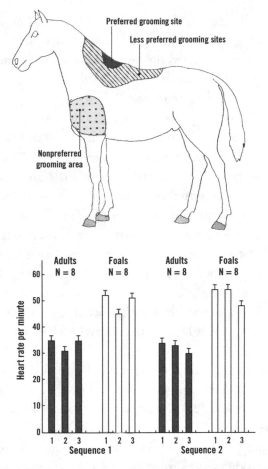

Figure 11.3 Top: preferred grooming sites, minimum preferred site, and tested nonpreferred grooming site. Source: Feh and de Mazières 1993, 1193. **Bottom: significant decrease in heart rate in adults and foals by scratching sequence.** Source: Feh and de Mazières 1993, 1193.

ing at the same frequency (two per second) at which horses groom each other, while the other researcher dictated numbers of heartbeats through a stethoscope into a tape recorder every 30 seconds. The researchers began the experiment after having confirmed that the horse was resting. They recorded its heart rate without any stimulation for three minutes, then during the next three minutes while the horse was scratched on the preferred grooming site, and finally for three minutes during which the horse was scratched on the nonpreferred grooming site (sequence 1). A few hours later, or the next day, the researchers repeated the procedure with the same horse, but reversed the order in which the two sites were scratched, thus scratching the preferred site last (sequence 2).

At the end of this experiment, in addition to significantly higher heart rate in foals than in adults (in general $p < 0.001$ for the different situations), the researchers found (see figure 11.3) a significant decrease in heart rate ($p < 0.001$) in both foals and adults during stimulation of the preferred site, followed by an equally significant increase during stimulation of the nonpreferred grooming site. In contrast, there was no significant difference in heart rate ($p > 0.05$) between the period preceding scratching and that during which the horse was scratched on the nonpreferred grooming site. During sequence 2, scratching the horses at this site immediately after the resting period did not result in a significant decrease in heart rate; however, it did ($p < 0.001$) when the researchers began scratching the preferred site.

According to the authors, imitation of grooming resulted in a substantial decrease in cardiac rate of 11.4 percent on average in the eight adult horses and 13.5 percent on average in the eight foals for the preferred site, whereas the same stimulation applied to the nonpreferred grooming site had no effect, which suggests that grooming the preferred site did have a calming effect on the groomed individual.

Finally, in discussing why the calming effect is limited to a particular site on the body, the authors find it likely that the parasympathetic part of the autonomous nervous system is involved, noting that the ganglion stellatum, which is the most important ganglion in the horse, is close to the preferred grooming site and that it is known to be linked to major efferent cardiac pathways. The authors also note in concluding that the preferred site that they had identified corresponds to that used by veterinarians who use acupressure to calm nervous horses (Giniaux 1998).

Simona Normando and colleagues (2003) subsequently replicated this experiment with several modifications: They used an equine radio frequency counter to record heart rate measurements every 5 seconds; a population of sixteen geldings aged three to twelve years, regularly exercised; four scratching sites; and a Latin square design to determine the sequence of the stimulated sites. The four sites chosen were the base of the neck (at 10 centimeters around the withers and at 10 centimeters at the start of the mane), the withers, the shoulder (centered on the scapulohumeral joint), and the hip (just below the ventral iliac spine, *spina iliaca ventralis*).

Their results confirmed the main conclusion of the preceding study, that is, a net decrease in heart rate due to the calming effect of the scratching on the withers as well as, to a lesser extent, the base of the neck. The researchers also noted, however, a certain decrease in heart rate at the shoulder and the hip, whereas they expected to find stability there, or even an increase in the case of the shoulder, which is the ready target of bites or bite threats (Waring 2003). On this subject, the researchers proposed that the result might be due to conditioning, as the horses they used were accustomed to being patted with the flat of the hand in those places in a positive context, and such an act might, by association, be taken as a reward.

Viewed from the therapeutic perspective of massage, which in traditional Chinese medicine constitutes a form of acupressure that is relaxing, and considering the observations of Feh and de Mazières (1993) on the withers-grooming effect, Sebastian McBride and colleagues (McBride, Hemmings, and Robinson 2004) undertook to examine whether effleurage (a massage technique) of any one of six different body sites could reduce stress in the horse. The sites chosen were the withers, midneck, croup, second thigh, the forearm, and poll and ears. The first three were considered by Stephen Budiansky (1997) as the preferred sites commonly used in social grooming, and the three others as nonpreferred sites of social grooming (see figure 11.4).

It is generally thought that a preliminary period of massage is necessary to induce an effect, and consequently, it was applied to the ten horses (various ages, sexes, and breeds) used in this study for seven days. At the same time, they were being trained to carry a heart

Figure 11.4 The six sites of effleurage used by McBride and colleagues. Source: McBride et al. 2004, 77.

rate monitor (an equine radio frequency counter that recorded the heart rate every 5 seconds).

For a given site, the resting heart rate was recorded before the massage started, with the horse at rest during five minutes, the mean of the measurements giving a pre value; during the massage, which also lasted five minutes (mean during value); and finally for five consecutive minutes after the massage (mean post value). Throughout the experiment, an observer scored the behavioral response of the horse being massaged on a scale of 1 to 5 (1: highly negative, for example, restless behavior, attempts at biting; 2: negative response; 3: indifferent; 4: positive; 5: highly positive, for example, rubbing against the masseur, somnolent-type behavior), according to a list of behavioral score criteria. This procedure was repeated for each of the ten horses and for each of the six massage sites. The order of the sites according to which the horses were massaged was chosen at random, with one constraint: The sequence chosen was different for each of the ten horses.

The withers and midneck massages resulted in the strongest decrease in heart rate and also the highest (positive) behavioral response scores. These effects were increasingly less significant for

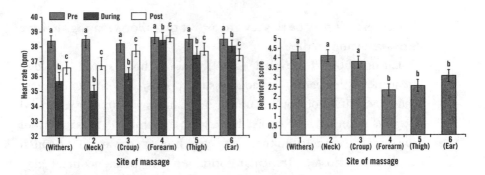

Figure 11.5 Left: mean heart rate during the three experimental phases for each site. Right: mean behavioral scores for each site. Source: McBride et al. 2004, 79, 80.

the croup, second thigh, ear, and forearm. Thus, it was indeed the preferred sites of social grooming that showed the greatest physiological and behavioral effects, compared with the sites where no social grooming was observed.

Overall, taking the mean of the measurements for all three sequences pre, during, and post, also shows a general trend toward a greater decrease (4.3 percent) in heart rate in the during phase than in the post phase (2.6 percent). However, two sites were an exception to this trend: the forearm, for which massage in any event did not result in a significant change in heart rate, and the poll and ear area, for which the significant decrease in heart rate continued beyond the end of the massage (see figure 11.5).

The authors remark that, in general, mutual grooming fills several functions: hygiene (cleaning, extraction of ectoparasites, and prevention of infections); distensive (the restraint, prevention, or diversion of potential aggression); and affiliative (establishing and maintaining bonds). However, compared with primates, for example, whose mutual grooming involves the entire body and thus appears especially to be related to hygiene, in ungulates, where the practice is more body-region specific, it may serve to emphasize the social aspect. However, the authors add, like other behaviors oriented toward a goal, mutual grooming appears to entail a hedonic reward element designed to maintain the behavior both in the shorter and longer term. It may be this reward aspect that allows massage to have stress-reducing qualities.

Citing Donald Kendall (1989), the authors state that low-frequency stimulation at acupressure points, identified as such, causes activation of small-diameter nerve fibers within the peripheral nervous system. These nerve fibers synapse within the dorsal horn of the spinal cord and from there activate sensory pathways in the spinal cord, the brainstem (reticular formation and periaqueductal gray area), and the hypothalamus. A large part of this activation is opioid-mediated, accompanied by activation of the serotonergic neurons in the raphe nuclei section of the reticular formation. "It is," they write, "this opioid mechanism that is considered responsible for the analgesic effects of acupuncture and the pleasurable or 'well-being' sensation of acupressure or massage" (80).[2] They add that the neural mechanisms at play are also consistent with a decrease in heart rate, as the relative decreases observed (for a single site, or between different sites) of this rate may reflect the differences in stimulation of the periaqueductal gray area. Finally, they suggest that the prolonged effect of stimulation of the ear area might be related to the observation that some areas within the ear are primary acupuncture points for treating pain. They conclude that, from a practical point of view, massage may be used to induce a more relaxed, calm state in the horse, under certain low to medium stressful situations.

Simona Normando and colleagues (2007), however, later reported an experiment carried out with twenty-seven horses, twelve of whom had been exhibiting stereotypies for at least two months (two of the animals showed oral stereotypies, seven locomotor stereotypies, and three others both types), whereas fifteen other horses, which served as a control group, showed no stereotypies at all. Stimulation consisted of scratching and a muscle massage performed with the ends of the fingers on the back, along the vertebral column, in particular the withers. Heart rate was measured with the aid of a stethoscope.

---

2. In this same vein, note that the desensitizing action of nose twitching also appears to be related to release of opioids, beta-endorphins (Lagerweij et al. 1984; Schelp 2000).

Table 11.1   HR (mean ± SD in beats per minute) in stereotyping (ST) and control (NST) horses during premassage and massage phases and percentage variation between the two phases.

	Premassage HR	Massage HR	% Variation
ST	40.7 ± 4.5	43.9 ± 4.3	7.9 ± 3.4
NST	40.9 ± 5.2	38.2 ± 4.7	−6.6 ± 4.0

Source: Normando et al. 2007, 103.

While the horse was in its usual box stall, the experimenter waited two minutes after entering, to give the horse time to adjust to his presence. He then measured the horse's heart rate for two consecutive minutes (taking the mean value of the first and second minutes) and finally imitating allogrooming for seven minutes, without interruption, measuring the heart rate separately during the sixth and seventh minutes. The results were surprising. For the horses presenting no stereotypies, the researchers did indeed observe a decrease in heart rate. However, for the horses exhibiting stereotypies, the heart rate increased, contrary to expectations (see table 11.1).

The authors note the existence of other studies (Minero et al. 1999; Bachmann et al. 2003) reporting changes in heart rate in horses exhibiting stereotypies compared with those who do not. However, while suggesting a different motivational state in the two populations, the authors state that they found no satisfactory explanation for the increase in heart rate and call for further research.

Tactile Stimulation and Interspecific Social Relationships

It turns out that procuring, or sharing, food plays an important role in establishing social relationships in the animal kingdom, especially in horses.

The social component of mutual grooming in horses and the observation of a calming effect of scratching on a horse's withers suggest, among other things, that tactile stimulation in itself may also help to establish an attachment relationship between a human and a horse, in

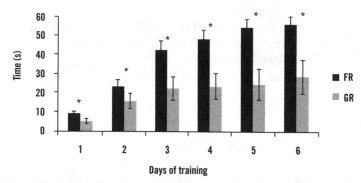

Figure 11.6 **Maximum duration of immobility on order based on reward type. FR: food reward. GR: grooming reward.** Source: Sankey et al. 2010a, 2.

the form of a primary positive reinforcement. Accordingly, Carol Sankey and colleagues (2010a) compared the effect of a food reward and a tactile reward in the course of training and in the evolution of the relationship between horses and humans, using two groups of ten horses each, rewarded by one or the other method.

The procedure consisted of individually training the horses to stand still for increasingly long periods of time, from 5 to 60 seconds, following the verbal command "Stay!" When the horse obeyed the command, the experimenter gave it a small piece of carrot in the case of the food reward and vigorously scratched its withers three times in the case of the tactile reward. Each horse was exposed to six daily training sessions, each lasting five minutes. To evaluate the evolution of their relationship with a human, the horses were also subjected to an "immobile human test" before and after the training: For five minutes, the horse was left free to approach the experimenter, who was positioned in the center of the testing space. The latency time for an approach of less then 0.50 meters was recorded, as well as the time the horse spent inside this limit.

As the results show, the two types of rewards produced marked differences (see figure 11.6). For example, the horses rewarded by food remained still longer after they received the command than the horses rewarded by scratching. Moreover, whereas almost all of the first group of horses succeeded in remaining still for a minute at the end of six days of training, four of the second group of horses did

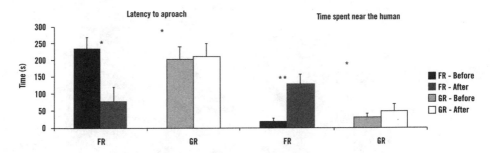

Figure 11.7 Approach latency and time spent near the experimenter based on reward type. Source: Sankey et al. 2010a, 3.

not. Thus, although the food reward did appear to be a primary rein-forcement, the tactile reward was not. Naturally, these findings do not rule out that tactile stimulation could also function as secondary reinforcement, through association with a primary reinforcement, for example, food.

As for relationships to humans, it is clear that, although the food reward had a significant effect in reducing the approach latency and in resulting in markedly increased time spent in close proximity to the person, scratching had practically no effect (see figure 11.7). Consequently, for horses as for humans, food appears to be a key element in the process of attachment, which the authors underscore amusingly in concluding that saying "The way to a man's heart is through his stomach" applies just as equally to our friend the horse.

In contrast, as the authors also note, in the horse "tactile human contact, even when it imitates natural intraspecific interactions, is not necessarily perceived in a positive way and certainly does not suffice to create attachment" (Sankey et al. 2010a, 3). Aside from the fact that the signaling of attachment in humans by social interac-tions and physical contact is not necessarily mirrored in all species, it would be hazardous to blithely transpose intraspecific relationships to those between humans and animals.

As has been shown in the preceding chapters, the horse inhabits its own world, . . . and we ours!

CONCLUSION

A note prior to wrapping up. In the preceding pages, I have touched both at length and yet very little on learning. At length, in the sense that I have (necessarily) frequently discussed and analyzed learning as a means of exploring different cognitive processes and the representations they entail. Very little, in the sense that I have only occasionally mentioned work relating to the study of learning as such or its application in the context of what has become known as equitation science. This material has in fact been fairly exhaustively dealt with elsewhere, as I indicated in the first chapter. Consequently, it seemed to me there was no reason to revisit it except where explicitly warranted, for example, in the discussion of feeding behavior.

Research on the cognitive ethology of the horse, carried out over the two last decades, and in particular very recently, has significantly advanced our understanding of the way in which the mind of the horse works. However, missing pieces of the current picture also highlight gaps still to be filled and, accordingly, indicate the road ahead.

To elaborate these points, and shuffling somewhat the order of the chapters, permit me here to make a few remarks on what is known about equine perceptual modalities as well as on the cognitive capacities that their study has revealed. Next, I will extend the discussion to other cognitive processes and suggest research approaches to questions that thus far have been largely neglected.

1. Knowledge of Perceptual Modalities as Such

The discussion of studies relevant to the different perceptual modalities of the horse clearly illustrates the central point of the introduction, which is that the degree of current knowledge varies depending on the modality. Without repeating material already covered in detail, this heterogeneity can be characterized as follows.

Diverse studies on visual perception carried out roughly since the end of the 1990s have revealed much about its characteristics, although the question of object perception in particular merits going into it more deeply, and little is really known about the perception of movement. Earlier uncertainties, for example, regarding stereoscopic vision, have now been resolved. The knotty issue of color perception has largely been sorted thanks to the recognition and characterization of equine dichromatic vision. The challenge is no longer to discover whether the horse differentiates this or that color from gray but rather to understand how its perception is organized within the light spectrum, an endeavor that today appears to be well on track.

Unlike visual perception, what is currently known about equine auditory perception was established in the decade spanning 1980–1990. Auditory acuity and the localization of sounds are generally well understood, although work on the latter takes into account only the azimuthal plane.

Equine chemical perception, on the other hand, is still largely virgin territory. This state of affairs is not so much a reflection of the lack of work in the area; rather, the phenomena involved are complex and require knowledge of how the underlying neuroanatomical mechanisms of mammals developed in general as well as their functional evolution. This knowledge is at least developing. By way of a "simple" illustration, consider, for example, the debate of recent years on the respective roles of the principal olfactory system (with the olfactory epithelium) and the accessory olfactory system (with the vomeronasal organ).

At the same time, it is less obvious why equine tactile perception continues to elude characterization, given that it is so important in the daily life of horses, whether free ranging or domestic. On the face of it, nothing would seem to hinder psychophysical investigations to determine its broad outlines. We can only hope to see progress in this area in the not-too-distant future.

2. Perception: A Window on the Mental World of the Horse

As I mentioned in the chapter introducing the study of perception, it constitutes a dynamic process that constructs the world in which the individual lives. The brain, remember, perceives the world by transforming the data supplied by the senses, in a context where perception and action are intimately linked. And, not surprisingly, perception acts to reveal the workings of various cognitive capacities, beyond apprehension of perceptual mechanisms *sensu stricto*.

For example, research on physical cognition has uncovered the capacity of the horse to associate an object and its image and the animal's probable limits in terms of mental imagery for recognizing objects. Similarly, investigation into social cognition has aided understanding of how the horse is able both to categorize according to social information transmitted by the vocalizations of its conspecifics and to recognize individuals within or between species.

Finally, study of different perceptual modalities has helped to answer the question of lateralization and hemispheric brain specialization, which also comes up in the context of cognition and emotion.

3. Evolution of Knowledge Related to Other General Cognitive Processes

More broadly, research on the cognitive ethology of the horse has contributed to better understanding of fundamental cognitive processes, although the last word on the subject remains to be written. The same is true especially of the equine capacity for representation, categorization, and even formation of concepts. Likewise, the animal's remarkable long-term memory, which previously had only anecdotal support, is now proven.

Yet, it is also clear that certain paths of investigation are beginning to generate more questions than answers. Take, for example, research on short-term visual memory, which—contradictory results notwithstanding—has raised questions about the ecological validity of the available experimental paradigms.

Another case is the range of study topics that fall within the broad field of human-horse relationships, which is attracting ever greater interest. Thus, the question of the capacity for facial discrimination (and *a fortiori* its cortical corollaries) is barely sketched in the horse, whereas in other animals, for instance, sheep, it has produced a substantial literature. In the same vein, whether the horse takes into account cues supplied by humans (posture, body or gaze orientation, pointing gestures, and so on) requires additional studies, though several have already been done.

4. Looking to the Horizon: Fields of Research to Be Encouraged or to Discover

As already noted, immense fields of research have only begun to be tapped or still await exploration. Let me summarize some of them.

Although, as remarked above, visual short-term memory in the horse has been studied, the result of which not least has been to highlight the need for further work, to date the question of the permanence of the object has been totally neglected.

That horses construct a concept of relative size has been demonstrated, but the more general capacity of the animal to form abstract concepts is unknown. For example, what is the horse's ability to represent time? Taking into account the conditions in which many domesticated horses are kept, the answer to such a question would be of great interest.

Research into feeding strategies has focused on the interaction of time and space, the assumption being that, in the absence of recent information, horses choose their feeding sites according to the long-term average richness of the available sites. But how does a horse represent its home range? How does it structure the space? How does it find its way around it? Thus far, none of these questions have been tackled in any systematic fashion.

Finally, in the area of social cognition, the apparent closeness of relationships between individual horses and humans naturally stimulates inquiry. In this context, central issues like individual recognition—not just intra- but also interspecies—have certainly been the subject of recent advances but remain largely unexplored. Moreover, with reference to whether horses take into account cues provided by humans, the recent suggestion, in particular, that horses might possibly respond in the direction of human attention (body or gaze orientation) when the human is familiar offers exciting perspectives on the nature of the cognitive processes involved and their exploration.

Knowledge in these areas will no doubt increase significantly in the coming years. The cognitive ethology of the horse, as this book affirms, is no longer terra incognita. Questions that yesterday had no real answer, and thus could only be speculated about, have now either been satisfactorily resolved or, at least, have been reformulated or elaborated in a way that makes them amenable to new avenues of research. The achievements of recent years prove that, around the world, momentum has been established that augurs well for the future.

REFERENCES

Adachi, I., H. Kuwahata, and K. Fujita. 2007. Dogs recall their owner's face on hearing the owner's voice. *Animal Cognition* 10: 17–21.

Adelson, E. H. 2000. Lightness perception and lightness illusions. In *The New Cognitive Neurosciences,* 2nd ed., edited by M. S. Gazzaniga. Cambridge, MA: MIT Press, 339–51.

Ahrendt, L. P., J. W. Christensen, and J. Ladewig. 2012. The ability of horses to learn an instrumental task through social observation. *Applied Animal Behaviour Science* 139: 105–13.

Altman, J. 1962. Are new neurons formed in the brains of adult mammals? *Science* 135: 1127–28.

Anderson, T. M., B. W. Pickett, J. C. Heird, and E. L. Squires. 1996. Effect of blocking vision and olfaction on sexual responses of haltered and loose stallions. *Journal of Equine Veterinary Science* 16: 254–61.

Antell, S. E., and D. P. Keating. 1983. Perception of numerical invariance in neonates. *Child Development* 54: 695–701.

Anthony, D. W. 1996. Bridling horse power: The domestication of the horse. In *Horses through Time,* edited by S. L. Olsen. Boulder, CO: Robert Rinehart, 57–82.

Archer, M. 1971. Preliminary studies on the palatability of grasses, legumes, and herbs to horses. *Veterinary Record* 89: 236–40.

———. 1973. The species preferences of grazing horses. *Journal of the British Grassland Society* 28: 123–28.

Aubin, T., and P. Jouventin. 1998. Cocktail-party effect in king penguin colonies. *Proceedings of the Royal Society of London B* 265: 1665–73.

Austin, N. P., and L. J. Rogers. 2007. Asymmetry of flight and escape turning responses in horses. *Laterality* 12: 464–74.

————. 2012. Limb preferences and lateralization of aggression, reactivity, and vigilance in feral horses (*Equus caballus*). *Animal Behaviour* 83: 239–47.

Bachmann, I., P. Bernasconi, R. Herrmann, M. A. Weishaupt, and M. Stauffacher. 2003. Behavioural and physiological responses to an acute stressor in crib-biting and control horses. *Applied Animal Behaviour Science* 82: 297–311.

Baer, K. L., G. D. Potter, T. H. Friend, and B. V. Beaver. 1983. Observation effects on learning in horses. *Applied Animal Ethology* 11: 123–29.

Bagot, J.-D. 2002. *Information, sensation et perception*. Paris: Armand Colin.

Bailey, R. E., and J. A. Gillaspy. 2005. Operant psychology goes to the fair: Marian and Keller Breland in the popular press, 1947–1966. *Behavior Analyst* 28: 143–59.

Baillargeon, R. 2004. Infants' reasoning about hidden objects: Evidence for event-general and event-specific expectations. *Developmental Science* 7: 391–424.

Baker, A. E. M., and B. H. Crawford. 1986. Observational learning in horses. *Applied Animal Behaviour Science* 15: 7–13.

Baragli, P., A. Gazzano, F. Martelli, and C. Sighieri. 2009. How do horses appraise humans' actions? A brief note over a practical way to assess stimulus perception. *Journal of Equine Veterinary Science* 29: 739–42.

Baragli, P., C. Mariti, L. Petri, F. De Giorgio, and C. Sighieri. 2011a. Does attention make the difference? Horses' response to human stimulus after 2 different training strategies. *Journal of Veterinary Behavior* 6: 31–38.

Baragli, P., E. Paoletti, V. Vitale, C. Sighieri, and A. R. Reddon. 2011b. Detour behaviour in horses (*Equus caballus*). *Journal of Ethology* 29: 227–34.

Baragli, P., V. Vitale, E. Paoletti, M. Mengoli, and C. Sighieri. 2011c. Encoding the object position for assessment of short term spatial memory in horses (*Equus caballus*). *International Journal of Comparative Psychology* 24: 284–91.

Barasa, A. 1960. Forma, grandezza e densitá dei neuroni della corteccia cerebrale in mammiferi di grandezza corporea differente. *Cell and Tissue Research* 53: 69–89.

Barbet, I., and J. Fagot. 2002. Perception of the corridor illusion by baboons (*Papio papio*). *Behavioural Brain Research* 132: 111–15.

Barfield, C. H., Z. Tang-Martinez, and J. M. Trainer. 1994. Domestic calves (*Bos taurus*) recognize their own mothers by auditory cues. *Ethology* 97: 257–64.

Barnett, K. C., S. M. Crispin, J. D. Lavach, and A. C. Matthews. 2004. *Equine Ophthalmology: An Atlas and Text*. 2nd ed. London: Saunders.

Barone, R. 1984. *Splanchnologie I, appareil digestif et appareil respiratoire*. Vol. 3 of *Anatomie comparée des mammifères domestiques*. Paris: Vigot Frères.

Barone, R., and R. Bortolami. 2004. *Neurologie I, système nerveux central*. Vol. 6 of *Anatomie comparée des mammifères domestiques*. Paris: Vigot Frères.

Barton, R. A., and P. H. Harvey. 2000. Mosaic evolution of brain structure in mammals. *Nature* 405: 1055–58.

Bartoš, L., J. Bartošová, and L. Starostová. 2008. Position of the head is not associated with changes in horse vision. *Equine Veterinary Journal* 40: 599–601.

Basile, M., S. Boivin, A. Boutin, C. Blois-Heulin, M. Hausberger, and A. Lemasson. 2009. Socially dependent auditory laterality in domestic horses (*Equus caballus*). *Animal Cognition* 12: 611–19.

Baucher, F. 1843. *Dictionnaire raisonné d'équitation*. Brussels: Société Nationale pour la propagation des bons livres.

————. 1850. *Méthode d'équitation*. Paris: J. Dumaine. Reissued by Péronnas, France: Éditions du Bastion, 1990.

————. 1852. *A Method of Horsemanship Founded upon New Principles*. Anonymous translator. Philadelphia: Hart, Cary, and Hart.

Baumont, R., S. Prache, M. Meuret, and P. Morand-Fehr. 2000. How forage characteristics influence behaviour and intake in small ruminants: A review. *Livestock Production Science* 61 (1): 15–28.

Baxi, K. N., K. M. Dorries, and H. L. Eisthen. 2006. Is the vomeronasal system really specialized for detecting pheromones? *Trends in Neurosciences* 29: 1–7.

Békésy, G. von. 1947. The variation of phase along the basilar membrane with sinusoidal vibrations. *Journal of the Acoustical Society of America* 19: 452–60.

—— ——. 1956. Current status of theories of hearing. *Science* 123: 779–83.

Berger, J. 1986. *Wild Horses of the Great Basin: Social Competition and Population Size*. Chicago: University of Chicago Press.

Berthoz, A. 2000. *The Brain's Sense of Movement*. Cambridge, MA: Harvard University Press.

Biederman, I. 1987. Recognition-by-components: A theory of human image understanding. *Psychological Review* 94: 115–47.

Binet, A. 1911. Nouvelles recherches sur la mesure du niveau intellectuel chez les enfants d'école. *L'Année Psychologique* 19: 145–201.

Binet, A., and T. Simon. 1905. Application des méthodes nouvelles au diagnostic des enfants normaux et anormaux d'hospice et d'école primaire. *L'Année Psychologique* 13: 245–336.

Birke, L., J. Hockenhull, E. Creighton, L. Pinno, J. Mee, and D. Mills. 2011. Horses' responses to variation in human approach. *Applied Animal Behaviour Science* 134: 56–63.

Bisulco, S., and B. Slotnick. 2003. Olfactory discrimination of short chain fatty acids in rats with large bilateral lesions of the olfactory bulbs. *Chemical Senses* 28: 361–70.

Blackmore, T. L., T. M. Foster, C. E. Sumpter, and W. Temple. 2008. An investigation of colour discrimination with horses (*Equus caballus*). *Behavioural Processes* 78: 387–96.

Blake, H. 1975. *Talking with Horses*. London: Souvenir.

————. 1977. *Thinking with Horses*. London: Souvenir.

Bliss, T. V., and T. Lomo. 1973. Long-lasting potentiation of synaptic transmission in the dentate area of the anaesthetized rabbit following stimulation of the perforant path. *Journal of Physiology* 232: 331–56.

Blodgett, H. C. 1929. The effect of the introduction of reward upon maze performance of rats. *University of California Publications in Psychology* 4: 113–34.

Boeglin, J. 2003. La vision des couleurs. In *Perception et Réalité*, edited by A. Delorme and M. Flückiger. Brussels: De Boeck Université, 105–26.

Bonde, M., and D. Goodwin. 1999. Behaviour of stabled horses when presented with different odours. *Equine Veterinary Journal* 60 (S28): 1–2.

Bonnardel, V. 2005. Colour naming and categorization in inherited colour vision deficiencies. *Visual Neuroscience* 26: 637–43.

Bonnet, C. 2003. Les sens chimiques. In *Perception et Réalité*, edited by A. Delorme and M. Flückiger. Brussels: De Boeck Université, 173–96.

Bonnet, C., R. Ghiglione, and J.-F. Richard. 1989. *Traité de psychologie cognitive, tome 1: Perception, action, langage.* Paris: Dunod.

Bovet, D., and J. Vauclair. 2000. Picture recognition in animals and humans. *Behavioural Brain Research* 109: 143–65.

Bovet, P. 2003. L'audition. In *Perception et Réalité*, edited by A. Delorme and M. Flückiger. Brussels: De Boeck Université, 127–49.

Bowker, R. M., A. M. Brewer, K. B. Vex, L. A. Guida, K. E. Linder, I. M. Sonea, and A. W. Stinson. 1993. Sensory receptors in the equine foot. *American Journal of Veterinary Research* 54: 1840–44.

Boyd, L., and R. Keiper. 2005. Behavioural ecology of feral horses. In *The Domestic Horse: The Evolution, Development, and Management of Its Behaviour*, edited by D. Mills and S. McDonnell. Cambridge, UK: Cambridge University Press, 55–82.

Brainard, D. H. 1997. The psychophysics toolbox. *Spatial Vision* 10: 433–36.

Brand, G., J. L. Millot, M. Saffaux, and N. Morand-Villeneuve. 2002. Lateralization in human nasal chemoreception: Differences in bilateral electrodermal responses related to olfactory and trigeminal stimuli. *Behavioural Brain Research* 133: 205–10.

Bravo, M. R., R. Blake, and S. Morrison. 1988. Cats see subjective contours. *Vision Research* 28: 861–65.

Breland, K., and M. Breland. 1951. A field of applied animal psychology. *American Psychologist* 6: 202–4.

———. 1961. The misbehavior of organisms. *American Psychologist* 16: 681–84.

Bremer, C. D., J. B. Pittenger, R. Warren, and J. J. Jenkins. 1977. An illusion of auditory saltation similar to the cutaneous "rabbit." *American Journal of Psychology* 90: 645–54.

Brennan, P. A., and F. Zufall. 2006. Pheromonal communication in vertebrates. *Nature* 444: 308–15.

Breslin, P. A., and A. C. Spector. 2008. Mammalian taste perception. *Current Biology* 18: R148–55.

Briant, C., A. Bouakkaz, Y. Gaude, I. Couty, D. Guillaume, J. M. Yvon, A. Touchard, A. Najjar, S. Ben Said, S. Ezzar, B. Benaoun, M. Ezzaouia, Y. Maurin, O. Rampin, B. Nielsen, and M. Magistrini. 2012. Sur la piste de l'odeur d'oestrus . . . *IFCE (Institut français du cheval et de l'équitation) 38ème Journée de la Recherche Equine*: 47–54.

Briant, C., Y. Gaudé, B. Bruneau, J. M. Yvon, D. Guillaume, and A. Bouakkaz. 2010. Olfaction is not absolutely necessary for detection of the estrous mare by the stallion. *Animal Reproduction Science*: S120–22.

Broca, P. 1888. *Mémoires d'anthropologie.* Paris: Reinwald.

Brooks, C. J., and S. Harris. 2008. Directed movement and orientation across a large natural landscape by zebras, *Equus burchelli antiquorum. Animal Behaviour* 76: 277–85.

Brown, C. H., T. Schessler, D. Moody, and W. Stebbins. 1982. Vertical and horizontal sound localization in primates. *Journal of the Acoustical Society of America* 72: 1804–11.

Brown, P. K., and G. Wald. 1964. Visual pigments in single rods and cones of the human retina. *Science* 144: 45–52.

Brown, W. M. 2001. Natural selection of mammalian brain components. *Trends in Ecology and Evolution* 16: 471–73.

Brownell, W. E., C. R. Bader, D. Bertrand, and Y. de Ribaupierre. 1985. Evoked mechanical responses of isolated cochlear outer hair cells. *Science* 227: 194–96.

Buchbauer, G., L. Jirovetz, W. Jäger, H. Dietrich, and C. Plank. 1991. Aromatherapy: Evidence for sedative effects of the essential oil of lavender after inhalation. *Zeitschrift für Naturforschung C* 46: 1067–72.

Buck, L., and R. Axel. 1991. A novel multigene family may encode odorant receptors: A molecular basis for odor recognition. *Cell* 65: 175–87.

Budiansky, S. 1997. *The Nature of Horses: Exploring Equine Evolution, Intelligence, and Behavior.* New York: Free Press.

Buffon, G.-L. Leclerc, Comte de. 1781. *Natural History: General and Particular.* Vol. 3. Trans. W. Smellie. London: J. S. Barr. Available through New Jersey City University: http://faculty.njcu.edu/fmoran/subjects3.htm.

Burgaud, F. 2001. Les vices d'écurie: Approche épidémiologique et comportementale. *Haras Nationaux, 27ème Journée de la Recherche Equine*: 55–63.

Busnel, R.-G. 1963. On certain aspects in animal acoustic signals. In *Acoustic Behaviour of Animals*, edited by R. G. Busnel. Amsterdam: Elsevier, 69–111.

Butler, R. A. 1986. The bandwidth effect on monaural and binaural localization. *Hearing Research* 21: 67–73.

Buytendijk, F. J. J. 1965. *L'homme et l'animal: Essai de psychologie comparée.* Paris: Gallimard Idées NRF. First published 1958.

Cairns, M. C., J. J. Cooper, H. P. B. Davidson, and D. S. Mills. 2002. Association in horses of orosensory characteristics of foods with their post ingestive consequences. *Animal Science* 75: 257–66.

Calas, A., J.-F. Perrin, C. Plas, and P. Vanneste. 1997. *Précis de Physiologie.* Vélizy, France: Doin Éditeurs.

Carlier, C. 2002. *Le Crocodile, le cheval, l'homme: Trois cerveaux, trois lois naturelles.* Paris: François-Xavier de Guibert.

Carroll, J., J. N. Ver Hoeve, M. Neitz, C. J. Murphy, and J. Neitz. 2001. Photopigment basis for dichromatic color vision in the horse. *Journal of Vision* 1: 80–87.

Cartier, M. 1993. Considérations sur l'histoire du harnachement et de l'équitation en Chine. *Anthropozoologica* 18: 29–43.

Chandrashekar, J., M. A. Hoon, N. J. Ryba, and C. S. Zuker. 2006. The receptors and cells for mammalian taste. *Nature* 444: 288–94.

Changeux, J. P. 1985. *Neuronal Man: The Biology of the Mind.* New York: Pantheon.

Changeux, J. P., P. Courrège, and A. Danchin. 1973. A theory of the epigenesis of neuronal networks by selective stabilization of synapses. *Proceedings of the National Academy of Sciences of the U.S.A.* 70: 2974–78.

Chapouthier, G. 2001. *L'homme en mosaïque.* Paris: Odile Jacob.

Chauchard, P. 1956. Poids du cerveau et poids du corps. *L'Année Psychologique* 56: 101–5.

Cherry, E. C. 1953. Some experiments on the recognition of speech, with one and with two ears. *Journal of the Acoustical Society of America* 25: 975–79.

Chneiweiss, H. 2002. Les astrocytes contrôlent la neurogenèse dans le système nerveux central adulte. *Médecine/Sciences* 18 (11): 1065–66.

Christensen, J. W., L. J. Keeling, and B. L. Nielsen. 2005. Responses of horses to novel visual, olfactory, and auditory stimuli. *Applied Animal Behaviour Science* 93: 53–65.

Christensen, J. W., and M. Rungren. 2008. Predator odour per se does not frighten domestic horses. *Applied Animal Behaviour Science* 112: 136–45.

Christopher, M. 1970. *Seers, Psychics, and ESP*. New York: Thomas Y. Crowell.

Claparède, E. 1912. Les chevaux savants d'Elberfeld. *Archives de Psychologie* (Geneva) 12: 263–304.

———. 1913a. Les chevaux savants d'Elberfeld. *Bulletin de la Société Française de Philosophie*, séance du 13 mars 1913: 115–34.

———. 1913b. Encore les chevaux d'Elberfeld. *Archives de Psychologie* (Geneva) 13 244–84.

Clarke, D., H. Whitney, G. Sutton, and R. Robert. 2013. Detection and learning of floral electric fields by bumblebees. *Science* 340: 66–69.

Clarke, J. V., C. J. Nicol, R. Jones, and P. D. McGreevy. 1996. Effects of observational learning on food selection in horses. *Applied Animal Behaviour Science* 50: 177–84.

Cleeremans, A. 2000. *Hans le malin: Les sciences cognitives à la croisée des chemins*. Research symposium in cognitive science, Université Libre de Bruxelles. http://srsc.ulb.ac.be/axcWWW/papers/pdf/01-Hans.pdf.

Clément, F., E. Bassecoulart-Zitt, M. Hausberger, M. F. Bouissou, and X. Boivin. 2002. La recherche sur le comportement du cheval en France et à l'étranger. In *Poulain à Venir, Poulain en Devenir*. Paris: Éditions Belin, 91–113.

Collings, V. B. 1974. Human taste response as a function of locus of stimulation on the tongue and soft palate. *Perception and Psychophysics* 16: 169–74.

Cooke, B., and E. Ernst. 2000. Aromatherapy: A systematic review. *British Journal of General Practice* 50: 493–96.

Corballis M. C. 2009. The evolution and genetics of cerebral asymmetry. *Philosophical Transactions of the Royal Society B* 364: 867–79.

Coulon, M., D. L. Deputte, Y. Heyman, and C. Baudoin. 2007. Discrimination visuelle de leur espèce chez les bovins. In *Le social dans tous ses états. 41ème Colloque de la SFECA*, Université Paris 13, Villetaneuse, France, 16.

Crasto, C., L. Marenco, P. L. Miller, and G. S. Shepherd. 2002. Olfactory receptor database: A metadata-driven automated population from sources of gene and protein sequences. *Nucleic Acids Research* 1: 354–60.

Croney, C. C., and R. C. Newberry. 2007. Group size and cognitive processes. *Applied Animal Behaviour Science* 103: 215–28.

Crowell-Davis, S. L., and A. B. Caudle. 1989. Coprophagy by foals: Recognition of maternal faeces. *Applied Animal Behaviour Science* 24: 267–72.

Crowell-Davis, S. L., and K. A. Houpt. 1985. Coprophagy by foals: Effect of age and possible functions. *Equine Veterinary Journal* 17: 17–19.

Crowell-Davis, S. L., and J. W. Weeks. 2005. Maternal behaviour and mare-foal interaction. In *The Domestic Horse: The Evolution, Development, and Management of Its Behaviour*, edited by D. S. Mills and S. M. McDonnell. Cambridge, UK: Cambridge University Press, 126–38.

Cummings, J. F., and A. de Lahunta. 1969. An experimental study of the retinal projections in the horse and sheep. *Annals of the New York Academy of Sciences* 167: 293–318.

Dallos, P., and D. Harris. 1978. Properties of auditory nerve responses in absence of outer hair cells. *Journal of Neurophysiology* 41: 365–83.

Darwin, C. 1871. *The Descent of Man, and Selection in Relation to Sex*. 2 vols. London: John Murray.

Davidson, R. J. 2004. Well-being and affective style: Neural substrates and biobehavioural correlates. *Philosophical Transactions of the Royal Society B* 359: 1395–1411.

de Boyer des Roches, A., V. Durier, M.-A. Richard-Yris, C. Blois-Heulin, M. Ezzaouïa, M. Hausberger, and S. Henry. 2011. Differential outcomes of unilateral interferences at birth. *Biology Letters* 7: 177–80.

de Boyer des Roches, A., M. A. Richard-Yris, S. Henry, M. Ezzaouïa, and M. Hausberger. 2008. Laterality and emotions: Visual laterality in the domestic horse (*Equus caballus*) differs with objects' emotional value. *Physiology and Behavior* 94: 487–90

Delacampagne, C. 2000. Howard Gardner: L'intelligence au pluriel. *La Recherche* 337: 109–11.

Delage, Y. 1916. Etude des réactions idéationnelles. *L'Année Biologique:* 378.

Delorme, A., and J. Lajoie. 2003. La perception de la troisième dimension et de la taille. In *Perception et Réalité*, edited by A. Delorme amd M. Flückiger. Brussels: De Boeck Universiré, 275–309.

Despret, V. 2004. *Hans, le cheval qui savait compter.* Paris: Les Empêcheurs de penser en rond/Le Seuil.

Devenport, J. A., M. R. Patterson, and L. D. Devenport. 2005. Dynamic averaging and foraging decisions in horses (*Equus caballus*). *Journal of Comparative Psychology* 119: 352–58.

de Winter, W., and C. E. Oxnard. 2001. Evolutionary radiations and convergences in the structural organization of mammalian brains. *Nature,* 409, 710–14.

Dobzhanski, P. 1973. Nothing in biology makes sense except in the light of evolution. *American Biology Teacher* 35: 125–29.

Dominy, N. J., J.-C. Svenning, and W.-H. Li. 2003. Historical contingency in the evolution of primate color vision. *Journal of Human Evolution* 44: 25–45.

Doré, F. Y., and P. Mercier. 1992. *Les fondements de l'apprentissage et de la cognition.* Montreal: Gaëtan Morin.

Dougherty, D. M., and P. Lewis. 1992. Matching by horses on several concurrent variable-interval schedules. *Behavioural Processes* 26: 69–76.

Døving, K. B., and D. Trotier. 1998. Structure and function of the vomeronasal organ. *Journal of Experimental Biology* 201: 2913–25.

Duncan, P. 1992. *Horses and Grasses: The Nutritional Ecology of Equids and Their Impact on the Camargue.* New York: Springer.

Duncan, P., and N. Vigne. 1979. The effect of group size in horses on the rate of attacks by blood-sucking flies. *Animal Behaviour* 27: 623–25.

Eblé, Lt. Col. 1981. Intelligent ou pas? *L'information Hippique* 281: 48–49.

Eckert, R., D. Randall, W. Burggren, and K. French, 1999. *Physiologie animale. Mécanismes et adaptation.* Brussels: De Boeck Université.

Ehret, G. 1987. Left hemisphere advantage in the mouse brain for recognizing ultrasonic communication calls. *Nature* 325: 249–51.

Ehrenhofer, M. C., C. A. Deeg, S. Reese, H.-G. Liebich, M. Stangassinger, and B. Kaspers. 2002. Normal structure and. age-related changes of the equine retina. *Veterinary Ophthalmology,* 5, 39–47.

Eibl-Eibesfeldt, I. 1970. *Ethology: The Biology of Behavior.* Trans. E. Klinghammer. New York: Holt, Rinehart and Winston.

Engen, T. 1982. *The Perception of Odors*. New York: Academic Press.

Erickson, R. P. 2008. A study of the science of taste: On the origins and influence of the core ideas. *Behavioural and Brain Sciences* 41: 59–105.

Evans, K. E., and P. D. McGreevy. 2006. Conformation of the equine skull: A morphometric study. *Anatomia, Histologia, Embryologia* 35: 221–27.

———. 2007. The distribution of ganglion cells in the equine retina and its relationship to skull morphology. *Anatomia, Histologia, Embryologia* 36: 151–56.

Fagot, J., E. Wasserman, and M. Young. 2004. Catégorisation d'objets visuels et concepts relationnels chez l'animal. In *L'éthologie cognitive*, edited by J. Vauclair and M. Kreutzer. Paris: Éditions Ophrys, Éditions de la Maison des Sciences de l'Homme, 117–36.

Fantz, R. L. 1958. Pattern vision in young infants. *Psychological Record* 8: 43–47.

Farmer, K., K. Krueger, and R. Byrne. 2010. Visual laterality in the domestic horse (*Equus caballus*) interacting with humans. *Animal Cognition* 13: 229–38.

Farrall, H., and M. C. Handscombe. 1990. Follow-up report of a case of surgical aphakia with an analysis of equine visual function. *Equine Veterinary Journal* 22 (S10): 91–93.

Faurion, A. 1993. Chacun ses goûts. In *La gourmandise. Délices d'un pêché*, edited by C. N'Diaye. Paris: Autrement, 149.

Fechner, G. 1860. *Elemente der Psychophysik*. Leipzig, Germany: Breitkopf and Härtel.

Feh, C. 1999. Alliances and reproductive success in Camargue stallions. *Animal Behaviour, 57*, 705–13.

———. 2001. Alliance between stallions are more than just multimale groups: Reply to Linklater and Cameron. *Animal Behaviour* 61 (5): F27–30.

———. 2005. Relationships and communication in socially natural horse herds. In *The Domestic Horse: The Evolution, Development, and Management of Its Behaviour*, edited by D. S. Mills and S. M. McDonnell. Cambridge, UK: Cambridge University Press, 83–93.

Feh, C., and J. de Mazières. 1993. Grooming at a preferred site reduces heart rate in horses. *Animal Behaviour* 46: 1191–94.

Feist, J. D. 1971. *Behavior of feral horses in the Pryor Mountain Wild Horse Range*. MSc thesis. University of Michigan, Ann Arbor, Michigan.

Feist, J. D., and D. R. McCullough. 1976. Behavior patterns and communication in feral horses. *Zeitschrift für Tierpsychoogie* 41: 337–71.

Ferguson, C. E., H. F. Kleinman, and J. Browning. 2013. Effect of lavender aromatherapy on acute-stressed horses. *Journal of Equine Veterinary Science* 33: 67–69.

Fillis, J. 1902. *Breaking and Riding*. New York: Scribner.

Fiset, S., C. Beaulieu, and F. Landry. 2003. Duration of dogs' working memory in search of disappearing objects. *Animal Cognition* 6: 1–10.

Fiset, S., and F. Y. Doré. 2006. Duration of cats' working memory for disappearing objects. *Animal Cognition* 9: 62–70.

Fiske Godfrey, J. 1979. *How Horses Learn: Equine Psychology Applied to Training*. Brattleboro, VT: Stephen Green Press.

Flannery, B. 1997. Relational discrimination learning in horses. *Applied Animal Behaviour Science* 54: 267–80.

Fleurance, G., M.-A. Leblanc, and P. Duncan. 2004. Le comportement des équidés en liberté. *Haras Nationaux, 30ème journée de la Recherche Equine*: 101–13.

Floyd, A., and R. Mansmann. 2007. *Equine Podiatry*. London: Saunders.

Francis-Smith, K., and D. G. Wood-Gush. 1977. Coprophagia as seen in thoroughbred foals. *Equine Veterinary Journal* 9: 155–57.

Francois, J., L. Wouters, V. Victoria-Troncoso, A. deRouck, and A. van Gerven. 1980. Morphometric and electrophysiological study of the horse. *Ophthalmologica* 181: 340–49.

Fraser, A. F. 1992. *The Behaviour of the Horse*. Wallingford, UK: CAB International.

Friend, T. H. 2001. A review of recent research on the transportation of horses. *Journal of Animal Science* 79 (E Suppl.): E32–40.

Fujita, K. 1997. Perception of the Ponzo illusion by rhesus monkeys, chimpanzees, and humans: Similarity and difference in three primate species. *Perception and Psychophysics* 59: 284–92.

Fujita, K., D. S. Blough, and P. M. Blough. 1991. Pigeons see the Ponzo illusion. *Animal Learning and Behavior* 19: 283–93.

Fureix, C., P. Jego, C. Sankey, and M. Hausberger. 2009. How horses (*Equus caballus*) see the world: Humans as significant "objects." *Animal Cognition* 12: 643–54.

Gaillard, D., P. Passilly-Degrace, F. Laugerette, and P. Besnard. 2006. Sur la piste du "goût du gras." *Oléagineux, Corps Gras, Lipides* 13: 309–14.

Gall, F. J., and J. C. Spurzheim. 1810–1819. *Anatomie et physiologie du système nerveux en général et du cerveau en particulier, avec des observations sur la possibilité de reconnaître plusieurs dispositions intellectuelles et morales de l'homme et des animaux par la configuration de leur tête*. Paris: Schoell.

Gallup, G. G. Jr. 1970. Chimpanzees: Self-recognition. *Science* 167: 86–87.

Garcia, J., and R. A. Koelling. 1966. Relation cue to consequence in avoiding learning. *Psychonomic Science* 4: 123–24.

Gardner, H. 1985. *Frames of Mind: The Theory of Multiple Intelligences*. New York: Basic Books.

Gayon, J. 2000. History of the concept of allometry. *American Zoologist* 40: 748–58.

Gazzaniga, M. S., R. B. Ivry, and G. R. Mangun. 1998. *Cognitive Neuroscience: The Biology of the Mind*. New York: Norton.

Geisbauer, G., U. Griebel, A. Schimd, and B. Timney. 2004. Brightness discrimination and neutral test point in the horse. *Canadian Journal of Zoology* 82: 660–70.

Geldard, F., and C. Sherrick. 1972. The cutaneous "Rabbit": A perceptual illusion. *Science* 178: 178–79.

Gervais, J., P. Livet, and A. Tête. 1992. *La représentation animale*. Nancy, France: Presses Universitaires.

Ghirlanda, S., and G. Vallortigara. 2004. The evolution of brain lateralization: A game-theoretical analysis of population structure. *Proceedings of the Royal Society of London B* 271: 853–57.

Gilbert, A. N., and C. J. Wysocki. 1992. Hand preference and age in the United States. *Neuropsychologia* 30: 601–8.

Ginane, C., Baumont, R., Lassalas, J., and Petit, M. 2002. Feeding behaviour and intake of heifers fed on hays of various quality, offered alone or in a choice situation. *Animal Research* 51: 177–88.

Giniaux, D. 1998. *Healing Hands: A Treatise on First Aid Equine Acupressure*. Cleveland Heights, OH: Xenophon.

Goldfiem, J. de. 1974. *Hippologie. II. La Psychologie du cheval.* Paris: L'Officiel de l'Artisanat Rural.

Goodwin, D., H. P. Davidson, and P. Harris. 2002. Foraging enrichment for stabled horses: Effects on behaviour and selection. *Equine Veterinary Journal* 34: 686–91.

———. 2005. Selection and acceptance of flavors in concentrate diets for stabled horses. *Applied Animal Behaviour Science* 95: 223–32.

Gould, E., and C. G. Gross. 2002. Neurogenesis in adult mammals: Some progress and problems. *Journal of Neuroscience* 22: 619–23.

Gould, S. J. 1981. *The mismeasure of man.* New York: Norton.

Govardovskii, V. I., N. Fyhrquist, T. Reuter, D. G. Kuzmin, and K. Donner. 1999. In search of the visual pigment template. *Visual Neuroscience* 17: 509–28.

Grandin, T. 2006. *Thinking in Pictures and Other Reports from My Life with Autism.* New York: Vintage. First published 1995 by Doubleday, New York.

Grandin, T., and C. Johnson. 2005. *Animals in Translation.* New York: Scribner.

Grau, J. W. 2002. Learning and memory without a brain. In *The Cognitive Animal: Empirical and Theoretical Perspectives on Animal Cognition,* edited by M. Bekoff, C. Allen, and M. Burghardt. Cambridge, MA: MIT Press, 77–87.

Griffin, D. 1984. *Animal Thinking.* Cambridge, MA: Harvard University Press.

Grzimek, B. 1944. Das Erkennen von Menschen durch Pferde. *Zeitschrift für Tierpsychologie* 6: 110–26.

———. 1949. Gedächtnisversuche mit Pferden. *Zeitschrift für Tierpsychologie* 6: 445–54.

———. 1952. Versuche über das Farbsehen von pflanzenessern. *Zeitschrift für Tierpsychologie* 9: 23–39.

Guillaume, P. 1942. Analyse bibliographique 375. B. Grzimek. *L'Année Psychologique* 43: 460.

———. 1979. *Psychologie de la Forme.* Paris: Flammarion.

Guo, X., and S. Sugita. 2000. Topography of ganglion cells in the retina of the horse. *Journal of Veterinary Medical Science* 62: 1145–50.

Hahn, C. 2004. Behavior and the brain. In *Equine Behaviour: A Guide for Veterinarians and Equine Scientists,* edited by P. McGreevy. Edinburgh: Saunders, 55–84.

Hall, C. 2007. The impact of visual perception on equine learning. *Behavioural Processes* 76: 29–33.

Hall, C. A., H. J. Cassaday, and A. M. Derrington. 2003. The effect of stimulus height on visual discrimination in horses. *Journal of Animal Science* 81: 1715–20.

Hall, C. A., H. J. Cassaday, C. J. Vincent, and A. M. Derrington. 2005. The selection of coloured stimuli by the horse (*Equus caballus*). Poster presented at BSAS Conference (Applying Equine Science: Research into Business), Royal Agricultural College, Cirencester, UK. Short paper in proceedings.

———. 2006. Cone excitation ratios correlate with color discrimination performance in the horse (*Equus caballus*). *Journal of Comparative Psychology* 120: 438–48.

Hanggi, E. B. 1999a. Categorization learning in horses (*Equus caballus*). *Journal of Comparative Psychology* 113: 243–52.

———. 1999b. Interocular transfer of learning in horses (*Equus caballus*). *Journal of Equine Veterinary Science* 19: 518–23.

———. 2001. Can horses recognize pictures? *Proceedings of the Third International Conference of Cognitive Science,* Beijing, China, 52–56.

———. 2003. Discrimination learning based on relative size concepts in horses (*Equus caballus*). *Applied Animal Behaviour Science* 83: 201–13.

———. 2005. The thinking horse: Cognition and perception reviewed. In *AAEP 51st Annual Convention Proceedings*, Seattle, Washington, December 3–7. www.ivis.org/proceedings/AEEP/2005/toc. asp.

———. 2006. Equine cognition and perception: Understanding the horse. In *Diversity of Cognition: Evolution, Development, Domestication, and Pathology*, edited by K. Fujita and S. Itakura. Kyoto, Japan: Kyoto University Press, 98–118.

———. 2010a. Short-term memory testing in domestic horses: Experimental design plays a role. *Journal of Equine Veterinary Science* 30: 618–24.

———. 2010b. Rotated object recognition in four domestic horses (*Equus caballus*). *Journal of Equine Veterinary Science* 30: 175–86.

Hanggi, E. B., and J. F. Ingersoll. 2009a. Long-term memory for categories and concepts in horses (*Equus caballus*). *Animal Cognition* 12: 451–62.

———. 2009b. Stimulus discrimination by horses under scotopic conditions. *Behavioural Processes* 82: 45–50.

———. 2012. Lateral vision in horses: A behavioral investigation. *Behavioural Processes* 91: 70–76.

Hanggi, E. B., J. Ingersoll, and T. L. Waggoner. 2007. Color vision in horses: Deficiencies identified using a pseudoisochromatic plate test. *Journal of Comparative Psychology* 121: 65–72.

Hansen, R. M. 1976. Foods of free-roaming horses in southern New Mexico. *Journal of Range Management* 29: 437.

Hare, B., and M. Tomasello. 2005. Human-like social skills in dogs? *Trends in Cognitive Sciences* 9: 439–44.

Harman, A. M. 1998. The way they see. *Hoofbeats* 19: 44–46.

Harman, A. M., S. Moore, R. Hoskins, and P. Keller. 1999. Horse vision and an explanation for the visual behaviour originally explained by the 'ramp retina.' *Equine Veterinary Journal* 31: 384–90.

Hartmann, E., E. Søndergaard, and L. J. Keeling. 2012. Keeping horses in groups: A review. *Applied Animal Behaviour Science* 136: 77–87.

Hausberger, M., C. Bruderer, N. Le Scolan, and J.-S. Pierre. 2004. Interplay between environmental and genetic factors in temperament/personality traits in horses (*Equus caballus*). *Journal of Comparative Psychology* 118: 434–45.

Hausberger, M., C. Fureix, M. Bourjade, S. Wessel-Robert, and M.-A. Richard-Yris. 2012. On the significance of adult play: What does social play tell us about adult horse welfare? *Naturwissenschaften* 99: 291–302.

Hausberger, M., S. Henry, C. Larose, and M.-A. Richard-Yris. 2007. First suckling: A crucial event for mother-young attachment? *Journal of Comparative Psychology* 121: 109–12.

Hausberger, M., and C. Muller. 2002 A brief note on some possible factors involved in the reactions of horses to humans. *Applied Animal Behaviour Science* 76: 339–44.

Hausberger, M., C. Muller, and C. Lunel. 2011. Does work affect personality? A study in horses. *PLoS ONE* 6: e1469.

Hausberger, M., and M. A. Richard-Yris. 2005. Individual differences in the domestic horse, origins, development, and stability. In *The Domestic Horse: The Origin,*

Development, and Management of Its Behaviour, edited by D. S. Mills and S. M. McDonnell. Cambridge, MA: Cambridge University Press, 33–52.

Hausberger, M., H. Roche, S. Henry, and E. Visser. 2008. A review of the human-horse relationship. *Applied Animal Behaviour Science* 109: 1–24.

Hauser, M. D., and K. Anderson. 1994. Functional lateralization for auditory temporal processing in adult, but not infant rhesus monkeys: Field experiments. *Proceedings of the National Academy of Sciences of the U.S.A.* 91: 3946–48.

Hauser, M. D., S. Carey, and L. B. Hauser. 2000. Spontaneous number representation in semi-free ranging rhesus monkeys. *Proceedings of the Royal Society of London B* 267: 829–33.

Hawken, P., C. Fiol, and D. B. Blache. 2011. Behavioral reactivity to psychosocial stress determines the effects of lavender oil on anxiety in sheep. *Journal of Animal Science* 89 (E Suppl. 1): 10.

Healy, S. D., and C. Rowe. 2007. A critique of comparative studies of brain size. *Proceedings of the Royal Society of London B* 274: 453–64.

Hebel, R. 1976. Distribution of retinal ganglion cells in five mammalian species (pig, sheep, ox, horse, dog). *Anatomy and Embryology* 150: 45–51.

Hediger, H. 1955. *Studies of the psychology and behaviour of captive animals in zoos and circuses.* Trans. G. Sircom. London: Butterworth's Scientific.

———. 1981. The Clever Hans phenomenon from an animal psychologist's point of view. In *The Clever Hans Phenomenon: Communication with Horses, Whales, Apes, and People,* edited by T. Sebeok and R. Rosenthal. Vol. 364 of *Annals of the New York Academy of Sciences,* 1–17. New York: The New York Academy of Sciences.

Heffner, H. E. 1998. Auditory awareness in animals. *Applied Animal Behaviour Science* 57: 259–68.

Heffner, H. E., and R. S. Heffner. 1982. Hearing in the elephant (*Elephas maximus*): Absolute sensitivity, frequency discrimination, and sound localization. *Journal of Comparative and Physiological Psychology,* 96, 926–944

———. 1983a. Sound localization and high frequency hearing in horses. *Journal of Acoustical Society of America* 73: S42.

———. 1983b. The hearing ability of horses. *Equine Practice* 5: 27–32.

———. 1984. Sound localization in large mammals: Localization of complex sounds by horses. *Behavioral Neuroscience* 98: 541–55.

———. 1992a. Auditory perception. In *Farm Animals and the Environment,* edited by C. Philips and D. Piggins. Wallingford, UK: CAB International, 159–84.

———. 1998. Hearing. In *Comparative Psychology: A Handbook,* edited by G. Greenberg and M. M. Haraway. New York: Garland, 290–303.

———. 2003. Audition. In *Handbook of Research Methods in Experimental Psychology,* edited by S. F. Davis. Malden, MA: Blackwell, 413–40.

———. 2007. Hearing ranges of laboratory animals. *Journal of the American Association for Laboratory Animal Science* 46: 11–13.

———. 2008. High-frequency hearing. In *Handbook of the Senses: Audition,* edited by P. Dallos, D. Oertel, and R. Hoy. New York: Elsevier, 55–60.

Heffner, H. E., and R. B. Masterton. 1980. Hearing in glires: Domestic rabbit, cotton rat, feral house mouse, and kangaroo rat. *Journal of the Acoustical Society of America* 68 (6): 1584–99.

Heffner, R. S. 1986. Localization of tones by horses: Use of binaural cues and the role of the superior olivary complex. *Behavioral Neuroscience* 100: 93–103.

———. 1989. Sound localization, use of binaural cues, and the superior olivary complex in pigs. *Brain, Behavior, and Evolution* 33: 248–58.

———. 1997. Comparative study of sound localization in mammals and its anatomical correlates. *Acta Otolaryngologica* 532: 46–53.

———. 2001. Audiograms of five species of rodents: Implications for the evolution of hearing and the perception of pitch. *Hearing Research* 157: 138–52.

Heffner, R. S., and H. E. Heffner. 1983c. Hearing in large mammals: Horses (*Equus caballus*) and cattle (*Bos taurus*). *Behavioral Neuroscience* 97: 299–309.

———. 1992b. Evolution of sound localization in mammals. In *The Evolutionary Biology of Hearing*, edited by D. B. Webster, R. R. Fay, and A. N. Popper. New York: Springer, 691–715.

———. 1992c. Visual factors in sound localization in mammals. *Journal of Comparative Neurology* 317: 219–32.

Heffner, R. S., and R. B. Masterton. 1990. Sound localization in mammals: Brain-stem mechanisms. In *Basic Mechanisms,* edited by M. A. Berkley and W. C. Stebbins. Vol. 1 of *Comparative Perception.* New York: Wiley, 285–314.

Heleski, C. R., and R. Anthony. 2012. Science alone is not always enough: The importance of ethical assessment for a more comprehensive view of equine welfare. *Journal of Veterinary Behavior* 7: 169–78.

Heleski, C., L. Bauson, and N. Bello. 2008. Evaluating the addition of positive reinforcement for learning a frightening task: A pilot study with horses. *Journal of Applied Animal Welfare Science* 11: 213–22.

Helmholtz, H. von. 1867. *Handbuch der physiologischen Optik.* Leipzig, Germany: Voss.

Hemmi, J. M. 1999. Dichromatic colour vision in an Australian Marsupial, the tammar wallaby. *Journal of Comparative Physiology A* 185: 509–15.

Henry, S., D. Hemery, M.-A. Richard, and M. Hausberger. 2005. Human-mare relationships and behaviour of foals toward humans. *Applied Animal Behaviour Science* 93: 341–62.

Henry, S., M.-A. Richard-Yris, and M. Hausberger. 2006. Influence of various early human-foal interferences on subsequent human-foal relationship. *Developmental Psychobiology* 48: 712–18.

Henry, S., M.-A. Richard-Yris, S. Tordjman, and M. Hausberger. 2009. Neonatal handling affects durably bonding and social development. *PLoS ONE* 4 (4): e5216.

Henry, S., A. Zanella, C. Sankey, M.-A. Richard-Yris, A. Marko, and M. Hausberger. 2012. Adults may be used to alleviate weaning stress in domestic foals (*Equus caballus*). *Physiology and Behavior* 106: 428–36.

Hering, E. 1878. *Zur Lehre vom Lichtsinn.* Vienna: C. Gerold's Sohn.

Heuzé, P. 1928. *La plaisanterie des animaux calculateurs.* Paris: Les Éditions de France.

Heyers, D., M. Manns, H. Luksch, O. Güntürkün, and H. Mouritsen. 2007. A visual pathway links brain structures active during magnetic compass orientation in migratory birds. *PLoS ONE* 9: e937.

Hollard, V. D., and J. D. Delius. 1982. Rotational invariance in visual pattern recognition by pigeons and humans. *Science* 218: 804–6.

Holley, A. 2006a. Système olfactif et neurobiologie. *Terrain* 47: 107–22.

———. 2006b. *Le cerveau gourmand.* Paris: Odile Jacob.

———. 2007. Les secrets de l'odorat. *Cerveau et Psycho* 21: 50–55.

Hook, M. A. 2004. The evolution of lateralized motor functions. In *Comparative Vertebrate Cognition*, edited by L. J. Rogers and G. Kaplan. New York: Kluwer Academic/Plenum, 325–70.

Hopkins, W. D., J. L. Russell, C. Cantalupo, and H. Freeman. 2005. Factors influencing the prevalence and handedness for throwing in captive chimpanzees (*Pan troglodytes*). *Journal of Comparative Psychology* 119: 363–70.

Hothersall, B., P. Harris, L. Sortoft, and C. J. Nicol. 2010. Discrimination between conspecific odour samples in the horse (*Equus caballus*). *Applied Animal Behaviour Science* 126: 37–44.

Houpt, K. A., and L. Guida. 1984. Flehmen. *Equine Practice* 6: 32–35.

Houpt, K. A., and D. M. Zahorik. 1981. Species differences in feeding strategies, food hazards, and the ability to learn food aversions. In *Foraging Behavior,* edited by A. C. Kamil and T. D. Sargent. New York: Garland, 289–310.

Houpt, K. A., D. M. Zahorik, and J. A. Swartzman-Andert. 1990. Taste aversion learning in horses. *Journal of Animal Science* 68: 2340–44.

Howery, L. D., D. W. Bailey, and E. A. Laca. 1999. Impact of spatial memory on habitat use. In *Grazing Behavior of Livestock and Wildlife*, edited by K. L. Lunchbaugh, K. D. Sanders, and J. C. Mosley. Moscow: University of Idaho, 91–100.

Huang, A. L., X. Chen, M. A. Hoon, J. Chandrashekar, W. Guo, D. Tränkner, N. J. P. Ryba, and C. S. Zuker. 2006. The cells and logic for mammalian sour taste detection. *Nature* 442: 934–38.

Hubel, D. 1995. *Eye, Brain, and Vision.* http://hubel.med.harvard.edu/book. First published 1988 by W. H. Freeman, New York.

Hughes, A. 1977. The topography of vision of contrasting life style: Comparative optics and retinal organization. In *The Visual System in Vertebrates,* edited by F. Cresitelli. Vol. 7/5 of *Handbook of Sensory Physiology.* Berlin: Springer, 613–57.

Hummel, J. E., and I. Biederman. 1992. Dynamic binding in a neural network for shape recognition. *Psychological Review* 99: 480–517.

Imbert, M. 2006. *Traité du cerveau.* Paris: Odile Jacob.

Innes, L., and S. McBride. 2008. Negative versus positive reinforcement: An evaluation of training strategies for rehabilitated horses. *Applied Animal Behaviour Science* 112: 357–68.

Jacobs, G. H. 1981. *Comparative Color Vision.* New York: Academic Press.

Jacquot, L., J. Monnin, and G. Brand. 2004. Influence of nasal trigeminal stimuli on olfactory sensitivity. *Comptes Rendus de l'Académie des Sciences, Biologies* 327: 305–11.

Janet, P. 1936. *L'Intelligence avant le langage.* Paris: Bibliothèque de philosophie scientifique, Flammarion. http://classiques.uqac.ca/classiques/janet_pierre/intelligence_langage/intelligence.html.

Jansen, T., P. Forster, M. Levine, H. Oelke, M. Hurles, C. Renfrew, J. Weber, and K. Olek. 2002. Mitochondrial DNA and the origins of the domestic horse. *Proceedings of the National Academy of Sciences of the U.S.A.* 99: 10905–10.

Janzen, D. H. 1978. How do horses find their way home? *Biotropica* 10 (3): 240.

Jeannerod, M. 2002. *La nature de l'esprit*. Paris: Odile Jacob.

Jennings, R. C. 1998. A philosophical consideration of awareness. *Applied Animal Behaviour Science* 57: 201–11.

Jerison, H. J. 1973. *Evolution of the Brain and Intelligence*. New York: Academic Press.

Johnston, R. E., and T. A. Bullock. 2001. Individual recognition by use of odors in golden hamsters: The nature of individual representations. *Animal Behaviour* 61: 545–57.

Jones, C. D., D. Osorio, and R. Baddeley. 2001. Colour categorisation by domestic chicks. *Proceedings of the Royal Society of London B* 268: 2077–84.

Jouvent, R. 2009. *Le cerveau magicien. De la réalité au plaisir psychique*. Paris: Odile Jacob.

Julesz, B. 1971. *Foundations of Cyclopean Perception*. Chicago: University of Chicago Press.

Kainer, R. A., and T. O. McCracken. 1998. *Horse Anatomy: A Coloring Atlas*. Loveland, CO: Alpine Publications.

Kaissling, K.-E., and E. Priesner. 1970. Die Riechschwelle des Seidenspinners. *Naturwissenschaften* 57: 23–28.

Kare, M. R. 1971. Comparative study of taste. In *Chemical Senses: Taste*, edited by I. M. Beidler. Vol. 4, part 2, of *Handbook of Sensory Physiology*. New York: Springer, 278–92.

Kare, M. R., W. C. Pond, and J. Campbell. 1965. Observations on taste reactions in pigs. *Animal Behaviour* 13: 265–69.

Karlson, P., and M. Luscher. 1959. Pheromones: A new term for a class of biologically active substances. *Nature* 153: 55–56.

Kaseda, Y., and A. Khalil. 1996. Harem size and reproductive success in stallions in Misaki feral horses. *Applied Animal Behaviour Science* 78: 163–73.

Keiper, R. 1985. *The Assateague Ponies*. Centreville, MD: Tidewater Publishers.

Keiper, R., and J. Berger. 1982. Refuge-seeking and pest avoidance by feral horses in desert and island environments. *Applied Animal Ethology* 9: 111–20.

Kelber, A., and L. S. V. Roth. 2006. Nocturnal colour vision—not as rare as we might think. *Journal of Experimental Biology* 209: 781–88.

Kelber, A., M. Vorobyev, and D. Osorio. 2003. Animal colour vision—behavioural tests and physiological concepts. *Biological Reviews* 78: 81–118.

Keller, M., M. J. Baum, O. Brock, P. A. Brennan, and J. Bakker. 2009. Both main and accessory olfactory systems contribute to mate recognition and sexual behavior. *Behavioural Brain Research* 200: 268–76.

Kendall, D. E. 1989. A scientific model for acupuncture. *American Journal of Acupuncture* 17: 251–68, 343–60.

Kendrick, K. M., K. Atkins, M. R. Hinton, K. D. Broad, C. Fabre-Nys, and B. Keverne. 1995. Facial and vocal discrimination in sheep. *Animal Behaviour* 49: 1665–76.

Kendrick, K. M., A. P. Da Costa, A. E. Leigh, M. R. Hinton, and J. W. Peirce. 2001. Sheep don't forget a face. *Nature* 414: 165–66.

Kendrick, K. M., A. Leigh, and J. Peirce. 2001. Behavioural and neural correlates of mental imagery in sheep using face recognition paradigms. *Animal Welfare* 10 (S1): 89–101.

Kent, J. P. 1987. Experiments on the relationship between the hen and chick (*Gallus gallus*): The role of the auditory mode in recognition and the effects of maternal separation. *Behaviour* 102: 1–14.

Kiley, M. 1972. The vocalizations of ungulates, their causation and function. *Zeitschrift für Tierpsychologie* 31: 171–222.

Kimchi, T., J. Xu, and C. Dulac. 2007. A functional circuit underlying male sexual behaviour in the female mouse brain. *Nature* 448: 1009–14.

Klingel, H. 1967. Soziale Organisation und Verhalten freilebender Steppenzebras. *Zeitschrift für Tierpsychologie* 24: 580–624.

———. 1969. The social organization and population ecology of the plains zebra. *Zoologica Africana* 4: 249–63.

———. 1978. La vie sociale des zèbres et des antilopes. *La Recherche* 9: 112–20.

Koba, Y., C. Goto, N. Horikawa, K. Tsuchiya, and H. Tanida. 2004. Ability of ponies to discriminate among colors and among similarly dressed people. In *Proceedings of the 38th International Congress of the International Society of Applied Ethology.* Helsinki, Finland: ISAE, 202.

Koestler, A. 1967. *The Ghost in the Machine.* New York: Macmillan.

Köhler, W. 1925. *The Mentality of Apes.* London: Kegan Paul, Trench.

Komiya, M., A. Sugiyama, K. Tanabe, T. Uchino, and T. Takeuchi. 2009. Evaluation of the effect of topical application of lavender oil on autonomic nerve activity in dogs. *American Journal of Veterinary Research* 70: 764–69.

König, H. E., H. Wisdorf, A. Probst, R. Macher, S. Voss, and E. Polsterer. 2005. Considerations about the function of the mimic muscles and the vomeronasal organ of horses during the Flehmen reaction. *Pferdeheilkunde* 21: 297–300.

Kostov, D. I. 2007. Vomeronasal organ in domestic animals. *Bulgarian Journal of Veterinary Medicine* 10: 53–57.

Krall, K. 1912. *Denkende Tiere. Beiträge zur Tierseelenkunde auf Grund eigener Versuche. Der Kluge Hans und meine Pferde Muhamed und Zarif.* Leipzig, Germany: Friedrich Engelmann.

Kreutzer, M., and J. Vauclair. 2004. La cognition animale au carrefour de l'éthologie et de la psychologie. In *L'éthologie cognitive,* edited by J. Vauclair and M. Kreutzer. Paris: Éditions Ophrys, Éditions de la Maison des Sciences de l'Homme, 1–19.

Krueger, K., and B. Flauger. 2007. Social learning in horses from a novel perspective. *Behavioural Processes* 76: 37–39.

———. 2011. Olfactory recognition of individual competitors by means of faeces in horse (*Equus caballus*). *Animal Cognition* 14: 245–57.

Krueger, K., B. Flauger, K. Farmer, and K. Maros. 2011. Horses (*Equus caballus*) use human local enhancement cues and adjust to human attention. *Animal Cognition* 14: 187–201.

Kruska, D. 1988. Effects of domestication on brain structure and behaviour in mammals. *Human Evolution,* 3, 473–85.

Kumar, P., J. F. Timoney, H. H. Southgate, and A. S. Sheoran. 2000. Light and scanning electron microscopic studies of the nasal turbinates of the horse. *Anatomia, histologia, embryologia* 29: 103–9.

Kusunose, R., and A. Yamanobe. 2002. The effect of training schedule on learned tasks in yearling horses. *Applied Animal Behaviour Science* 78: 225–33.

Lachman, R., J. L. Lachman, and E. C. Butterfield. 1979. *Cognitive Psychology and Information Processing.* Hillsdale, NJ: Lawrence Erlbaum Associates.

Lagerweij, E., P. C. Nelis, V. M. Wiegant, and J. M. van Ree. 1984. The twitch in horses: A variant of acupuncture. *Science* 4667: 1172–74.

Lajoie, J., and A. Delorme. 2003. Fonctions et processus visuels. In *Perception et Réalité*, edited by A. Delorme and M. Flückiger. Brussels: De Boeck Université, 71–103.

Lalande, A. 2006. *Vocabulaire technique et critique de la philosophie*. Paris: Quadrige, Presses Universitaires de France.

Lampe, J. F., and J. Andre. 2012. Cross-modal recognition of human individuals in domestic horses (*Equus caballus*). *Animal Cognition* 15: 623–30.

Land, E. H. 1977. The retinex theory of color vision. *Scientific American*, December, 108–128.

Land, E. H. 1983. Recent advances in retinex theory and some implications for cortical computations: Color vision and the natural image. *Proceedings of the National Academy of Sciences of the U.S.A.* 80: 5163–69.

Land, E. H., and J. J. McCann. 1971. Lightness and retinex theory. *Journal of the Optical Society of America* 61: 1–11.

Langevin, P., E. Rabaud, H. Laugier, A. Marcelin, and I. Meyerson. 1923. Rapport au sujet des phénomènes produits par le médium J. Guzik. *L'Année Psychologique* 24: 664–72.

Lansade, L., M. Bertrand, and M.-F. Bouissou. 2005. Effects of neonatal handling on subsequent manageability, reactivity, and learning ability of foals. *Applied Animal Behaviour Science* 92: 143–58.

Lansade, L., and M. F. Bouissou. 2008. Reactivity to humans: A temperament trait of horses which is stable across time and situations. *Applied Animal Behaviour Science* 114: 492–508.

Lansade, L., M. F. Bouissou, and X. Boivin. 2007. Temperament in preweanling horses: Development of reactions to humans and novelty, and startle responses. *Developmental Psychobiology* 49: 401–513.

Lansade, L., M. F. Bouissou, and H. W. Erhard. 2008a. Reactivity to isolation and association with conspecifics: A temperament trait stable across time and situations. *Applied Animal Behaviour Science* 109: 355–73.

———. 2008b. Fearfulness in horses: A temperament trait stable across time and situations. *Applied Animal Behaviour Science* 115: 182–200.

Lansade, L., G. Pichard, and M. Leconte. 2008. Sensory sensitivities: Components of a horse's temperament dimension. *Applied Animal Behaviour Science* 114: 534–53.

Lansade L., and F. Simon. 2010. Horses' learning performances are under the influence of several temperamental dimensions. *Applied Animal Behaviour Science* 125: 30–37.

Lapicque, L., G. Dumas, H. Piéron, and H. Laugier. 1922. Rapport sur des expériences de contrôle relatives aux phénomènes dits ectoplasmiques. *L'Année Psychologique,* 23: 604–12.

Larose, C., M. A. Richard-Yris, M. Hausberger, and L. J. Rogers. 2006. Laterality of horses associated with emotionality in novel situations. *Laterality* 11: 355–67.

Laska, M., A. Seibt, and A. Weber. 2000. "Microsmatic" primates revisited: Olfactory sensitivity in the squirrel monkey. *Chemical Senses* 25: 47–53.

Lassalle, J.-M. 2004. Apprentissage, adaptation et cognition. In *L'éthologie cognitive*, edited by J. Vauclair and M. Kreutzer. Paris: Éditions Ophrys, Éditions de la Maison des Sciences de l'Homme, 49–75.

Latto, R. 1988. Making decisions about the conscious experiences of animals. *Behavioural and Political Animal Studies* 1: 7–16.

Lautrey, J. 2004. Hauts potentiels et talents: La position actuelle du problème. *Psychologie Française* 49: 219–32.

Lavoie, M. P. 2007. Évaluation de la sensibilité rétinienne dans le but d'élucider le dérèglement chimique à l'origine de la dépression saisonnière hivernale et les mécanismes biologiques de la luminothérapie. Ph.D. thesis. Université Laval, Quebec.

Lazaris, S. 2005. Considérations sur l'apparition de l'étrier: Contribution à l'histoire du cheval dans l'Antiquité tardive. In *Les équidés dans le monde méditerranéen antique,* edited by A. Gardeisen. Lattes, France: Monographies d'Archéologie Méditerranéenne, 275–88.

Leblanc, M.-A. 1984. *Le cheval. Comportement, vie sociale, relations avec l'environnement.* Montreal: Les Éditions de l'Homme.

Leblanc, M.-A., and M. F. Bouissou. 1981. Mise au point d'une épreuve destinée à l'étude de la reconnaissance du jeune par la mère chez le cheval. *Biology of Behaviour* 6: 283–90.

Leblanc, M.-A., M. F. Bouissou, and F. Chéhu. 2004. *Cheval qui es-tu? L'éthologie du cheval, du comportement naturel à la vie domestique.* Paris: Éditions Belin.

Leblanc M.-A., and P. Duncan. 2007. Can studies of cognitive abilities and of life in the wild really help us to understand equine learning? *Behavioural Processes* 76: 49–52.

Le Bon, G. 1923. *L'équitation actuelle et ses principes. Recherches expérimentales.* 4th ed. with a preface by Lieutenant-Colonel Blacque-Belair. Paris: Flammarion.

Leclerc, V., P. MacLeod, and B. Schaal. 2002. Le goût. *La Recherche* 349: 54–57.

Lefèvre-Balleydier, A., P. MacLeod, and A. Holley. 2006. L'odorat. *La Recherche,* 393, 91–94.

Lehman, H. 1998. Animal awareness. *Applied Animal Behaviour Science* 57: 315–25.

Leliveld, L., J. Langbein, and B. Puppe. 2013. The emergence of emotional lateralization: Evidence in non-human vertebrates and implications for farm animals. *Applied Animal Behaviour Science* 145: 1–14.

Le Maléfan, P. 2004. La psychopathologie confrontée aux fantômes. L'épisode de la villa Carmen. *Psychologie et Histoire* 5: 1–19.

Lemasson, A., A. Boutin, S. Boivin, C. Blois-Heulin, and M. Hausberger. 2009. Horse (*Equus caballus*) whinnies: A source of social information. *Animal Cognition* 12: 693–704.

Lemasson, M., and P.-M. Lledo. 2003. Le cerveau adulte: Un perpétuel chantier! *Médecine/Sciences* 19: 664–66.

Leonard, J. A. 2002. Ancient DNA evidence for Old World origin of New World dogs. *Science* 298: 1613.

Leopold, D. A., and G. Rhodes. 2010. A comparative view of face perception. *Journal of Comparative Psychology* 124: 233–51.

Lesimple, C., C. Sankey, M. A. Richard, and M. Hausberger. 2012. Do horses expect humans to solve their problems? *Frontiers in Psychology* 3 (306): 1–4.

Lettvin, J. Y., H. R. Maturana, W. S. McCulloch, and W. H. Pitts. 1959. What the frog's eye tells the frog's brain. *Proceedings of the Institute of Radio Engineers* 47: 1940–51.

Levine, J. M., G. J. Levine, A. G. Hoffman, and G. R. Bratton. 2008. Comparative anatomy of the horse, ox, and dog: The brain and associated vessels. *Equine Compendium on Continuing Education for the Practicing Veterinarian* 3: 153–63.

Levine, M. A. 2005. Domestication and early history of the horse. In *The Domestic Horse: The Evolution, Development, and Management of Its Behaviour*, edited by D. Mills and S. McDonnell. Cambridge, UK: Cambridge University Press, 5–22.

Lévy, F., M. Keller, and P. Poindron. 2004. Olfactory regulation of maternal behavior in mammals. *Hormones and Behavior* 46: 284–302.

L'Hotte, A.-F. 1997. *The Quest for Lightness in Equitation*. Trans. H. Nelson. London: J. A. Allen. First published 1906 by Plon-Nourrit et cie, Paris.

Liberles, S. D., and L. B. Buck. 2006. A second class of chemosensory receptors in the olfactory epithelium. *Nature* 442: 645–50.

Lima, S. L., and L. M. Dill. 1990. Behavioural decisions made under the risk of predation: A review and prospectus. *Canadian Journal of Zoology* 68: 619–40.

Lindberg, A. C., A. Kelland, and C. J. Nicol. 1999. Effects of observational learning on acquisition of an operant response in horses. *Applied Animal Behaviour Science* 61: 187–99.

Linklater, W. L., and E. Z. Cameron. 2000. Distinguishing cooperation from cohabitation: The feral horse case study. *Animal Behaviour* 59: F17–21.

Linklater, W. L., E. Z. Cameron, K. J. Stafford, and C. J. Veltman. 2000. Social and spatial structure and range use by Kaimanawa wild horses. *New Zealand Journal of Ecology* 24: 139–52.

Lledo, P.-M., and G. Gheusi. 2006. Neurogenèse adulte: Aspects fondamentaux et potentiels thérapeutiques. *Bulletin de l'Académie Nationale de Médecine* 190: 385–402.

Lledo, P.-M., and J.-D. Vincent. 1999. Odeurs. *Médecine/Sciences* 15: 1211–18.

Lorenz, K. Z. 1981. *The Foundations of Ethology*. New York: Springer. First published 1978 by Springer-Verlag/Wien, New York.

MacLean, P. D. 1962. New findings relevant to the evolution of psychosexual functions of the brain. *Journal of Nervous and Mental Disorders* 135: 289–301.

———. 1989. *The Triune Brain in Evolution: Role in Paleocerebral Functions*. New York: Plenum.

MacLeod, P., and F. Sauvageot. 1986. L'audition. In *Bases neurophysiologiques de l'evaluation sensorielle des produits alimentaires*, edited by P. MacLeod, F. Sauvageot, and G. Chevalier. Les cahiers de l'ENSBANA, 5, 21–45

Macuda, T. 2000. Equine color vision. Ph.D. dissertation. University of Western Ontario, London, Ontario.

Macuda, T., and B. Timney. 1998. The spectral sensitivity of the horse. *Investigative Ophthalmology and Visual Science* (Suppl. 35): 2210.

Macuda, T., and B. Timney. 1999. Luminance and chromatic discrimination in the horse *(Equus caballus)*. *Behavioural Processes* 41: 301–7.

Maeterlinck, M. 1914. *The Unknown Guest*. Trans. A. Teixeira de Mattos. London: Methuen.

Mair, T., S. Love, J. Schumacher, and E. Watson. 1998. *Equine Medicine, Surgery, and Reproduction*. Philadelphia: Saunders.

Marinier, S. L., and A. J. Alexander. 1994. The use of a maze in testing learning and memory in the horse. *Applied Animal Behaviour Science* 39: 177–82.

———. 1995. Coprophagy as an avenue for foals of the domestic horse to learn food preferences from their dams. *Journal of Theoretical Biology* 173: 121–24.

Marinier, S. L., A. J. Alexander, and G. H. Waring. 1988. Flehmen behaviour in the domestic horse: Discrimination of conspecific odors. *Applied Animal Behaviour Science* 19: 227–37.

Marks, W. B., W. H. Dobelle, and E. F. MacNichol. 1964. Visual pigments of single primate cones. *Science* 143: 1181–82.

Marmin, N. 2001. Métapsychique et psychologie en France (1880–1940). *Revue d'histoire des sciences humaines* 1 (4): 145–71.

Maros, K., M. Gácsi, and Á. Miklósi. 2008. Comprehension of human pointing gestures in horses (*Equus caballus*). *Animal Cognition* 11: 457–66.

Marshall, D. A., and D. G. Moulton. 1981. Olfactory sensitivity to alpha-ionone in humans and dogs. *Chemical Senses* 6: 53–61.

Martinengo Cesaresco, E. 1906. *The Psychology and Training of the Horse*. New York: Charles Scribner's Sons.

Martínez-García, F., J. Martínez-Ricós, C. Agustín-Pavón, J. Martínez-Hernández, A. Novejarque, and E. Lanuza. 2009. Refining the dual olfactory hypothesis: Pheromone reward and odour experience. *Behavioural Brain Research* 200: 277–86.

Masland, R. H. 2001. The fundamental plan of the retina. *Nature Neuroscience* 4: 877–86.

Massad, C. M., Hubbard, M., and Newtson, D. 1979. Selective perception of events. *Journal of Experimental Social Psychology*, 15, 513–32.

Masterton, R. B., and I. T. Diamond. 1973. Hearing: Central neural mechanisms. In *Biology of Perceptual Systems*, edited by E. C. Carterette and M. P. Friedman. Vol. 3 of *Handbook of Perception*. New York: Academic Press, 407–48.

McAfee, L. M., D. S. Mills, and J. J. Cooper. 2002. The use of mirrors for the control of stereotypic weaving behaviour in the stabled horse. *Applied Animal Behaviour Science* 78: 159–73.

McBride, S. D., A. J. Hemmings, and K. Robinson. 2004. A preliminary study on the effect of massage to reduce stress in the horse. *Journal of Equine Vetterinary Science* 24: 76–81.

McCall, C. A. 1990. A review of learning behavior in horses and its application in horse training. *Journal of Animal Science* 68: 75–81.

———. 2007. Making equine learning research applicable to training procedures. *Behavioural Processes* 76: 27–28.

McCall, C. A., S. Hall, W. H. McElhenney, and K. A. Cummins. 2006. Evaluation and comparison of four methods of ranking horses based on reactivity. *Applied Animal Behaviour Science* 96: 115–27.

McCall, C. A., M. A. Salters, and S. M. Simpson. 1993. Relationship between number of conditioning trials per training session and avoidance learning in horses. *Applied Animal Behaviour Science* 36: 291–99.

McCort, W. D. 1984. Behavior of feral horses and ponies. *Journal of Animal Science* 58: 493–99.

McDonald, B. J., and A. K. Warren-Smith. 2010. Mare and foal recognition after a prolonged period of separation. *Journal of Veterinary Behavior*, 5, 215.

McDonnell, S. M. 2003. *A Practical Guide to Horse Behavior: The Equid Ethogram*. Lexington, KY: Eclipse Press.

McDonnell, S. M., and J. C. S. Haviland. 1995. Agonistic ethogram of the equid bachelor band. *Applied Animal Behaviour Science* 43: 147–88.

McFarland, D. 1985. *Animal Behavior: Psychobiology, Ethology, and Evolution.* Menlo Park, CA: Benjamin/Cummings.

———. 1986. *Problems of Animal Behaviour.* New York: Wiley.

———. 2006. *Dictionary of Animal Behaviour.* Oxford, UK: Oxford University Press.

McGreevy, P. D. 2004. *Equine Behaviour: A Guide to Veterinarians and Equine Scientists.* Edinburgh: Saunders.

McGreevy, P. D., A. Harman, A. N. McLean, and L. A. Hawson. 2010. Over-flexing the horse's neck: A modern equestrian obsession? *Journal of Veterinary Behavior* 5: 180–86.

McGreevy, P. D., and A. N. McLean. 2010. *Equitation Science.* Chichester, UK: Wiley-Blackwell.

McGreevy, P. D., and L. J. Rogers. 2005. Motor and sensory laterality in thoroughbred horses. *Applied Animal Behaviour Science* 92: 337–52.

McGreevy, P. D., and P. C. Thomson. 2006. Differences in motor laterality between breeds of performance horse. *Applied Animal Behaviour Science* 99: 183–90.

McGurk, H., and J. MacDonald. 1976. Hearing lips and seeing voices. *Nature* 264: 746–48.

McKinley, J., and T. D. Sambrook. 2000. Use of human-given cues by domestic dogs (*Canis familiaris*) and horses (*Equus caballus*). *Animal Cognition* 3: 13–22.

McLean, A. N. 2001. Cognitive abilities—the result of selective pressures on food acquisition? *Applied Animal Behaviour Science* 71: 241–58.

———. 2004. Short-term spatial memory in the domestic horse. *Applied Animal Behaviour Science* 85: 93–105.

Meehan, J. 1904. The Berlin "thinking" horse. *Nature* 70: 602–3.

Meierhenrich, U. J., J. Golebiowski, X. Fernandez, and D. Cabrol-Bass. 2005. De la molécule à l'odeur. Les bases moléculaires des premières étapes de l'olfaction. *L'actualité chimique* 289: 29–40.

Menault, E. 1869. *The Intelligence of Animals.* Anonymous translator. New York: Scribner.

Meuret, M., C. Viaux, and J. Chadoeuf. 1994. Land heterogeneity stimulates intake rates during grazing trips. *Annales de Zootechnie* 43: 296.

Miller, G. 2003. The cognitive revolution: A historical perspective. *Trends in Cognitive Science* 7: 141–44.

Miller, P. E. 2001. Vision in animals: What do dogs and cats see? *Proceedings of the 25th Annual Waltham/Ohio State University Symposium for the Treatment of Small Animal Diseases: Small Animal Ophthalmology.* Edited by D. A. Wilkie and A. Olaerts. Waltham USA. www.vin.com/OSUWaltham/2001.

Miller, R. 1981. Male aggression, dominance, and breeding behavior in Red Desert feral horses. *Zeitschrift für Tierpsychologie* 57: 340–51.

Miller, R., and R. H. Denniston. 1979. Interband dominance in feral horses, *Zeitschrift für Tierpsychologie* 51: 41–47.

Miller, R. M. 1991. *Imprint Training of the Newborn Foal.* Colorado Springs, CO: Western Horseman.

Miller, P. E., and C. J. Murphy. 1995. Vision in dogs. *Journal of the Veterinary Medical Association* 207: 1623–34.

Mills, D., and S. McDonnell, eds. 2005. *The Domestic Horse: The Evolution, Development, and Management of Its Behaviour.* Cambridge, UK: Cambridge University Press.

Mills, D. S. 1998. Personality and individual differences in the horse, their significance, use, and measurement. *Equine Veterinary Journal* 30 (S27): 10–13.

Mills, D. S., and M. Riezebos. 2005. The role of the image of a conspecific in the regulation of stereotypic head movements in the horse. *Applied Animal Behaviour Science* 91: 155–65.

Mills, D. S., K. D. Taylor, and J. J. Cooper. 2005. Weaving, headshaking, cribbing, and other stereotypies. *51st Annual Convention of the American Association of Equine Practitioners,* Seattle, Washington. Ithaca, NY: International Veterinary Information Service. www.ivis.org.

Minero, M., E. Canali, V. Ferrante, M. Verga, and F. O. Ödberg. 1999. Heart rate and behavioural responses of crib-biting horses to two acute stressors. *Veterinary Record* 145: 430–33.

Monard, A. M., and P. Duncan. 1996. Consequences of natal dispersal in female horses. *Animal Behaviour* 52: 565–79.

Morgan, C. L. 1894. *Introduction to Comparative Psychology.* London: Scott.

———. 1903. *Introduction to Comparative Psychology.* Rev. 2nd ed. London: Scott.

Morgan, M. J., A. Adam, and J. D. Mollon. 1992. Dichromats detect colour-camouflaged objects that are not detected by trichromats. *Proceedings of the Royal Society of London B* 248: 291–95.

Morris, P. H., A. Gale, and K. Duffy. 2002. Can judges agree on the personality of horses? *Personality and Individual Differences* 33: 67–81.

Moss, C. F., and C. E. Carr. 2002. Comparative psychology of audition. In *Biological Psychology*, edited by R. Nelson and M. Gallagher. Vol. 3 of *Handbook of Psychology*. Hoboken, NJ: Wiley, 71–107.

Muller, D. 2004. La plasticité des fonctions et structures syntaxiques. *Epileptologie* 2: 2–6.

Murphy, J., and S. Arkins, eds. 2007. Equine learning behaviour. *Behavioural Processes* 76: 1–60.

Murphy, J., C. Hall, and S. Arkins. 2009. What horses and humans see: A comparative review. *International Journal of Zoology* 2009: 1–15. www.hindawi.com/journals/ijz/2009/721798.

Murphy, J., A. Sutherland, and S. Arkins. 2005. Idiosyncratic motor laterality in the horse. *Applied Animal Behaviour Science,* 91, 297–310.

Neisser, U. 1968. The processes of vision. *Scientific American*, September, 204–14.

Neisser, U., G. Boodoo, T. J. Bouchard, A. W. Boykin, N. Brody, J. Cecis, D. F. Halpern, J. C. Loehlin, R. Perloff, R. J. Sternberg, and S. Urbina. 1996. Intelligence: Knowns and unknowns. *American Psychologist* 51 (2): 77–101.

Nicol, C. J. 2002. Equine learning: Progress and suggestions for future research. *Applied Animal Behaviour Science* 78: 193–208.

———. 2005. Learning abilities in the horse. *The Domestic Horse: The Evolution, Development, and Management of Its Behaviour*, edited by D. Mills and S. McDonnell. Cambridge, UK: Cambridge University Press, 169–83.

Nieder, A. 2002. Seeing more than meets the eye: Processing of illusory contours in animals. *Journal of Comparative Physiology A* 188: 249–60.

Nieder, A. and H. Wagner. 1999. Perception and neural coding of subjective contours in the owl. *Nature Neuroscience* 2: 660–63.

Noreika, A. 1998. Ophthalmoscopy of eye fundus in horses. Abstract. *Veterinarija ir Zootechnika* 5: 27.

Normando, S., A. Haverbeke, L. Meers, F. O. Ödberg, M. Ibanez Talegon, and G. Bono. 2003. Effects of manual imitation of grooming on riding horses' heart rate in different environmental situations. *Veterinary Research Communications* 27: 615–17.

Normando, S., C. Trevisan, O. Bonetti, and G. Bono. 2007. A note on heart rate response to massage in stereotyping and non stereotyping horses. *Journal of Animal and Veterinary Advances* 6: 101–4.

Nottebohm, F. 1981. A brain for all seasons: Cyclical anatomical changes in song control nuclei of the canary brain. *Science* 214: 1368–70.

Ödberg, F. O. 1974. Some aspects of the acoustic expression in horses. In *Ethologie und Ökologie bei der Haustierhaltung,* edited by K. Zeeb. Darmstadt, Germany: KTBL, 89–105.

———. 1978. A study of the hearing ability of horses. *Equine Veterinary Journal* 10: 82–84.

Okano, M., S. Fujimaki, and K. Onishi. 1998. The structures of the vomeronasal complex in the horse. *Journal of Bioresource Sciences* 1: 19–26.

O'Keefe, J., and L. Nadel. 1978. *The Hippocampus as a Cognitive Map.* Oxford, UK: Oxford University Press.

Ollivier, F. J., D. A. Samuelson, D. E. Brooks, P. A. Lewis, M. E. Kallberg, and A. M. Komáromy. 2004. Comparative morphology of the tapetum lucidum (among selected species). *Veterinary Ophthalmology* 7: 11–22.

Olsen, S. L., ed. 1996. Horse hunters of the ice age. In *Horses through Time.* Lanham, MD: Roberts Rinehart, 35–56.

Outram, A. K, N. A. Stear, R. Bendrey, S. Olsen, A. Kaspartov, V. Zaibert, N. Thorpe, and R. P. Evershed. 2009. The earliest horse harnessing and milking. *Science* 323: 1332–35.

Pasternak, T., and W. H. Merigan. 1980. Movement detection by cats: Invariance with direction and target configuration. *Journal of Comparative Physiology and Psychology* 94: 943–52.

Payne, K., W. R. Langbauer Jr., and E. Thomas. 1986. Infrasonic calls of the Asian elephant (*Elephas maximus*). *Behavioral Ecology and Sociobiology* 18: 297–301.

Peeples, D. R. and D. Y. Teller. 1975. Color vision and brightness discrimination in two-month-old human infants. *Science* 189: 1102–03.

Pérez-Barberia, F. J., and I. J. Gordon. 2005. Gregariousness increases brain size in ungulates. *Oecologia* 145: 41–52.

Pérez-Barberia, F. J., S. Shultz, and R. I. M. Dunbar. 2007. Evidence for coevolution of sociality and relative brain size in three orders of mammals. *Evolution* 61: 2811–21.

Pernollet, J.-C., G. Sanz, and L. Briand. 2006. Les récepteurs des molécules odorantes et le codage olfactif. *Comptes Rendus de l'Académie des Sciences, Biologies* 329: 679–90.

Pfister, J. A. 1999. Behavioral strategies for coping with poisonous plants. In *Grazing Behavior of Livestock and Wildlife,* edited by K. L. Launchbaugh, J. C. Mosley, and K. D. Sanders. Idaho Forest, Wildlife, and Range Experiment Station, Station Bulletin 70. Moscow: University of Idaho, 45–59.

Pfister, J. A., B. L. Stegelmeier, C. D. Cheney, and D. R. Gardner. 2007. Effect of previous locoweed (*Astragalus* and *Oxytropis* species) intoxication on conditioned taste aversions in horses and sheep. *Journal of Animal Science* 85: 1836–41.

Pfister, J. A., B. L. Stegelmeier, C. D. Cheney, M. H. Ralphs, and D. R. Gardner. 2002. Conditioning taste aversions to locoweed (*Oxytropis sericea*) in horses. *Journal of Animal Science* 80: 79–83.

Pfungst, O. 2000. *Clever Hans (The Horse of Mr. Von Osten): A Contribution to Experimental Animal and Human Psychology.* London: Thoemmes Continuum. First published 1911 by Henry Holt, New York. Trans. C. L. Rahn.

Pick, D. F., G. Lovell, S. Brown, and D. Dail 1994. Equine color perception revisited. *Applied Animal Behaviour Science,* 42, 61–65.

Piéron, H. 1913. Le problème des animaux pensants. *L'Année Psychologique* 20: 218–38.

Piggins, D., and C. J. C. Phillips. 1998. Awareness in domesticated animals—concepts and definitions. *Applied Animal Behaviour Science* 57: 181–200.

Pihlström, H., M. Fortelius, S. Hemilä, R. Forsman, and T. Reuter. 2005. Scaling of mammalian ethmoid bones can predict olfactory organ size and performance. *Proceedings of the Royal Society of London B* 272: 957–62.

Pinker, S. 1997. *How the Mind Works.* New York: Norton.

Plas, R. 2004. Comment la psychologie expérimentale française est-elle devenue cognitive? *La Revue pour l'histoire du CNRS* 10. http://histoire-cnrs.revues.org/586.

Pond, R. L., M. J. Darre, P. M. Scheifele, and D. G. Browning. 2010. Characterization of equine vocalization. *Journal of Veterinary Behavior* 5: 7–12.

Porter, J., B. Craven, R. M. Khan, S. J. Chang, I. Kang, B. Judkewitz, J. Volpe, G. Settles, and N. Sobel. 2007. Mechanisms of scent-tracking in humans. *Nature Neuroscience* 10: 27–29.

Portmann, M. 1992. L'oreille, structure et fonction. *Journal de Physique IV* 2: C1–13, C1–18.

Pretterer, G., H. Bubna-Littiz, G. Windischbauer, C. Gabler, and U. Griebel. 2004. Brightness discrimination in the dog. *Journal of Vision* 4: 241–49.

Privat, A. 1988. La synaptogénèse: Faits et perspectives. *Cahiers du Mouvement Universel de la Responsabilité Scientifique* 13: 19–36.

Proops, L., and K. McComb. 2012. Cross-modal individual recognition in domestic horses (*Equus caballus*) extends to familiar humans. *Proceedings of the Royal Society of London B* 279: 3131–38.

———. 2010. The use of human-given cues by domestic horses (*Equus caballus*). *Animal Cognition* 13: 197–205.

Proops, L., K. McComb, and D. Reby. 2009. Cross-modal individual recognition in domestic horses (*Equus caballus*). *Proceedings of the National Academy of Sciences of the U.S.A.* 106: 947–51.

Proops, L., M. Walton, and K. McComb. 2010. The use of human-given cues by domestic horses, *Equus caballus,* during an object choice task. *Animal Behaviour* 709: 1205–9.

Prost-Squarcioni, C. 2006. Histologie de la peau et des follicules pileux. *Médecine/ Sciences* 22: 131–37.

Proust, J. 2000. L'animal intentionnel. In *Les animaux pensent-ils? Terrain* 34: 23–36.

Pryor, K. W. 2008. *Don't Shoot the Dog: The New Art of Teaching and Training.* Rev. ed. Reading, UK: Ringpress.

Pryor, K. W., R. Haag, and J. O'Reilly. 1969. The creative porpoise: Training for novel behavior. *Journal of the Experimental Analysis of Behavior* 12: 653–61.

Purves, D., G. J. Augustine, D. Fitzpatrick, W. C. Hall, A.-S. Lamantia, J. O. McNamara, and S. M. Williams, eds. 2004. *Neuroscience.* Sunderland, MA: Sinauer Associates.

Putman, R. J., R. M. Pratt, J. R. Ekins, and P. J. Edwards. 1987. Food and feeding behaviour of cattle and ponies in the New Forest Hampshire. *Journal of Applied Ecology* 24: 369–80.

Quaranta, A., M. Siniscalchi, and G. Vallortigara. 2007. Asymmetric tail-wagging responses by dogs to different emotive stimuli. *Current Biology* 17 (6): R199–201.

Quignon, P., E. Kirkness, E. Cadieu, N. Touleimat, R. Guyon, C. Renier, C. Hitte, C. André, C. Fraser, and F. Galibert. 2003. Comparison of the canine and human olfactory receptor gene repertoires. *Genome Biology* 4 (R80): 1–9.

Ramachandran, V. S., and D. Rogers-Ramachandran. 2009. Seeing in stereo: Illusions of depth. *Scientific American Mind*, July/August, 20–22.

Randall, R. P., W. A. Schurg, and D. C. Church. 1978. Response of horses to sweet, salty, sour, and bitter solutions. *Journal of Animal Science* 47: 51–55.

Renck, J.-L., and V. Servais. 2002. *L'éthologie. Histoire naturelle du comportement.* Paris: Seuil, Points Sciences.

Restrepo, D., J. Arellano, A. M. Oliva, M. L. Schaefer, and W. Lin. 2004. Emerging views on the distinct but related roles of the main and accessory olfactory systems in responsiveness to chemosensory signals in mice. *Hormones and Behavior* 46: 247–56.

Reznikova, Z. 2007. *Animal Intelligence: From Individual to Social Cognition.* Cambridge, UK: Cambridge University Press.

Rhine, J. B., and L. E. Rhine. 1929a. An investigation of a "mind reading" horse. *Journal of Abnormal and Social Psychology* 23: 449–66.

———. 1929b. Second report on Lady, the "mind reading" horse. *Journal of Abnormal and Social Psychology* 24: 287–92.

Ringo, J. L., R. W. Doty, S. Demeter, and P. Y. Simard. 1994. Time is of the essence: A conjecture that hemispheric specialization arises from inter-hemispheric conduction delay. *Cerebral Cortex* 4: 331–43.

Risold, P.-Y. 2008. Avez-vous un "cerveau reptilien"? *Cerveau et Psycho* 29: 66–70.

Roberts, S. M. 1992. Equine vision and optics. *Veterinary Clinics of North America. Equine Practice* 8: 451–57.

Rogers, L. J. 2002. Lateralization in vertebrates: Its early evolution, general pattern, and development. *Advances in the Study of Behavior* 31: 107–61.

Rogers, L. J., and R. J. Andrew, eds. 2002. *Comparative Vertebrate Lateralization.* New York: Cambridge University Press.

Rogers, L. J., P. Zucca, and G. Vallortigara. 2004. Advantage of having a lateralized brain. *Proceedings of the Royal Society of London B* 271: S420–22.

Romanes, G. J. 1884. *Animal Intelligence.* New York: Appleton.

Rosolen, S., and F. Rigaudière. 2003. Exploration de la fonction visuelle sensorielle chez l'animal. Première partie: Rappels anatomo-physiologiques. *Bulletin de l'Académie Vétérinaire de France* 156: 15–24.

Roth, G., and U. Dicke. 2005. Evolution of the brain and intelligence. *Trends in Cognitive Sciences* 9: 250–57.

Roth, L. S. V., A. Balkenius, and A. Kelber. 2007. Colour perception in a dichromat. *Journal of Experimental Biology* 210: 2795–800.

———. 2008. The absolute threshold of colour vision in the horse. *PLoS ONE* 3 (11): e3711.

Royet, J. P., and J. Plailly. 2004. Lateralization of olfactory processes. *Chemical Senses* 29: 731–45.

Rubenstein, D., and M. Hack. 1992. Horse signals: The sounds and scents of fury. *Evolutionary Ecology* 6: 254–60.

Rubin, L., C. Oppegard, and H. F. Hintz. 1980. The effect of varying the temporal distribution of conditioning trials on equine learning behaviour. *Journal of Animal Science* 50: 1184–87.

Rugani, R., L., Fontanari, E. Simoni, L. Regolin, and G. Vallortigara. 2009. Arithmetic in newborn chicks. *Proceedings of the Royal Society of London B* 276 (1666): 2451–60.

Rushton, W. A. H. 1972. Pigments and signals in colour vision. *Journal of Physiology, London* 220: 1–31P.

Rutberg, A., and R. Keiper. 1993. Proximate causes of natal dispersal in feral ponies: Some sex differences. *Animal Behaviour* 46: 969–75.

Rzoska, J. 1953. Bait shyness: A study in rat behavior. *British Journal of Animal Behaviour* 1: 128–35.

Sacks, O. 1993. An anthropologist on Mars. *The New Yorker*, December 27, 106–125.

Salter, R. E., and R. J. Hudson. 1982. Social organization of feral horses in western Canada. *Applied Animal Ethology* 8: 207–23.

Salter, R. E., and D. J. Pluth. 1980. Determinants of mineral lick utilization by feral horses. *Northwest Science* 54: 109–18.

Sandmann, D., B. B. Boycott, and L. Peichl. 1996. Blue-cone horizontal cells in the retinae of horses and other equidae. *Journal of Neuroscience* 16: 3381–96.

Sankey, C., S. Henry, N. André, M.-A. Richard-Yris, and M. Hausberger. 2011a. Do horses have a concept of person? *PLoS ONE* 6 (3): e18331.

Sankey, C., S. Henry, C. Clouard, M. Richard-Yris, and M. Hausberger. 2011b. Asymmetry of behavioral responses to a human approach in young naïve vs. trained horses. *Physiology and Behavior* 104: 464–68.

Sankey, C., S. Henry, A. Gorecka-Bruzda, M.-A. Richard-Yris, and M. Hausberger. 2010a. The way to a man's heart is through his stomach: What about horses? *PLoS ONE* 5 (11): e15446.

Sankey, C., M.-A. Richard-Yris, S. Henry, C. Fureix, F. Nassur, and M. Hausberger. 2010b. Reinforcement as a mediator of the perception of humans by horses (*Equus caballus*). *Animal Cognition* 13: 753–64.

Sankey, C., M.-A. Richard-Yris, H. Leroy, S. Henry, and M. Hausberger. 2010c. Positive interactions lead to lasting positive memories in horses, *Equus caballus*. *Animal Behaviour* 79: 869–75.

Sappington, B. F., and L. Goldman. 1994. Discrimination learning and concept formation in the Arabian horse. *Journal of Animal Science* 72: 3080–87.

Saslow, C. A. 1999. Factors affecting stimulus visibility for horses. *Applied Animal Behaviour Science* 61: 273–84.

———. 2002. Understanding the perceptual world of horses. *Applied Animal Behaviour Science* 75: 209–24.

Saurel, E. 1964. *Pratique de l'équitation d'après les Maîtres français*. Paris: Flammarion.

Schaal, B., and R. D. Porter. 1990. Microsmatic humans revisited: The generation and perception of chemical signals. *Advances in the Study of Behavior* 20: 135–99.

Schelp, D. 2000. Studies on ethological and physiological parameters concerning the mode of action of the twitch in horses and its possible relevance for animal protection. Ph.D. dissertation. University of Munich. Abstract: http://library.vetmed.fu-berlin.de/diss-abstracts/115867.html.

Schino, G., S. Scucchi, D. Maestripieri, and P. G. Turillazzi. 1988. Allogrooming as a tension-reduction mechanism: A behavioral approach. *American Journal of Primatology* 16: 43–50

Schwarzlose, R. F., C. I. Baker, and N. Kanwisher. 2005. Separate face and body selectivity on the fusiform gyrus. *Journal of Neuroscience* 25: 11055–59.

Scorolli, A. L., A. C. Lopez Cazorla, and L. A. Tejera. 2006. Unusual mass mortality of feral horses during a violent rainstorm in Parque Provincial Tornquit, Argentina. *Mastozoologia Neotropical* 13: 255–58.

Scott, D. W., and W. H. Miller Jr. 2003. *Equine dermatology.* St. Louis, MO: Saunders.

Scott, K. 2004. The sweet and the bitter of mammalian taste. *Current Opinion in Neurobiology* 14: 423–27.

Sebeok, T. 1981. The ultimate enigma of "Clever Hans": The union of nature and culture. In *The Clever Hans Phenomenon: Communication with Horses, Whales, Apes, and People,* edited by T. Sebeok and R. Rosenthal. Vol. 364 of *Annals of the New York Academy of Sciences.* New York: The New York Academy of Sciences, 199–205.

Sebeok, T., and R. Rosenthal, eds. 1981. *The Clever Hans Phenomenon: Communication with Horses, Whales, Apes, and People.* Vol. 364 of *Annals of the New York Academy of Sciences.* New York: The New York Academy of Sciences.

Seyfarth, R. M., and D. L. Cheney. 2009. Seeing who we hear and hearing who we see. *Proceedings of the National Academy of Sciences of the U.S.A.* 106: 669–70.

Shams, L., Y. Kamitani, and S. Shimojo. 2000. What you see is what you hear. *Nature* 408 (6814): 788.

Sharpe, L. T., A. Stockman, H. Jägle, and J. Nathans. 1999. Opsin genes, cone photopigments, and color vision. In *Color Vision: From Genes to Perception,* edited by C. K. N. Gegenfurtner and L. T. Sharpe. Cambridge, UK: Cambridge University Press, 3–51.

Shepherd, G. M. 2004. The human sense of smell: Are we better than we think? *PLoS Biology* 5: 572–75.

Shepard, R. N., and J. Metzler. 1971. Mental rotation of three-dimensional objects. *Science* 171: 701–3.

Shultz, S., and R. I. M. Dunbar. 2006. Both social and ecological factors predict ungulate brain size. *Proceedings of the Royal Society of London B* 273: 207–15.

Simunovic, M. P., B. C. Regan, and J. D. Mollon. 2001. Is color vision deficiency an advantage under scotopic conditions? *Investigative Ophthalmology and Visual Science* 42: 3357–64.

Siniscalchi, M., A. Quaranta, and L. J. Rogers. 2008. Hemispheric specialization in dogs for processing different acoustic stimuli. *PLoS ONE* 3 (10): e3349.

Sivak, J. G., and D. B. Allen. 1975. An evaluation of the ramp retina on the horse eye. *Vision Research* 15: 1353–56.

Skinner, B. F. 1960. Pigeons in a pelican. *American Psychologist* 15: 28–37.

Slater, C., and S. Dymond. 2011. Using differential reinforcement to improve equine welfare: Shaping appropriate truck loading and feet handling. *Behavioural Processes* 86: 329–39.

Smith, S., and L. Goldman. 1999. Color discrimination in horses. *Applied Animal Behaviour Science* 62: 13–25.

Smith, T. D., and K. P. Bhatnagar. 2004. Microsmatic primates: Reconsidering how and when size matters. *Anatomical Record* 279B: 24–31.

Søndergaard, E., and J. Jago. 2010. The effect of early handling of foals on their reaction to handling, humans and novelty, and the foal-mare relationship. *Applied Animal Behavior Science* 123: 93–100.

Spelke, E. S. 1979. Perceiving bimodally specified events in infancy. *Developmental Psychology* 15: 626–36.

———. 1985. Preferential looking methods as tools for the study of cognition in infancy. In *Measurement of Audition and Vision in the First Year of Postnatal Life,* edited by G. Gottleib and N. Krasnegor. Norwood, NJ: Ablex, 323–63.

———. 2000. Core knowledge. *American Psychologist* 1: 1233–43.

Spelke, E. S., and K. D. Kinzler. 2007. Core knowledge. *Developmental Science* 10: 89–96.

Spetch, M. L., and A. Friedman. 2006. Comparative cognition of object recognition. *Comparative Cognition and Behavior Reviews* 1: 12–35.

Stahlbaum, C. C., and K. A. Houpt. 1989. The role of the flehmen response in the behavioral repertoire of the stallion. *Physiology and Behavior* 45: 1207–14.

Sternberg, R. J., and D. K. Detterman, eds. 1986. *What Is Intelligence? Contemporary Viewpoints on Its Nature and Definition.* Norwood, NJ: Ablex.

Stevens, S. S., and Newman, E. B. 1936. On the nature of aural harmonics. *Proceedings of the National Academy of Sciences of the U.S.A.* 22: 668–72.

Stoffel-Willame, M., and Y. Stoffel-Willame. 1999. Horses of the Namib. *Africa, Environment and Wildlife* 7 (1): 58–67.

Stone, S. M. 2010. Human facial discrimination in horses: Can they tell us apart? *Animal Cognition,* 13, 51–61.

Stréri, A. 2003. Perception. Psychology. In *Dictionary of Cognitive Science: Neuroscience, Psychology, Artificial Intelligence, Linguistics, and Philosophy,* edited by O. Houdé, D. Kayser, O. Koenig, J. Proust, and F. Rastier. New York: Psychology Press.

Strutt, J. W. (Lord Rayleigh). 1907. On our perception of sound direction. *Philosophical Magazine* 13: 214–32.

Tanida, H., Y. Koba, N. Horikawa, and T. Nagai. 2005. Ability of ponies to discriminate between photographs of two similarly dressed people. In *Proceedings of the 39th International Congress of the International Society of Applied Ethology.* ISAE: Kanagawa, Japan, 85.

Tanzarella, S. 2006. *Perception et communication chez les animaux.* Brussels: De Boeck Université.

Tasaki, I. 1954. Nerve impulses in individual auditory nerve fibers of guinea pig. *Journal of Neurophysiology* 17: 97–122.

Tate, A. J., H. Fischer, A. E. Leigh, and K. M. Kendrick. 2006. Behavioural and neurophysiological evidence for face identity and face emotion processing in animals. *Philosophical Transactions of the Royal Society B* 361: 2155–72.

REFERENCES

Taube, J. S., R. I. Muller, and J. B. J. Ranck. 1990. Head direction cells recorded from the postsubiculum in freely moving rats. *Journal of Neuroscience* 10: 420–47.

Terlouw, E. M. C., A. Boissy, and P. Blinet. 1998. Behavioural responses of cattle to the odours of blood and urine from conspecifics and to the odour of faeces from carnivores. *Applied Animal Behaviour Science* 57: 9–21.

Terrace, H. S. 1984. Animal cognition. In *Animal Cognition*, edited by H. L. Roitblat, T. G. Bever, and H. S. Terrace. Hillsdale, NJ: Lawrence Erlbaum Associates, 7–28.

Theriault, E., and J. Diamond. 1988. Nociceptive cutaneous stimuli evoke localized contractions in a skeletal muscle. *Journal of Neurophysiology* 60: 446–62.

Thorndike, E. L. 1898. Animal intelligence: An experimental study of the associative processes in animals. *Psychological Review Monograph Supplements* 2 (8): 1–109.

Thurstone, L. L. 1940. Current issues in factor analysis. *Psychological Bulletin* 37: 189–236.

Tibbets, E. A., and J. Dale. 2007. Individual recognition: It is good to be different. *Trends in Ecology and Evolution* 22: 529–37.

Timney, B. 2009. Photopic spectral sensitivity and wavelength discrimination in the horse (*Equus caballus*). In *Proceedings of the International Equine Science Meeting 2008*, edited by K. Kruger. Wald, Germany: Xenophon, 58.

Timney, B., and K. Keil. 1992. Visual acuity in the horse. *Vision Research* 32: 2289–93.

———. 1996. Horses are sensitive to pictorial depth cues. *Perception* 25: 1121–28.

———. 1999. Local and global stereopsis in the horse. *Vision Research* 39: 1861–67.

Timney, B., and T. Macuda. 2001. Vision and hearing in horses. *Journal of the American Veterinary Medical Association* 218: 1567–74.

Timney, B., and B. Wright. 2007. Le monde visuel du cheval. Ontario, Canada: Ministry of Agriculture, Food, and Rural Affairs. www.omafra.gov.on.ca/french/livestock/horses/facts/info-visualworld.htm.

Tinbergen, N. 1963. On aims and methods in ethology. *Zeitschrift für Tierpsychologie* 20: 410–33.

Tolman, E. C. 1948. Cognitive maps in rats and men. *Psychological Review* 55: 189–208.

Tolman, E. C., and C. H. Honzik. 1930. Insight in rats. *University of California Publications in Psychology* 4: 215–32.

Tomchik, S. M., S. Berg, J. W. Kim, N. Chaudhari, and S. D. Roper. 2007. Breadth of tuning and taste coding in mammalian taste buds. *Journal of Neuroscience* 27: 10840–48.

Treisman, M. 1975. Predation and the evolution of gregariousness. I. Models of concealment and evasion. *Animal Behaviour* 23: 779–800.

Turner, J. W., A. Perkins, and J. F. Kirkpatrick. 1981. Elimination marking behavior in feral horses. *Canadian Journal of Zoology* 59: 1561.

Tyler S. J. 1972. The behaviour and social organization of the New Forest Ponies. *Animal Behaviour Monographs* 5: 85–196.

Uller, C., and J. Lewis. 2009. Horses (*Equus caballus*) select the greater of two quantities in small numerical contrasts. *Animal Cognition* 12: 733–38.

Vallortigara, G. 2000. Comparative neuropsychology of the dual brain: A stroll through animals' left and right perceptual worlds. *Brain and Language* 73: 189–219.

Vallortigara, G., and A. Bisazza. 2002. How ancient is brain lateralization? In *Comparative Vertebrate Lateralization: The Evolution of Brain Lateralization*, edited by L. J. Rogers and R. J. Andrew. Cambridge, UK: Cambridge University Press, 9–69.

Vallortigara, G., and L. J. Rogers. 2005. Survival with an asymmetrical brain: Advantages and disadvantages of cerebral lateralization. *Behavioral and Brain Sciences* 28: 575–89.

Vallortigara, G., A. Snyder, G. Kaplan, P. Bateson, N. S. Clayton, and L. J. Rogers. 2008. Are animals autistic savants? *PLoS Biology* 6 (e42): 208–14.

VanDierendonck, M. C., and B. M. Spruijt. 2012. Coping in groups of domestic horses— review from a social and neurobiological perspective. *Applied Animal Behaviour Science* 138: 194–202.

van Hateren, J. H., M. V. Srinivasan, and P. B. Wait. 1990. Pattern recognition in bees: Orientation discrimination. *Journal of Comparative Physiology* 167: 649–54.

van Rillaer, J. 1980. *Les illusions de la psychanalyse.* Brussels: Mardaga.

Vauclair, J. 1992. Psychologie cognitive et représentations animales. In *La Représentation animale*, edited by J. Gervet, P. Livet, and A. Tête. Nancy, France: Presses de l'Université de Nancy, 127–42.

———. 1995. *L'intelligence de l'animal.* Paris: Seuil/Points Science.

———. 1996. *Animal Cognition: An Introduction to Modern Comparative Psychology.* Cambridge, MA: Harvard University Press.

Vauclair, J., J. Fagot, and W. D. Hopkins. 1993. Rotation of mental images in baboons when the visual input is directed to the left cerebral hemisphere. *Psychological Science* 4: 99–103.

Vauclair, J., and M. Kreutzer, eds. 2004. *L'éthologie cognitive.* Paris: Éditions Ophrys, Éditions de la Maison des Sciences de l'Homme.

Vauclair, J., and P. Perret. 2003. The cognitive revolution in Europe: Taking the developmental perspective seriously. *Trends in Cognitive Sciences* 7: 284–85.

Verhulst, S., and F. W. Maes. 1998. Scotopic vision in colour-blinds. *Vision Research* 38: 3387–90.

Verrill, S., and S. McDonnell. 2008. Equal outcomes with and without human-to-horse eye contact when catching horses and ponies in an open pasture. *Journal of Equine Veterinary Science* 28: 309–12.

Vialar, P. 1982. *Cheval, mon bel ami.* Paris: Albin Michel.

Vincent, J. D. 2004. *Au commencement était l'émotion.* Séance solennelle de l'Académie des Sciences. www.academie-sciences.fr/conferences/seances_solennelles/pdf/ discours_Vincent_15_06_04. pdf.

Visser, E. K., C. G. van Reenen, J. T. N. van derWerf, M. B. H. Schilder, J. H. Knaap, A. Barneveld, and H. J. Blokhuis. 2002. Heart rate and heart rate variability during a novel object test and a handling test in young horses. *Physiology and Behavior* 76: 289–96.

von Frisch, K. 1955. *The Dancing Bees: An Account of the Life and Senses of the Honey Bee.* Trans. D. Ilse. New York: Harcourt, Brace.

von Uexküll, J. 1957. A stroll through the worlds of animals and men: A picture book of invisible worlds. In *Instinctive Behavior: The Development of a Modern Concept,* edited and translated by C. H. Schiller. New York: International Universities Press, 5–80. First published 1934 by Verlag von Julius Springer, Berlin.

Vulink, J. T. 2001. *Hungry Herds: Management of Temperate Lowland Wetlands by Grazing.* Lelystad, The Netherlands: Rijksuniversiteit Groningen.

Wachtler, T., U. Dohrmann, and R. Hertel. 2004. Modeling color percepts of dichromats. *Vision Resarch* 44: 2843–55.

Waggoner, T. L. 1994. *Color Vision Testing Made Easy Test.* Elgin, IL: Good-Lite.

Wakuri, H., K. Mutoh, H. Ichikawa, and B. Liu. 1995. Microscopic anatomy of the equine skin with special reference to the dermis. *Okajimas Folia Anatomica Japonica* 72: 177–83.

Walser, E. E. S. 1986. Recognition of the sow's voice by neonatal piglets. *Behaviour* 99: 177–88.

Walser, E. S., P. Hague, and E. Walters. 1981. Vocal recognition of recorded lambs voices by ewes of three breeds of sheep. *Behaviour* 78: 260–72.

Waring, G. H. 1971. Sounds of the horse (*Equus caballus*). *Bulletin of the Ecological Society of America* 52: 45.

———. 2003. *Horse Behaviour: The Behavioral Traits and Adaptations of Domestic and Wild Horses, Including Ponies.* 2nd ed. New York: Noyes.

Warmuth, V., A. Eriksson, M. A. Bower, G. Barker, E. Barrett, B. K. Hanks, S. Li, D. Limitashvili, M. Ochir-Goryaeva, G. V. Sizonov, V. Soyonov, and A. Manica. 2012. Reconstructing the origin and spread of horse domestication in the Eurasian steppe. *Proceedings of the National Academy of Sciences of the U.S.A.* 109:8202–6.

Warrant, E. J. 1999. Seeing better at night: Life style, eye design, and the optimum strategy of spatial and temporal summation. *Vision Research* 39: 1611–30.

Warren-Smith, A. K., and P. D. McGreevy. 2007. The use of blended positive and negative reinforcement in shaping the halt response of horses (*Equus caballus*). *Animal Welfare* 16: 481–88.

Watson, J. B. 1913. Psychology as the behaviorist views it. *Psychological Review* 20: 158–77.

Weary, D. M., G. Lawson, and B. K. Thompson. 1996. Sows show stronger responses to isolation calls of piglets associated with greater levels of piglet need. *Animal Behaviour* 52: 1247–53.

Weeks, J. W., S. L. Crowell-Davis, and G. Heusner. 2002. Preliminary study of the development of the flehmen response in *Equus caballus*. *Applied Animal Behaviour Science* 78: 329–34.

Wells, A. E. D., and D. Blache. 2008. Horses do not exhibit motor bias when their balance is challenged. *Animal* 2: 1645–50.

Welsh, D. A. 1973. The life of Sable Island's wild horse. *Canadian Nature* 2: 7–14.

———. 1975. Population, behavioural, and grazing ecology of the horses of Sable Island. Ph.D. dissertation. Dalhousie University, Halifax, Nova Scotia.

White, M. 1979. A new effect on perceived lightness. *Perception* 8: 413–16.

Williams, D. E., and B. J. Norris. 2007. Laterality in stride pattern preferences in racehorses. *Animal Behaviour* 74: 941–50.

Williams, M. 1976. *Horse Psychology.* London: J. A. Allen.

Wolff, A., and M. Hausberger. 1996. Learning and memorisation of two different tasks in horses, the effects of age, sex, and sire. *Applied Animal Behaviour Science* 46: 137–43.

Wolff, A., M. Hausberger, and N. Le Scolan. 1997. Experimental tests to assess emotivity in horses. *Behavioural Processes* 40: 209–21.

Wolski, T. R., K. A. Houpt, and R. Aronson. 1980. The role of the senses in mare-foal recognition. *Applied Animal Ethology* 6: 121–38.

Wolter, R. 1975. *L'alimentation du cheval.* Paris: Éditions Vigot Frères.

Wong, D., V. Buechner-Maxwell, and T. Manning. 2005. Equine skin: Structure, immunologic function, and methods of diagnosing disease. *Compendium, Continuing Education for Veterinarians* 27: 463–72.

Wouters, L., and A. De Moor. 1979. Ultrastructure of the pigment epithelium and the photoreceptors in the retina of the horse. *American Journal of Veterinary Research* 40: 1066–71.

Wouters, L., A. De Moor, and Y. Moens. 1980. Rod and cone components in the electro-retinogram of the horse. *Zentralblatt für Veterinärmedizin Reihe A* 27: 330–38.

Yeon, S. C. 2012. Acoustic communication in the domestic horse (*Equus caballus*). *Journal of Veterinary Behavior* 7: 179–85.

Yerkes, R. M. 1916a. *The Mental Life of Monkeys and Apes: A Study of Ideational Behavior.* Cambridge, MA: Henry Holt. Reprinted 1979 by Scholars' Facsimiles and Reprints, Delmar, NY.

———. 1916b. A new method of studying ideational and allied forms of behavior in man and other animals. *Proceedings of the National Academy of Sciences of the U.S.A.* 2: 631–33.

Yokoyama, S., and F. B. Radlwimmer. 1999. The molecular genetics of red and green color vision in mammals. *Genetics* 153: 919–32.

Young, T. 1802. Bakerian lecture: On the theory of light and colours. *Philosophical Transactions of the Royal Society of London* 92: 12–48.

Zarzo, M. 2007. The sense of smell: Molecular basis of odorant recognition. *Biological reviews of the Cambridge Philosophical Society* 82: 455–79.

Zayan, R. 1992. Représentation de la reconnaissance sociale chez l'animal. In *La Représentation animale*, edited by J. Gervet, P. Livet, and A. Tête. Nancy, France: Presses de l'Université de Nancy, 143–64.

———. 1994. Editor's preface. *Behavioural Processes* 33: 1–2.

Zazzo, R. 1979. Des enfants, des singes et des chiens devant le miroir. *Revue de Psychologie Appliquée* 29: 235–46.

———. 1993. Alfred Binet (1857–1911). *Prospects: The quarterly review of comparative education* 23: 101–12.

Zeki, S. 1990. Colour vision and functional specialization in the visual cortex. *Discussions in Neuroscience* 6: 8–64.

———. 2005. The construction of colours in the brain. *Actes du Colloque IRIS, Essence et sens des couleurs,* Nancy, France, November 9–10, 9.

ACKNOWLEDGMENTS

Among the people who helped and supported me during the preparation of this book, there are some who I would particularly like to thank.

First and foremost, I would like to express my gratitude to Patrick Duncan, Michel Kreutzer, and Jacques Vauclair. In addition to their friendship, over the years they have been an inexhaustible source of exchanges, reflections, and knowledge in the fields of behavioral ecology, biology of behavior, cognitive ethology, and comparative psychology.

I am also deeply indebted to eminent colleagues all over the world who over the span of this project responded immediately to requests for information. Among them, I must mention Alison Harman, Henry Heffner, Paul McGreevy, Andrew McLean, Leanne Proops, and especially Evelyn Hanggi and Brian Timney, who very kindly allowed me free access to work in progress.

Writing the French edition of this book also gave me the chance to appreciate the efficiency and enthusiasm of Lucie Blettery, who took charge of the project at Éditions Belin.

My unending appreciation, too, to Annick Weil-Barais for her warm welcome and our excellent interactions at the Psychology Laboratory at the University of Angers.

A big thank-you to Martine Hausberger, whose own work and that of her team are described in these pages, for doing me the kindness of writing the foreword to the French edition.

The English adaptation has been updated to take into account relevant work published since the book appeared in French.

Once again, people have been unstinting in their friendship. I would like to reiterate my thanks in particular to Patrick Duncan, Jacques Vauclair, and Brian Timney for their help, as well as Evelyn Hanggi, who was kind enough to contribute the preface to the present edition.

Michael Fisher, at Harvard University Press, has been a most attentive, effective, and considerate negotiating partner. He is also responsible for introducing me to Giselle Weiss, a remarkable translator in every respect. Her professional competence is equaled only by her gentle powers of persuasion in helping me to "see reason" for the benefit of my English audience.

Finally, my very warmest thanks to all the members of the team at Harvard University Press with whom I have been in contact for their professionalism and their high commitment to quality.

INDEX

Page references in *italics* indicate photographs and illustrations in the text.

Austin, Nicole, 182

Aversion, conditioned, 366–367

Awareness, in animals: 67–70

Axel, Richard, 336

Axons, 72–73, 136

Azimuth, 297–298, 301

Bachelors, bands of, 8, 10

Baddeley, Roland, 244–245

Bagot, Jean-Didier, 150

Balkenius, Anna, 154n13, 244–249, 262–265

Baragli, Paolo, 17, 95

Barasa, Antonio, 85

Barone, Robert, 71, 85–88, 146n

Barrey, Jean-Claude, 212, 212n8

Bartlett, Sir Frederic, 61

Bartoš, L., 141–142

Barton, R. A., 109

Bartošová, J., 141–142

Basal ganglia, 89

Basal layer, 370, 372

Basilar membrane: auditory, 283, 285–286; olfactory, 334–335

Basile, Muriel, 326

Baucher, François, 26–27

Behavioral ontogenesis, 12

Behaviorism, 48–50, 61; critique of, 51–54; neo-, 56

Berger, Joël, 21

Berthoz, Alain, 114, 121

Bhatnagar, Kunwar, 341

Biederman, Irving, 208

Binaural cues, 279–280, 298, 302–303

Binet, Alfred, 40, 42

Binocular vision fields, 127, 140–143, 142; depth perception and, 198–203; hearing and, 306; lateral bias and, 184–186

Bipolar cells, 135–136

Bitter taste, 359

Blache, Dominique, 95

Blackmore, Tania, 255–261

Blacque-Belaire, Henri, 25

Blake, Henry, 39, 39n

Blind spots, 127, 134

Blodgett, Hugh, 56–57

Blows, 310

Blue, recognition of, 223, 223, 228, 231, 232, 258, 258–260

Bodie (horse), 205–206

Boeglin, John, 157

Bonde, M., 346

Bonnardel, Valerie, 270

Bortolami, Ruggero, 71, 85–88, 146n

Bovet, Dalila, 204

Brain: color construction by, 151–159; localization, 108–109; meaning of, 71, 79–80; and mind (evolution), 99–112; structure, 108–109; triune theory of, 100–101; and visual perception, 132–134; weight/size, 103–105, 104, 105, 107, 110, 110n13

Brain, equine: brainstem of, 80, 92–94, 93; cerebellum of, 94; forebrain of, 80, 85–92; laterality and, 94–99; left lateral aspect of, 81, 83; main nerves of, 80; median section of, 82, 84; organization of, 79–85

Brain plasticity, 74–76

Brainstem, 80, 92–94, 93

Breland, Keller and Marian, 52–54

Breslin, Paul, 359, 361

Briant, Christine, 349

Brightness and color perception, 218–236, 259; dichromats and, 244

Broca, Paul, 340n

Brown, Suzanne, 225–226

Buck, Linda, 336

Budiansky, Stephen, 381